TRACE ELEMENT SPECIATION IN SURFACE WATERS
and Its Ecological Implications

NATO CONFERENCE SERIES

I	Ecology
II	Systems Science
III	Human Factors
IV	Marine Sciences
V	Air—Sea Interactions
VI	Materials Science

I ECOLOGY

Volume 1   Conservation of Threatened Plants
            edited by J. B. Simmons, R. I. Beyer, P. E. Brandham, G. Ll. Lucas,
            and V. T. H. Parry

Volume 2   Environmental Data Management
            edited by Carl H. Oppenheimer, Dorothy Oppenheimer, and William
            B. Brogden

Volume 3   The Breakdown and Restoration of Ecosystems
            edited by M. W. Holdgate and M. J. Woodman

Volume 4   Effects of Acid Precipitation on Terrestrial Ecosystems
            edited by T. C. Hutchinson and M. Havas

Volume 5   In Vitro Toxicity Testing of Environmental Agents: Current and Future
            Possibilities (Parts A and B)
            edited by Alan R. Kolber, Thomas K. Wong,
            Lester D. Grant, Robert S. DeWoskin, and Thomas J. Hughes

Volume 6   Trace Element Speciation in Surface Waters and
            Its Ecological Implications
            edited by Gary G. Leppard

TRACE ELEMENT SPECIATION IN SURFACE WATERS
and Its Ecological Implications

Edited by
Gary G. Leppard
National Water Research Institute
Burlington, Ontario, Canada

Published in cooperation with NATO Scientific Affairs Division
PLENUM PRESS · NEW YORK AND LONDON

Library of Congress Cataloging in Publication Data

NATO Advanced Research Workshop Program on Trace Element Speciation in Surface Waters and Its Ecological Implications (1981: Nervi, Italy)
Trace element speciation in surface waters and its ecological implications.

(NATO conference series. I, Ecology; v. 6)
"Proceedings of a NATO Advanced Research Workshop Program on Trace Element Speciation in Surface Waters and Its Ecological Implications, held November 2–4, 1981, in Nervi (Genoa), Italy"—Verso t.p.
Includes bibliographical references and index.
1. Trace elements in water—Environmental aspects—Congresses. 2. Trace elements in water—Analysis—Congresses. I. Leppard, Gary G. II. North Atlantic Treaty Organization. Scientific Affairs Division. III. Title. IV. Series.
QH545.T7N37 1981   574.5'222   83-2177
ISBN 0-306-41269-1

Proceedings of a NATO Advanced Research Workshop Program on Trace Element Speciation in Surface Waters and Its Ecological Implications, held November 2–4, 1981, in Nervi (Genoa), Italy.

©1983 Plenum Press, New York
A Division of Plenum Publishing Corporation
233 Spring Street, New York, N.Y. 10013

All rights reserved

No part of this book may be reproduced, stored in a retrieval system, or transmitted, in any form or by any means, electronic, mechanical, photocopying, microfilming, recording, or otherwise, without written permission from the Publisher

Printed in the United States of America

To my dear wife, Kristine
and
To my patient family

Preface

This volume contains papers presented at a NATO Advanced Research Workshop Programme entitled "Trace element speciation in surface waters and its ecological implications", held in Nervi (Genoa), Italy from November 2-4, 1981. It was co-sponsored by the Scientific Affairs Division of NATO, in conjunction with its Eco-Sciences Panel, and by the Associazione Italiana di Oceanologia e Limnologia. The organizing expenses and the expenses of the chairman and all invited participants were provided by the North Atlantic Treaty Organization. The scientific programme was planned by Dr. G.G. Leppard (Chief Organizer, Canada) together with Dr. O. Ravera (Eco-Sciences Panel, Italy), Dr. R. Mayer (Eco-Sciences Panel, Germany), Dr. R. Baudo (Chairman of the Local Organizing Committee, Italy), Dr. H. Muntau (Secretary of the Local Organizing Committee, Italy), and Dr. R. de Bernardi (President A.I.O.L., Italy). The final programme established by Dr. G.G. Leppard was based on one suggested by the NATO Eco-Sciences Panel and developed by Drs. R. Baudo, H. Muntau and O. Ravera.

Toxic elements, such as mercury, cadmium, lead and arsenic, are continually being released into surface waters where they may exist in trace quantities in various physico-chemical forms (chemical species) which disrupt ecological relationships and present society with health hazards. As a result of several catastrophic events caused by such toxic heavy metals (most notably the massive poisoning of humans at Minamata Bay in Japan), the attention of scientists, governments and the public alike was drawn to a pressing need to assess the impacts of toxic metals in natural waters. At the same time, serious eutrophication problems in the industrialized countries focussed attention on the need to understand the dynamic relationships between micronutrients, such as phosphorus, and the growth of aquatic biota. The consequent advent of more sensitive methods for measuring the low but biologically active amounts of individual trace elements in natural waters and aquatic organisms has carried us a long way in our understanding of trace element impacts on surface waters. However, such a simple approach to quantification is not enough.

There has long been evidence that it is not the total concentration of a given element which determines its impact, but rather it is the concentration of certain physico-chemical species. So, it has been necessary to couple the development of chemical analytical technology to a biological problem, that of ascertaining which trace element species are the important ones to focus on for detection and quantification. During the past decade, this attitude has gathered strength. The time is now ripe to facilitate direct exchanges between scientists of different disciplines working with this attitude. This NATO Advanced Research Workshop has been established to permit a limited number of scientific leaders and experts to come together for a short period to achieve the following general objectives:

(1) to exchange ideas at the frontiers of knowledge;
(2) to review and critically assess the state-of-the-art;
(3) to formulate recommendations for future research directions;
(4) to stimulate the formation of international and multidisciplinary contacts leading to more meaningful experiments.

To accomplish these general objectives, the Workshop was organized in three parts in which the themes were, in sequence:

(1) analytical approaches to the problem of trace element speciation;
(2) trace element species and general aspects of their interactions with aquatic life;
(3) new perspectives and future actions.

Then, a fourth theme was created, on site, to formulate concise statements of where we are and where we should be going. These statements in turn led to specific proposals for action.

As chief organizer and editor, I wish to record my gratitude to all participants who exchanged information so openly and made both the Workshop and its resultant book such a stimulating enterprise. I thank also the NATO Eco-Sciences Panel for its seminal role, the A.I.O.L. for its support role, the Local Organizing Committee for its energy, enthusiasm and perseverance, and the Local Recording Secretaries (Mrs. H. Muntau and Ms. L. Giannoni) for their continued helpfulness and efficiency. I thank also the Hotel Astor in Nervi for providing superb facilities and atmosphere, the mayor of Genoa for his role in hosting our social evening, Dr. M. Hotz for considerable valuable advice, and the many scientists from the La Spezia region who helped sponsor and enliven our post-Workshop tour of the major aquatic science laboratories at and near Fiascherino.

My own specific duties would not have been possible had it not

PREFACE

been for the excellent support received at my home base, the National Water Research Institute of the Canada Centre for Inland Waters at Burlington, Ontario. The Aquatic Ecology Division and the Staff Services Division were especially helpful. Ms. Mary C. Miscione of the Water Quality Branch of the Inland Waters Directorate at C.C.I.W. was indefatigable in providing administrative support and Ms. Dina Paolini of the Aquatic Ecology Division provided excellent service as my editorial assistant.

Gary G. Leppard

CONTENTS

1. Trace Element Speciation and the Quality of
   Surface Waters: An Introduction to the
   Scope for Research 1
   Gary G. Leppard

   ANALYTICAL APPROACHES TO THE PROBLEM OF
   TRACE ELEMENT SPECIATION

2. The Current Status of Trace Element
   Speciation Studies in Natural Waters 17
   G.E. Batley

3. Physical Separation Techniques in Trace
   Element Speciation Studies 37
   E. Steinnes

4. Voltammetric Studies on Trace Metal
   Speciation in Natural Waters. Part I: Methods . . . 49
   Pavel Valenta

5. Complex Formation in Solution and in
   Heterogeneous Systems. 71
   F.H. Frimmel

6. Direct Speciation Analysis of Molecular
   and Ionic Organometals 87
   Y.K. Chau and P.T.S. Wong

   TRACE ELEMENT SPECIES AND GENERAL ASPECTS OF
   THEIR INTERACTIONS WITH AQUATIC LIFE

7. Bioavailability, Trace Element Associations
   With Colloids and an Emerging Interest
   in Colloidal Organic Fibrils 105
   Gary G. Leppard and B. Kent Burnison

8. Biological Effects Under Laboratory
   Conditions 123
   C. Barghigiani, R. Ferrara, O. Ravera and A. Seritti

9. Feasibilities and Limits of Field
   Experiments to Study Ecological
   Implications of Heavy Metal Pollution. 137
   René Gächter and Jacques Urech

10. Biological Response to Trace Metals and
    Their Biochemical Effects. 159
    V. Albergoni and E. Piccinni

11. Biological Aspects of Trace Element
    Speciation in the Aquatic Environment. 177
    Maarten Smies

12. Biological Control of Trace Metal
    Equilibria in Surface Waters 195
    John P. Giesy

NEW PERSPECTIVES AND FUTURE ACTIONS

13. Voltammetric Studies on Trace Metal
    Speciation in Natural Waters
    Part II: Application and Conclusions
    for Chemical Oceanography and Chemical Limnology . . 211
    Hans Wolfgang Nürnberg

14. Neutron Activation Analysis Applied to
    Speciate Trace Elements in Freshwater. 231
    E. Orvini, M. Gallorini, and M. Speziali

15. Trace Element Speciation in Surface Waters:
    Interactions with Particulate Matter 245
    U. Förstner and W. Salomons

16. Is Analytically-Defined Chemical Speciation
    the Answer We Need to Understand
    Trace Element Transfer Along A
    Trophic Chain? 275
    Renato Baudo

17. Laboratory and Field Approaches to Environmental
    Effects Monitoring with Emphasis
    on Some Microbial-Heavy Metal Relations. 291
    Ralph F. Vaccaro

Index. 315

TRACE ELEMENT SPECIATION AND THE QUALITY OF SURFACE WATERS:

AN INTRODUCTION TO THE SCOPE FOR RESEARCH

        Gary G. Leppard

        National Water Research Institute
        Environment Canada
        Canada Centre for Inland Waters
        Burlington, Ontario, Canada. L7R 4A6

    Considerable research effort has been devoted by environmental scientists to measuring the concentrations of biologically-important trace elements in surface waters (McNeely et al., 1979; Florence and Batley, 1980; Förstner and Wittmann, 1981). This body of work has set the stage for a major thrust towards a profound understanding of the impact of trace elements on biota; it is now generating an increasing need to couple the development of chemical analytical technology to process-related biological problems. Current technology development strives to measure concentrations, for a given trace element, of those particular physico-chemical forms (or chemical species) which are biologically active. Concurrently, a new focus is being imposed on ecological impact studies, that of determining which active trace element species merit the most intensive research from the standpoint of environmental perturbation. From the point of view of environmental management, this major thrust should be directed towards the development of chemical speciation schemes which can be related directly to measures of bioavailability. With such a focus in mind, NATO supported my organizational role in the 1981 state-of-the-art Workshop which has led to this volume of the same title.

    It was felt in 1980 that such a workshop might well be premature; a consideration of several of the words in the title reveals why one would anticipate difficulty in dealing with the overall theme - "trace element speciation in surface waters and its ecological implications". Let us consider the term "surface waters". For proper coverage of our theme from a conceptual standpoint it was necessary to select research contributions dealing with lacustrine, riverine and estuarine waters as well as with marine coastal and oceanic waters. Generalizing from such a wide variety of natural water types is done only with caution. Let us consider also the term "trace element". The word "trace" refers to a specific quantitative limit which differs according to the scientific discipline of the expert one consults; the term "trace element" tends incorrectly to be used synonymously with the term "trace metal" as a result of there being a preponderance of analytical chemists who focus most or all of their methods development research on metals. This strong tendency is reflected in the contents of this volume, wherein even an essential non-metal such as phosphorous receives comparatively little attention, relative to some of the less troublesome toxic metals, despite its frequently powerful impact on many important aquatic processes. With regard to "ecological implications", one also encounters a tendency in need of correction. Considerable research on readily-grown laboratory organisms has been done to guide one to an understanding of how aquatic organisms are likely to behave in response to trace element perturbations. However, many important extensions of laboratory and theoretical research still need to be carried out on relevant aquatic organisms per se; this need too often goes unanswered. Hopefully, any future book built upon this volume will have a richer literature to draw from for background descriptions of physiological studies employing ecologically-significant aquatic biota. In particular, I would like to see detailed studies of the interplay between cells and the surrounding aquatic milieu wherein the respeciation of an applied trace element is followed through time.

Leaving aside for the moment the problem of what "speciation" really means in terms of bioavailability and how this linkage relates to the quality of surface waters, let us examine the conclusions of the Workshop. 1980 presented difficulties in integrating a variety of topics and disciplines into a coherent whole but 1981 produced three sets of recommendations on research areas of high promise. Table 1 lists scientific projects which are likely to lead to significant advances in understanding. They are long-term endeavors requiring a multidisciplinary team approach. Table 2 lists promising projects having a predominantly methodological nature. All seven projects in Table 1 and all of the five in Table 2 were endorsed by the participants present at the session on future actions (see Appendix 1 for a complete list of the participants). The future actions group also endorsed a list of thirteen trace elements worthy of intensive study (Table 3). An improved understanding of these particular elements is likely to lead to advances in chemical speciation methodology and in relating chemical speciation to biological phenomena affecting water quality. A guideline to the use of Table 3 is found in Appendix 2.

The discussions leading to the creation of Tables 1, 2 and 3 provide some interesting background and commentary on the endorsed research areas. A brief consideration of them in terms of the format of the Workshop yields ideas which provide a scope for specific research endeavors. Below is the format in abbreviated form:

Table 1. Research topics endorsed by * the Workshop - Problems

IMPORTANT PROBLEM-ORIENTED TOPICS**

---

1. To understand dynamic feedback interactions, between a trace element and aquatic biota, from the cellular level to the level of organisms.

2. To improve the quantitative aspect of the identification of "dissolved organic materials" - both by chemistry and by activity.

3. To study the speciation properties of major components of particulate matter as currently defined.

4. To delineate the transformation of trace element species within organisms.

5. To extend, in a systematic manner, speciation studies to all trace elements listed in Table 3, with a focus on surface water types of greatest concern to environmental management needs.

6. To ascertain the speciation of trace elements in interstitial (pore) waters.

7. To ascertain the real residence times for trace elements in important compartments of the biogeochemical cycles within aquatic ecosystems.

---

* A list of Workshop participants is found in Appendix 1.

** It is noted that analytical quality control is important for all of these topics - the reader is referred to Table 2.

Table 2. Research topics endorsed by * the Workshop – Methods

IMPORTANT METHODOLOGY-ORIENTED TOPICS

---

1. To develop speciation schemes which relate directly to bioavailability.
2. To establish the standardization of complexation capacity measurements.
3. To define an improved methodology for the separation of "dissolved" from "particulate" materials.
4. To provide a framework for analytical quality assurance and to develop appropriate reference materials.
5. To develop criteria for a proper selection of test organisms.

---

* A list of Workshop participants is found in Appendix 1.

Table 3. "Trace"* elements of primary interest** for speciation studies

RELEVANT ELEMENTS OF SURFACE WATERS TYPICALLY FOUND IN OBVIOUSLY TRACE AMOUNTS

Al	As	Cd	Cr	Cu
Fe	Hg	Mn	Pb	Zn

RELEVANT NUTRIENT ELEMENTS WHOSE BIOAVAILABLE SPECIES TEND TO BE FOUND IN LOW CONCENTRATIONS IN SURFACE WATERS

N	P	Si

---

* It is noted that the word "trace" is defined loosely and somewhat inconsistently by a multidisciplinary group. For the moment, this defect must persist, as illustrated by Leppard and Burnison in this volume. The elements listed here were chosen on the basis of their potential to provide new insight into speciation phenomena and speciation relationships to important biological phenomena.

** See Appendix 2.

Day one

Analytical approaches to the problem of trace element speciation

        Part one: the state of the art
                        physical separation techniques
                        voltammetric methods
        Chairman: Dr. H.W. Nürnberg
        Part two: complex formation
                        direct speciation analysis
        Chairman: Dr. P. Valenta

Discussion leader: Dr. H.W. Nürnberg

Day two

Trace element species and general aspects of their interactions with aquatic life

        Part one: bioavailability and fibrils
                        biological effects in the laboratory
                        field experiments
                        biological responses
        Chairman: Dr. E. Duursma
        Part two: biological consequences
                        biological control of equilibria
        Chairman: Dr. G.G. Leppard

Discussion leader: Dr. E. Duursma

Day three

New perspectives and future actions

        New perspectives: voltammetry application
                                    neutron activation analysis
                                    interactions with particles
                                    trace element trophic transfer
                                    environmental effects monitoring
        Chairman: Dr. H. Muntau
        Future actions: round table discussion
                                general recommendations
                                specific proposals for action
        Chairman: Dr. G.G. Leppard

Discussion leaders: rotated according to topic.

The discussion of day one produced several dominant questions which were formulated repeatedly in various ways according to the discipline of the inquirer. Some of these, as formulated by the editor, are as follows.

1) What research is necessary to establish a relationship between bioavailability and a chemical measure of labile species?
2) How can one speciate colloids in meaningful detail?
3) When will toxicity testing provide aquatic scientists with realistic levels as a baseline for proper interpretation of measured environmental levels?
4) How can we define operationally the concept of "particle" by a means superior to the current operation which employs a 0.45 μm filter?
5) With respeciation being a function of time, for what level of detail can we realistically strive in a speciation study involving a long separation stage or a period of storage?
6) What good features can we expect from a speciation plan based on lipid solubility?

The discussion of day one also produced several cautionary notes which appear to state the obvious but which bear emphasizing anyway:

1) chemical speciation schemes are designed to isolate classes of species - attempts at a "real" chemical speciation of a surface water sample would require intensive subfractionation and might have to be restricted to a class or classes already deemed to be important according to previously established information;
2) there are many "bioavailabilities" for a given chemical entity and not just one - bioavailability data for a given biological species, in a given ecological and/or physiological setting, do not necessarily relate to biota in general;
3) the term "biomethylation" tends to be incorrectly used, often resulting in a misleading impression - it has in the past been confused with chemical methylation and is readily mistaken for biologically-facilitated chemical methylation;
4) if one has the means, one should try one's best to discriminate between the binding of trace elements to colloids and the binding of them to other entities whose size range extends into the colloidal;
5) a chemical analysis should be restricted to a sample size and handling procedure appropriate to the analysis - one should be prepared to question whether or not one has done this;
6) biologists and chemists with a common interest tend to become totally disconnected - there is no real collaboration if there is no real communication;
7) the "interface" between biological cells and external water contains no magic - it is a structured zone of active materials whose behaviour obeys the laws of physics and chemistry.

Day two produced discussions probing an increasingly important basic theme - <u>integration</u> of the biological analyses related to trace element perturbation with the relevant chemical analyses of trace element speciation (and respeciation). It is clear that one does face considerable complexity in trying to comprehend the ecological implications of perturbed trace element speciation in surface waters. One must deal with many levels in the organization of aquatic life (ecosystems, communities, populations, organisms, organs and even finer levels down to the molecular) and one must deal also with many different simultaneous chemical measures to assess speciation changes (total element present, inorganic species, organic species, individual species of particular interest, and ligands which affect respeciation).

The aquatic scientist often must face two basic information needs which, in the light of constrained research support, are not compatible within the context of feasible experimental systems. There is a need to reduce ecological complexity to generalizations that one can work with in a practical way. There is also a need to describe the complexity of some natural systems to such a detailed extent that the mechanics of the driving processes can be delineated well enough to permit "highly-tuned" environmental manipulations and predictions. Since great depth and great breadth appear too much to ask for in a given project over the short term, one must compromise. This can be done effectively by a creative use of the technology and concepts already at hand. The secret to compromising for profit lies in asking the right questions, a capacity which increasingly requires a multidisciplinary framework.

The dominant topic of day two discussions was the interplay between living cells and their aquatic milieu. An understanding of this interplay is vital to future research on trace element effects as well as to research on trace element redistribution, respeciation, uptake and transformations. The cell surface layers must become considered more as a physiological system of major importance, and less as a magical black box, to all scientists engaged in the study of surface water processes. In relation to this topic, the colloids again revealed themselves to be worthy of further research. In the waters of some temperate lakes of various sizes and trophic states, one class of colloid is attracting a particular interest now, the colloidal organic fibril (Leppard et al., 1977; see also Leppard and Burnison, this volume). Its role in respeciation is under investigation; its presence in many lacustrine water types may explain a major difficulty in the use of filters to define the "particulate" compartment of many physical separation schemes. At even a finer level of "biological organization", we can also focus on a specific subject of particular interest, the class of molecules called metallothioneins (see Albergoni and Piccinni, this volume). Studies of the relationships of these molecules to cell surface events, detoxification processes, trace metal respeciation processes and trace metal redistribution phenomena are certain to lead to a deeper physiological understanding of trace element relations to biota.

Certain lines of questioning on "cell-milieu interplay" brought out a serious problem in oversimplification of the design of some past lacustrine and marine researches. Many experiments have consisted of adding a given trace element species to a laboratory vessel, or to an external enclosure, and then interpreting subsequent biological events in terms of the amount and species of added perturbant. One can imagine that the rate and the nature of the respeciation in the milieu could have profound effects on the biological responses. One can also imagine that factors affecting respeciation of an added perturbant were not always controlled well by past experimenters. Now that respeciation of added trace perturbants can be more readily monitored through time, we will be able to repeat and extend these earlier experiments in ways more amenable to proper interpretation.

The discussion of the final day session on "new perspectives" assessed the strengths and limitations of some currently powerful techniques, stressed the importance of particles and their surface relations with the aquatic milieu in general, reinforced the basic theme of day two, and elaborated upon the attitude of creative compromising in environmental monitoring. The individual chapters based on this last scientific session cover these subjects well when read in the context of earlier related chapters. The detailed discussion sections (one is present at the end of each chapter) provide insight into our thinking on perspectives in general. The session on "new perspectives" refined the framework, established earlier, for the discussion on "future actions" which led, in turn, to the endorsed lists of projects of high promise (Table 1, Table 2) and to the list of aquatic trace elements worthy of particular interest (Table 3).

This volume reviews and assesses critically the state-of-the-art. It establishes where we are in 1982 and indicates where science might profitably go in the near future. In addition to consulting the endorsed lists, one can gain a sense of how the various areas are developing by studying the degree to which each chapter tends to favorize certain contributions. Expert opinions, as opposed to bland generalizations, have been encouraged and speculation has been stimulated, sometimes to a high degree. As editor, it is my hope that the considerable organizational effort that has gone into this volume will stimulate its readers to consult more than just those chapters of immediate concern to their own specialty.

REFERENCES

Albergoni, V., and Piccinni, E., in: this volume.
Florence, T.M., and Batley, G.E., 1980, Chemical speciation in natural waters, CRC Crit. Rev. Anal. Chem., 9:219.
Förstner, U., and Wittmann, G., 1981, "Metal Pollution in the Aquatic Environment, Second Edition", Springer-Verlag, Berlin.

Leppard, G.G., Massalski, A., and Lean, D.R.S., 1977, Electron-opaque microscopic fibrils in lakes: their demonstration, their biological derivation and their potential significance in the redistribution of cations, Protoplasma, 92:289.

Leppard, G.G., and Burnison, B.K., in: this volume.

McNeely, R.N., Neimanis, V.P., and Dwyer, L., 1979, "Water Quality Sourcebook, a Guide to Water Quality Parameters", Environment Canada, Inland Waters Directorate, Water Quality Branch, Ottawa.

Appendix 1. The Workshop participants

A LIST OF THE PRINCIPAL SPEAKERS, CHAIRMEN, PRINCIPAL DISCUSSANTS AND CO-AUTHORS OF NATO WORKSHOP #9, HELD AT NERVI (GENOA), ITALY, IN 1981.

Dr. V. Albergoni
Istituto di Biologia Animale
Fisiologia Generale
Università di Padova
Via Loredan, 10
35100 Padova
Italy

Dr. C. Barghigiani
Consiglio Nazionale delle
  Ricerche
Istituto per lo Studio delle Pro-
  prietà Fisiche di Biomolecole
  e Cellule
Via S. Lorenzo, 24-26-28
56100, Pisa
Italy

Dr. G.E. Batley
Environmental Chemistry Group
Division of Energy Chemistry
CSIRO
Lucas Heights
New South Wales, 2232
Australia

Dr. R. Baudo
Consiglio Nazionale delle
  Ricerche
Istituto Italiano di Idrobiologia
28048, Pallanza (Novara)
Italy

Dr. M. Bernhard
Senior Expert
Environmental Protection Research
  Division
Comitato Nazionale per l'Energia
  Nucleare
19030, Fiascherino (La Spezia)
Italy

Dr. R. Boniforti
Comitato Nazionale per l'Energia
  Nucleare
19030, Fiascherino (La Spezia)
Italy

Dr. M. Branica
Centre for Marine Research
"Ruder Bošković" Institute
41001 Zagreb, Croatia
Yugoslavia

Dr. R. Breder
Stazione Marina del Centro di
  Ricerche
KFA Jülich
Via Byron 11
Fiascherino, (La Spezia)
Italy

Dr. D. Bregant
Consiglio Nazionale delle Ricerche
Istituto Talassografico
Viale R. Gessi, 2
Trieste
Italy

Dr. B.K. Burnison
Aquatic Ecology Division
National Water Research Institute
Environment Canada
C.C.I.W., P.O. Box 5050
Burlington, Ontario
Canada   L7R 4A6

Dr. Y.K. Chau
Environmental Contaminants
  Division
National Water Research Institute
Environment Canada
C.C.I.W., P.O. Box 5050
Burlington, Ontario
Canada   L7R 4A6

Dr. R. de Bernardi
Consiglio Nazionale delle
  Ricerche
Istituto Italiano di Idrobiologia
28048 Pallanza (Novara)
Italy

Dr. E.K. Duursma, Director
Delta Instituut Voor Hydrobio-
  logisch Onderzoek
Koninklijke Nederlandse Akademie
  van Wetenschappen
Vierstraat 28
4401 EA Yerseke (Zeeland)
The Netherlands

Dr. R. Ferrara
Consiglio Nazionale delle
  Ricerche
Istituto per lo Studio delle
  Proprietà Fisiche di Biomole-
  cole e Cellule
Via S. Lorenzo, 24-26-28
56100, Pisa
Italy

Dr. U. Förstner
(for recent address change, see
his chapter)

Dr. U. Förstner
Institut für Sedimentforschung
Universität Heidelberg
Im Neuenheimer Feld 236
Postfach 103020
D-6900 Heidelberg 1
West Germany

Dr. R. Frache
Istituto di Chimica
Viale Benedetto XXV 3
Università di Genova
Genova
Italy

Dr. F.H. Frimmel
Institut für Wasserchemie und
  Chemische Balneologie
Der Technischen Universität München
8 München 70
Marchioninistrasse 17
(Grosshadern)
West Germany

Dr. R. Gächter
E.A.W.A.G.
Seenforschungslaboratorium der
  EAWAG/ETH
CH-6047 Kastanienbaum
Switzerland

Dr. M. Gallorini
Consiglio Nazionale delle Ricerche
Centro di Radiochimica e Analisi
  per Attivazione
Viale Taramelli, 12
27100 Pavia
Italy

Dr. J.P. Giesy
Pesticide Research Centre and
  Department of Fisheries and
  Wildlife
Michigan State University
E. Lansing
Michigan 48824, U.S.A.

Dr. G.G. Leppard
Aquatic Ecology Division
National Water Research Institute
Environment Canada (C.C.I.W.)
P.O. Box 5050
Burlington, Ontario
Canada   L7R 4A6

Dr. H. Muntau
Department of Chemistry
Commission of the European
  Communities
J.R.C.
21020, Ispra (Varese)
Italy

Dr. H.W. Nürnberg
Direktor Am Institut für Chemie
  der
Kernforschungsanlage Jülich GmbH
Institut 4: Angewandte Physika-
  lische Chemie
Postfach 1913
D-5170 Jülich 1
West Germany

Dr. E. Orvini
Consiglio Nazionale delle
  Ricerche
Centro di Radiochimica e Analisi
  per Attivazione
Viale Taramelli, 12
27100 Pavia
Italy

Dr. E. Piccinni
Istituto di Biologia Animale
Fisiologia Generale
Università di Padova
Via Loredan, 10
35100 Padova
Italy

Dr. O. Ravera
Dept. of Physical & Natural Sciences
Commission of the European Communi-
  ties
J.R.C.
21020, Ispra (Varese)
Italy

Dr. W. Salomons
Delft Hydraulics Laboratory
Haren Branch
c/o Institute for Soil Fertility
P.O. Box 30003
9750 RA Haren (Gr)
The Netherlands

Dr. E. Schulte
Comitato Nazionale per L'Energia
  Nucleare
19030, Fiascherino (La Spezia)
Italy

Dr. A. Seritti
Consiglio Nazionale delle Ricerche
Istituto per lo Studio delle Pro-
  prietà Fisiche di Biomolecole
  e Cellule
Via S. Lorenzo, 24-26-28
56100, Pisa
Italy

Maarten Smies
Shell Internationale Research
  Maatschappij BV
Group Toxicology Division
P.O. Box 162
2501 AN Den Haag
The Netherlands

Dr. M. Speziali
Consiglio Nazionale delle Ricerche
Centro di Radiochimica e Analisi
  per Attivazione
Viale Taramelli, 12
27100 Pavia
Italy

Dr. E. Steinnes
Universitetet I Trondheim
NLHT-Kjemisk Institutt
Rosenborg (Trondheim)
7055, Dragvoll
Norway

Jacques Urech
E.A.W.A.G.
Seenforschungslaboratorium der
　EAWAG/ETH
CH-6047 Kastanienbaum
Switzerland

Dr. R.F. Vaccaro
Senior Scientist
Woods Hole Oceanographic Institution
Woods Hole, Massachusetts, 02543
U.S.A.

Dr. P. Valenta
Institut für Chemie der
Kernforschungsanlage Jülich GmbH
Institut 4: Angewandte Physika-
　lische Chemie
Postfach 1913
D-5170 Jülich 1
Germany

Dr. P.T.S. Wong
Great Lakes Biolimnology Laboratory
Canada Department of Fisheries
　and Oceans
Burlington, Ontario
Canada　　　L7R 4A6

Appendix 2. The trace elements of primary interest: a guideline for setting research priorities (as constructed by Dr. Eiliv Steinnes).

The speciation of all elements present in trace quantities in natural waters is of potential interest from a purely scientific point of view. Nevertheless, it is evident from the presentations and discussions at this Workshop that some elements are considered more important than others in this respect, and it seems desirable to arrive at a preferentially rated list of trace elements of primary interest in speciation studies.

Before selecting the elements, it is useful to consider what may be the most important goals of speciation work related to natural waters. In my opinion the major goals can be defined as follows.

I. Short-term scale

   The study of elements (and their chemical species) known to be harmful to man or to other organisms.

II. Long-term scale

   The improvement of our general understanding of natural processes.
   a. Geochemical and biogeochemical cycles
   b. Interactions between trace elements and organisms

On this basis I would recommend that the following elements be given the highest priority in speciation studies related to surface waters.

A. Potentially harmful elements:

   Pb, Cd, Hg, Cr, As, Cu

B. Nutritionally important elements:

   N, P, Mn, Zn, [Fe]

C. Major elements in the lithosphere that occur in trace concentrations in most natural waters:

   Si, Al, [Fe]

Some of these elements would deserve a high priority in more than one of the above groups, e.g., Fe.

THE CURRENT STATUS OF TRACE ELEMENT SPECIATION STUDIES

IN NATURAL WATERS

        G.E. Batley

        Analytical Chemistry Section
        CSIRO, Division of Energy Chemistry
        Lucas Heights, NSW
        Australia, 2232

INTRODUCTION

    Biochemists have long been aware that the assimilation by the human body of essential trace elements takes place in certain preferred chemical forms. Cobalamin, the glucose tolerance factor, and heme iron, for example, are the favoured forms of cobalt, chromium, and iron, respectively. Elements may be classified as either essential, such as Cu, Zn, Cr, Mo, V, Mn, Sn, Fe, Ni, Co, and Se or non-essential, Ag, Cd, Hg, Tl, Pb and As. Excesses of either class can be toxic, although in general, the non-essential elements are of greater toxicity, and as with bioavailability, this toxicity will be a function of chemical form.

    The range of trace elements found in natural waters will include heavy metals, the lanthanide and actinide elements, non-metallic elements such as As, Se and Te, and the micronutrient elements Si, P and N. Techniques for the speciation of the micronutrient elements are well-established, for the most part, and will not be dealt with here. The metallic elements which comprise the most important group of inorganic environmental contaminants will be the principle focus of this review.

    Interest in the measurement of metal speciation in aquatic ecosystems assumed importance in the late 1960's when, following the earlier mercury pollution incidents such as that at Minamata Bay, laboratory studies revealed the differing toxicities of methylated and inorganic mercury species and the possibility of chemical and biological transformations from one form to another. It was shown that the toxicity and fate of waterborne metal contaminants was dependent on chemical forms, and that the quantification of these forms would be more meaningful than measurements of total metal concentrations. This finding was echoed by those responsible for the supervision of water quality, at a time when methods for metal speciation measurement were still being evolved.

Now, some fifteen years further on, in reviewing the progress that has been achieved in metal speciation studies, it is evident that, despite the significant advances that have been made in measurement techniques, much remains to be done. The methods that have been developed do not provide an absolute breakdown of metal species; they are usually operationally-defined classifications. We are still a long way from formulating standard procedures. As will be shown, the approaches of research groups have been quite diversified; although the best that can be achieved might indeed be an operationally-defined speciation procedure, there must be additional refining of these procedures, and biological testing of their utility in making satisfactory classifications. It is to be hoped that meetings such as this will contribute to clarifying and coordinating the directions that this research should take.

TRACE ELEMENT SPECIATION: WHY AND WHEN?

In pursuing the problems of how to measure speciation, it is important that the questions of why and when to study speciation are carefully considered and not lost sight of, especially by those who advocate speciation measurements as part of regular monitoring programs.

The principal purpose of measuring metal species relates to their relative toxicities to aquatic biota. Even studies of metal transport are governed by this basic justification. It is important, therefore, to consider metal species in all phases of the aquatic system which these organisms may contact. These will include not only the dissolved metal species in water which passes a $0.45 \mu m$ membrane filter, but also colloidal and particulate fractions, interstitial waters and sediment phases.

In addition to this immediate purpose, a second and longer-term aim of speciation studies is to further an understanding of metal interactions in aquatic systems. Of greatest importance in this area will be studies of the concentration and transformations of metal species within and at the surface of organisms. Our current lack of knowledge in this area will be discussed in more detail in subsequent chapters.

With reference then to our principal aim of species toxicity research, the problem is to define toxicity limits, not only for metal species but even for total metal. Chronic and sub-acute effects are difficult to identify. Water quality standards are usually formulated on the basis of acute toxicity testing, yet sub-acute effects may be detectable at levels well below those set using application factors. It is fair to say that while the analytical chemist has progressed to the point where he can confidently measure total metal concentrations in the ng L^{-1} range in natural waters and can now quantify metal species at concentrations near 100 ng L^{-1}, toxicity testing has yet to evolve to a similar degree of sophistication. For example, chronic toxicity effects of heavy metals on some algae have been shown to be exhibited at lower metal concentrations if testing is carried out using a continuous flow system with constant renewal of nutrients and toxicants, rather than the usually accepted batch culture testing (Wong, unpublished results). In static tests, the metal binding ability of algal exudates is often overlooked (Van den Berg et al., 1979). Also, in the area of total metal testing, documenting the synergistic effects of certain mixed metal combinations is only now being realized. Closer examination of the effects of sub-acute metal concentrations on phytoplankton has revealed instances where certain smaller size fractions are eliminated while a slightly larger size fraction may multiply (Wong, unpublished results). Such changes are significant, given the important role of microalgae in the food chain, and should be considered when establishing "safe" metal limits.

In addition, while the analytical chemist has now recognised the problems of possible sources of metal contamination or losses, in many instances toxicity testing at parts per billion concentrations and below is being performed using procedures which were only acceptable for much higher concentrations. Researchers working with trace and ultratrace metal concentrations must be aware of the meticulous care that is required if meaningful results are to be obtained.

In studying the toxicity of chemical forms, progress has been limited. There are many examples where the toxicity of ionic metal has been shown to be reduced by the addition of organic chelators (Anderson and Morel, 1978; Gnassia-Barelli et al., 1978). Other approaches have included the use of calculations based on known equilibrium data to predict metal species in solution and relate this to observed toxic behaviour (Andrew et al., 1977; Jackson and Morgan, 1978). Attempts to look at the effect of natural chelators on the presence of metal in colloidal and particulate size fractions has yet to be satisfactorily addressed. A problem is that the metal concentrations at which one can readily identify toxicity are considerably in excess of 10 $\mu g\ L^{-1}$, which, for toxic metals such as cadmium, lead and copper, is well above the natural concentrations.

Since, in almost all instances, the concentrations at which chronic effects have been detected for even the most toxic metals, such as cadmium, are in excess of 1 $\mu g\ L^{-1}$ (Klapow and Lewis, 1979; Great Lakes Science Advisory Board, 1980), it can be argued

that there is little point in attempting to measure speciation where
the total metal concentrations are one to two orders of magnitude
below this. In these instances, speciation measurements become an
academic exercise, if not in many cases a practical impossibility.
The assumption of course is that these toxicity testing results are
in fact correct, and that safe levels represent no detectable effects
to the most susceptible natural organisms over an acceptably long
time interval. This area of speciation research needs to be more
fully investigated.

It should be noted that the above discussions have been restricted to metals in ionic and complexed forms which may be in either
soluble, adsorbed or particulate fractions. Organometallic compounds,
where by definition the metal is covalently bound to carbon atoms
of organic groups, have not been considered. Most common among this
class are the methylated species of mercury, lead, arsenic and tin,
which have been shown to exist in natural waters (Chau and Wong,
1981). Their chemical behaviour differs markedly from the other
metal species discussed in this paper. The low-molecular-weight
species are volatile and may be detected by gas chromatography or
related techniques. Their toxicities to algae are, in many instances,
significantly higher than free ionic metal (Chau and Wong, 1981).

APPROACHES TO THE MEASUREMENT OF METAL SPECIATION

The two basic approaches to the measurement of metal species
distributions in natural waters are (1) chemical modeling and (2)
experimental measurement. These have been discussed in considerable
detail in the recent review by Florence and Batley (1980), and will
be briefly summarized here together with advances since the writing
of that review. More detailed considerations of specific methods
will be given in other papers in this volume.

A third approach to metal speciation is the study of metal adsorption on selected substrates and metal complexation with chemically isolated "natural" chelators. These studies can provide useful
thermodynamic data for chemical models and add to the understanding
of natural systems.

CHEMICAL MODELING

The current status of chemical modeling of natural waters has
been reviewed in a recent symposium (Jenne, 1979), where a comparison of various chemical models pointed out many of the limitations of this approach. A predictive model of metal speciation,
using measured total metal concentrations and the concentrations
of other major, minor and trace species, will be only as good as
the data on which it is based. Problems arise in seawater, for

example, with the use of activity coefficients and our uncertainty of ionic interactions in aqueous electrolyte solutions. Secondly, a major problem is the reliability of equilibrium constants. For many reactions these constants are unknown, while for others more accurate constants are still being obtained (e.g., Sylva and Davidson,1979).Our incomplete knowledge and failure to account for all interactions is the major limitation of this approach. Most importantly, it is unlikely that one could ever account for the heterogeneous interaction of metal species with the mixed organic and inorganic colloidal and particulate phases which represent a major component of the total metal concentrations in most natural systems (Florence and Batley, 1980). Because of this, the use of modeling is currently restricted to artificial colloid-free solutions. Valuable information has, however, been obtained in this manner concerning the toxicity of simple inorganic and complexed metal forms (Andrew et al., 1977).

EXPERIMENTAL SEPARATION TECHNIQUES

With the exception of electrodeposition techniques, such as anodic stripping voltammetry (ASV), most measurement techniques are unable to selectively detect different metal species at natural levels in waters. Ion selective electrodes respond to free metal ion activity but at appreciably higher metal concentrations. A common approach to metal speciation has been, therefore, to make use of a separation procedure to isolate or concentrate selected species prior to their measurement by one of the standard techniques. A recognized limitation of such processes is the possibility that the pre-existing dynamic equilibria in solution may be disturbed by the removal of species, to the detriment of subsequent speciation measurements. It is important therefore to acknowledge the operationally-defined nature of any multistage separation process.

Size Fractionation

Size fractionation alone is a potentially useful classification since the most biologically-available metals are likely to be those in the smallest size categories, compared to those bound to high-molecular-weight organic macromolecules or colloidal species (Table 1). The simplest and most common size fractionation is the separation of particulate species by filtration through a 0.45 μm membrane filter. Even this operation may disturb solution equilibria (Sharp, 1973).

Ultrafiltration membranes permit the separation of the "dissolved" fraction into molecular weight fractions from 500 to 300,000, while other membranes are capable of subdividing the colloidal size range. Measurements of metals in the separated fractions are usually accompanied by dissolved organic carbon analyses to enable deductions as to the nature of the separated species (Giesy and

Table 1. Some Chemical Forms of Metals in Natural Waters

Chemical Form	Example	Approximate diameter, nm
Hydrated Metal ion	$M(H_2O_6)^{2+}$	1
Inorganic complexes	$M(H_2O)_5Cl^+$	1-2
	$M(OH)_2^0$	1-2
	$M_2(OH)_2CO_3$	2-3
Organic complexes	M-amino acid	2-4
	M-fatty acid	2-6
	M-fulvic acid	2-6
Colloidal species	M-clay	10-500
	$M-Fe_2O_3$	10-500
	$M-MnO_2$	10-500
	M-humic acid	10-500
	M-humic acid – Fe_2O_3	10-500
Particulate species	minerals, soils, clays, detritus	450

Briese, 1977). The technique is susceptible to contamination problems, which could restrict its application to solutions having higher metal and dissolved organic carbon contents than, for example, those in coastal seawater (Smith, 1976). Potential losses of metal by adsorption and blockage of membrane pores during filtration could also lead to errors in the measured size distributions (Beneš and Steinnes, 1974).

Similar fractionation can be achieved by gel-permeation chromatography, as used by Gjessing (1965) for dissolved organic matter in natural waters. Steinberg (1980) recently applied this technique to a natural river water. Metal concentrations were again correlated with dissolved organic carbon distributions. Gel chromatography has similarly been used to isolate high-molecular-weight metal complexes from seawater (Betz, 1979). This technique is not applicable to "natural" metal concentrations where a preconcentration is required.

For the separation of low-molecular-weight size fractions, dialysis offers a possible solution. Diffusion through dialysis membranes is slow, however, with the possibility of dissociation of metal complexes at the membrane surface. The problems of contamination due to the large membrane surface areas involved can be significant, as can adsorption of metals on clean membranes. In situ dialysis has been successfully employed by Beneš and Steinnes (1974). Hart and Davies (1981) used dialysis as a component stage in a trace metal speciation scheme.

Centrifugation is an alternative, little explored technique for the separation of colloidal species (Beneš and Steinnes, 1975).

Solvent Extraction

Separations based on the polarity of metal species can be achieved using solvent extraction. Organically-associated metal in seawater has been equated to metal extractable with chloroform by Slowey et al. (1967). Florence and Batley (1981) chose a 9:1 hexane-butanol mixture, having a dielectric constant similar to that of the lipid bilayer of the living cell membrane, to extract the equivalent of lipid-soluble metal species from natural waters.

Recently, high pressure liquid chromatography (HPLC), using a C-18 reversed-phase column and acetonitrile-water solvent mixtures, has been investigated by Lee (1981) for the separation of copper and cobalt associated with natural water organics. Elution profiles were obtained for both organics, using U.V. detection, and metals, detected by flameless atomic absorption spectrometry. A major problem with this potentially valuable technique could be the small sample size and the limited sensitivity attainable. Cassidy and Elchuck (1980) preconcentrated metal species using a cation-exchange resin cartridge prior to HPLC separation.

Resin Separations

The use of conventional ion exchange resins has been proposed for metal speciation studies (Filby et al., 1974). To achieve an effective breakdown, sequential use of both a cation and anion exchange column is required. Separations on the basis of charge, however, have little biological relevance, except perhaps where valency states of differing toxicities are being separated (Pankow et al., 1977). Even in this instance, involving chromium, it is doubtful whether complete separation of Cr (III) and Cr (VI) species can be achieved (Florence and Batley, 1980).

More meaningful separations can be obtained on the basis of metal complexation, using a chelating resin. Iminodiacetate groups on the resin Chelex-100 are able to bind ionic metal species, for a range of metals, in the pH range of natural waters. It has been demonstrated that the pore size of the resin is such that macromolecules and colloidal species are excluded from the internal surfaces of the resin beads and will not exchange, except perhaps on the much smaller external surfaces (Florence and Batley, 1980). In addition, kinetically labile metal complexes will be dissociated and retained by the resin.

The difference in exchange kinetics of metal complexes has been exploited by Figura and McDuffie (1980) using a 7-second column separation, and a 3-day batch equilibration, to differentiate "moderately labile" and "slowly labile" species. Chelating resin systems have also been incorporated in speciation schemes by Batley and Florence (1976) and Hart and Davies (1977b).

Polymeric adsorbents, such as the non-polar, polymeric, styrenedivinylbenzene adsorbent, XAD-2, have been used by Sugimura and coworkers (1978) to separate metal-organic complexes. Organic compounds with aromatic structures were shown to be retained by the resin, but those with aliphatic structures and inorganic compounds were not. Leonard and Crewe (1981) have, however, used this resin to extract straight-chain fatty acids from seawater, while Florence and Batley (1981) found that an equivalent resin, Bio-beads SM-2, adsorbed more than 70% of ionic lead and copper spikes from seawater at pH 8.1, and even more from distilled water at the same pH. Ionic metal adsorption did not occur at pH 4, which was used by the authors to separate lipid-soluble metal species.

Electrodeposition Techniques

The electrochemical preconcentration of metal species from natural waters prior to their measurement by chemical (potentiometric stripping), electrochemical (anodic stripping) or physical (electrothermal atomization) means, offers the potential of discriminating between certain electroactive and electroinactive metal forms (Batley, 1981). The deposited species, collectively called "labile", are those which are in labile equilibrium or dissociated within the time scale of the measurement. This fraction will therefore be dependent on pH, the selected deposition potential, and the electrode system and its conditions of operation. The techniques are restricted to metal ions which are electrochemically reducible to the metal, and for those metals commonly analyzed at natural levels this requires amalgam formation with mercury, either co-deposited with the metal in a film electrode (MFE), or in a hanging mercury drop electrode. The relative advantages of these systems are well known (Batley and Florence, 1974; Batley, 1981). Electroactive metals include Cu, Pb, Cd, Zn, Tl, Sb, Sn, Bi, Cr, Co, Mn, Ni, Ag, Hg, and Au, which fortuitously encompass most of the metals of environmental concern. Most studies to date have concentrated on anodic stripping voltammetry. However, this discussion will apply equally to the other electrodeposition techniques.

To obtain meaningful speciation results, careful consideration of operating conditions is required. Voltammetric measurements at natural pH values greater than 7 are insensitive to certain electroinactive ionic metal species, for example, hydroxycarbonates of lead and cadmium. Acidification to pH 2 may dissociate some organic complexes or release metals adsorbed on colloidal species. Measurements at such extreme pH values have been used in speciation studies (Baier, 1977). Buffering of the sample to values near pH 5 has been used as a compromise designed to detect all simple ionic species but not those in strong complexes or those strongly adsorbed. The effect on metal speciation of the acetate buffer addition, and of the 10^{-5} M mercury (II) ion required for MFE in situ deposition, have been discussed by Skogerboe et al. (1980). By its nature, however, ASV can only provide an operationally-defined measurement of metal species and cannot reflect true equilibrium concentrations. These operational conditions may yet provide data which is of biological significance (Young et al., 1979).

Other applications of voltammetry to speciation studies include the use of shifts in stripping peak potentials to determine conditional, metal-complex, stability constants (Raspor et al., 1978; Brown and Kowalski, 1979) and the variation of electrode rotation rate to detect complexes having different dissociation rate constants (Shuman and Michael, 1978). While these approaches can provide useful information for artificial solutions, they are of little value in studies of natural samples.

STUDIES OF THE ADSORPTION AND COMPLEXATION BEHAVIOUR OF METALS

Considerable information on the behaviour of metal ions in natural waters has been accumulated through studies of their adsorption and complexation behaviour. Because of the problems in obtaining reproducible data from natural adsorbants and chelators, these studies have generally involved reactions with what may be classified as homogeneous, chemically-purified matrices, using higher than natural metal concentrations.

The adsorption of heavy metals on silica, alumina, hydrated ferric oxide, illite and a range of other soil minerals has now been well documented (Florence and Batley, 1980). The relevance of these data to natural water systems is, however, uncertain since there is strong evidence that both colloidal and particulate matter in natural waters are heterogeneous mixtures of both organic and inorganic components (Davis and Leckie, 1978). These mixtures most often comprise organic coatings on an inorganic substrate. The role of natural organic matter in promoting or inhibiting metal adsorption is not well understood. Davis and Leckie (1978) attempted to model this system using simple complexing ligands and an amorphous iron oxide substrate. The surface orientation of binding groups was shown to be important. If adsorption occurred through the only-metal-binding groups in the organic molecule, metal adsorption was reduced, whereas if additional binding groups were available on the adsorbed molecule, metal uptake was enhanced. The implication is that in natural waters complex macromolecular organic ligands, such as humic and fulvic acids, will enhance trace metal adsorption.

Natural organics play an important role in stabilizing inorganic colloids in natural waters. The addition of humic acid to freshwater samples at pH 6.6 was shown to maintain hydrous iron oxide in a colloidal form, while in its absence flocculation occurred (Batley and Florence, unpublished results). More studies are needed in this area, preferably with natural substrates, to improve our understanding of metal partitioning and, in particular, how this will be affected by changes in pH, as might occur as a result of acid rain or acid mine drainage.

Similarly, the study of complexation with natural ligands has received much attention (Mantoura, 1979). A preoccupation with many workers has been the reaction with metals of soil-derived humic and fulvic acids, and estimates of stability constants have been obtained. In natural waters, these compounds are present as highly aggregated molecules containing hydroxyl, quinone and carboxylic acid binding groups. Separation processes often result in changes in the degree of aggregation (Wershaw and Pickney, 1977). Soil-derived humics differ both in origin and structure from humic acids of marine origin (Harvey et al., 1981). Marine fulvic and humic materials comprise between 25 and 50% of the organic carbon content of seawater and have been shown to interact with heavy metals in a complex manner (Piotrowicz et al., 1981).

The complexing ability of aquatic organisms and their secretion products has been demonstrated by a number of authors. McKnight and Morel (1980), for example, found evidence for copper complexation by siderophores from filamentous blue-green algae in seawater. Leppard and coworkers (1977) have suggested that organic fibrillar colloids, associated with algae and bacteria in lake waters, may be important binders of metals.

The ability of these and other natural molecules to bind metal is usually demonstrated by a measurement of complexing capacity, an operationally-defined parameter obtained by titration of the ligands with added metal (Florence and Batley, 1980). The interactions of metals with organic matter in natural waters are, however, very dynamic processes. Many ligands appear to be reaction products of simpler molecules released by organisms. Piotrowicz et al. (1981) have shown that once a seawater sample is taken into a closed container, natural productive and destructive equilibria slow and cease, meaning that the detection of complexed metal species will depend on how quickly a sample is analysed. As an example, zinc-fulvic acid interactions were shown to occur as part of a steady state cycle of less than 40 hours duration controlled by photooxidation and bacterial processes. A postulated cycle, where H_3L^I is a fulvic acid and H_3L^{II} a humic acid, is shown in Figure 1. These ligands have a biological source, by way of an intermediate L. Their complexes with metals are degraded at unknown rates, possibly involving light, through either biological activity or photolysis. Time constants may vary from hours to days depending on factors such as biological production rate, nutrient content of water, etc. These studies illustrate the

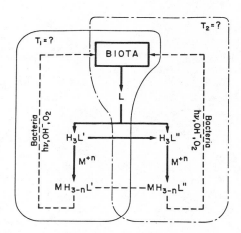

Figure 1. Postulated steady-state cycle of biological organisms, dissolved organic matter, and trace metals in the marine environment (Piotrowicz et al., 1981).

problems of extrapolating data based on synthetic complexing agents and of establishing a rapid speciation measuring scheme.

VALENCY-STATE SPECIATION

Where elements exist in more than one valency state, the separation of species in each state is a commonly-studied aspect of chemical speciation. These will include elements such as chromium, manganese, iron, cobalt, thallium, arsenic, antimony, tin and selenium (Florence and Batley, 1980). Such measurements are only important when one valency has a demonstrably greater toxicity than the other, as with chromium. Even then, more detailed measurements of the speciation of each valency state are necessary to permit a meaningful assessment of the results.

MEASUREMENT SCHEMES FOR HEAVY METAL SPECIATION

In a natural water system, we may expect metals to be present as simple hydrated ions, and in inorganic and/or organic complexed forms. Complexed species may be readily dissociated or inert, and of low molecular weight or in macromolecular colloidal aggregates. Metal may be adsorbed on inorganic or organic colloidal matter or associated in heterogeneous mixtures of both species. A number of measurement schemes have been proposed which attempt to classify dissolved metal species, in as many of these categories as possible, using combinations of the range of separation and measurement techniques previously outlined.

   Batley and Florence Scheme: This scheme (Batley and Florence, 1976) uses ASV to discriminate between labile and bound metal in a sample, before treatment, after passage through a chelating resin column, after U.V. irradiation, and after passage of the U.V.-irradiated sample through the chelating resin column. By this approach they were able to quantify seven classes of metal species.

   Irradiation of waters with U.V. light, for the decomposition of organic matter, has been used previously to release organically-bound metal (Foster and Morris, 1971). While irradiation of acidified samples will release all bound metal species (Batley and Farrar, 1978), it is assumed that, at natural pH values, inorganic colloidal species will not be solubilized and the only metal release will be that from organic ligands. This simple assumption is complicated in practice as a result of the heterogeneity of colloidal species. Thus, where an organic molecule binding a metal is stabilizing an inorganic colloid, destruction of the binding groups may release metal resulting in its desorption with a resultant flocculation of the colloid. This precipitate may carry with it some of the released metal. Experiments with artificially-prepared, humic acid-hydrous iron oxide-metal mixtures showed that metal was lost from solution after U.V. irradiation (Batley and Florence, unpublished results), although in measurements of natural river and estuarine waters, good metal balances were obtained suggesting no such losses.

   The chelating resin separation is used to remove ionic metal and metal which is exchangeable in weak complexes, while colloidal species will not be retained. Undoubtedly, some overlap of metal species will occur in these experimental classes, where complexes are partially dissociated, or some metal associated with colloidal species is exchanged. By its nature, the measured speciation is truly operationally-defined. Nevertheless, measurements obtained (Florence 1977; Batley and Gardner, 1978) have revealed an important, previously undocumented aspect of metal binding in waters, namely that despite predicted distributions, high percentages of copper, lead and cadmium are naturally present in colloidal forms. Even though it may not be possible to unequivocally define organic and inorganic colloidal forms, in coastal seawater this combined fraction accounted for 56% of total copper, 67% of total lead and 33% of total cadmium, while in river water, the respective percentages were 52, 24 and 5.

   The classifications afforded by this scheme are possibly more complex than is required to define biological availability, although to date this has not been demonstrated by toxicity testing. In addition, the analysis time of eight hours required per sample would preclude its application to routine analysis.

   Hart and Davies Scheme: This scheme does not attempt to separate exclusive classes, but instead defines filtrable, exchangeable and dialysable metal (Hart and Davies, 1977a, b). The exchangeable fraction is obtained by batch equilibration over 24 hours with Chelex-100. Dialysable metals are concentrated on a Chelex-100 column during a continuous 5-hour dialysis. Metals are analyzed by atomic absorption spectrometry after elution from the resin by nitric acid. The operations are relatively simple and the data from each separation, though not free from overlap, can be usefully evaluated.

<u>Figura and McDuffie Scheme</u>: Using Chelex-100, a breakdown is obtained on the basis of dissociation kinetics, giving inert, slowly labile, moderately labile and very labile fractions (Figura and McDuffie, 1980). The ASV-labile measurements are carried out at pH 6.3 on fractions after filtration, and resin separations, as shown in Figure 2. These are obtained at optimum plating potentials for each metal, to avoid reduction of some complexed species at the high cathodic potentials necessary for the simultaneous deposition of cadmium, lead and copper. The scheme does not provide mutually exclusive separations. The "very labile" fraction measured by ASV on the filtered sample will not necessarily be fully retained by the Chelex column. Batley and Florence (1976) showed that the proportion of this fraction not retained by Chelex-100 could be as high as 75% for some metals, giving erroneous "moderately-labile" numbers by difference. Nevertheless, the scheme is simple and it is likely that

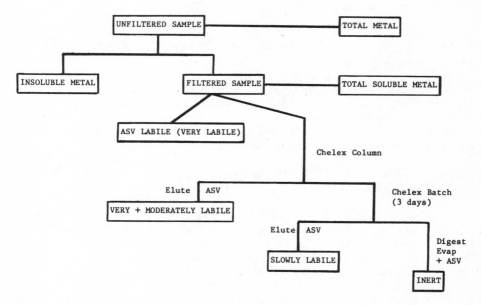

Figure 2. Lability Scheme for Metal Speciation (Figura and McDuffie, 1980).

this type of operational approach could be useful in relating to bioavailability.

Other schemes: The incorporation of XAD-2 resins in speciation schemes has been explored by Sugimura et al. (1978) and Montgomery and Santiago (1978), to provide breakdowns into organic and ionic (chelating-resin exchangeable) fractions. However, the selectivity of XAD-2 separations for organic species only is suspect.

A hypothetical scheme based on size fractionation has been proposed by Lee (1979) but has yet to be applied to natural waters. Harrison and Laxen (1980) preferred to measure total metal in ultrafiltered size fractions, followed by measurements on the total dissolved fraction of ASV-labile (at the natural pH) and Chelex-100 exchangeable metal, and of an acid ASV-labile fraction after 0.5% HNO_3 is added to the unfiltered sample. Hoffmann et al. (1981) developed a sequential ultrafiltration procedure with measurements of total metal, ASV-labile metal and dissolved organic carbon on each fraction. A mass balance procedure was used to compute concentrations in each size fraction.

The major limitations of the schemes so far discussed lie either in their complexity, or in their adaptability (in most cases undemonstrated) to ready interpretation in terms of bioavailability, acknowledging that this is the main purpose for pursuing speciation measurements. It is instructive then to examine the process of metal assimilation from the viewpoint of the organism.

The primary barrier to metal uptake by a living cell is the cell membrane. Whitfield and Turner (1979) have compared diffusion processes across a cell membrane with those occurring at an electrode surface, of either an ion selective electrode or a mercury film electrode for ASV, in attempting to define whether the cell responds to the thermodynamically-available (ion selective electrode) or electrochemically-available (ASV) metal fraction in solution. In the former case, availability is governed by an equilibrium distribution across the cell membrane, while in the latter the metal flux is kinetically limited. As with most modeling exercises, the complexities of the natural system and of the measurement techniques did not lend themselves to such a simplified approach.

Florence and Batley (1981) have considered routes for metal transport across a cell membrane and the ways in which these could be modeled by separation processes. A model of a cell membrane is shown in Figure 3. Mechanisms for the transport of ionic and molecular species across biological membranes have been the subject of detailed investigations in recent years (Selwyn and Dawson, 1977). Ion transport is believed to involve complexing molecules acting either as carriers or by forming ion-permeable pores or channels (Ovchinnikov, 1979). Phospholipids have demonstrated ionophoric properties (Green

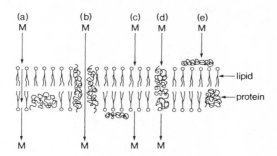

Figure 3. Possible interactions and transport routes for metal species, M, across a cell membrane.
a) direct lipid solubility
b) passage through an aqueous pore
c) lipid-mediated transport
d) protein-mediated transport
e) binding with surface proteins

et al., 1980); however, more important in the case of heavy metals is likely to be transport mediated by metal-binding proteins. Metallothioneins bind zinc, cadmium, copper and other heavy metals through thiol groups in the many cysteine residues that they contain. Their unique role in metal metabolism is not yet fully understood (Anon., 1980). In an attempt to model the interaction of metal species with these proteins, Florence and Batley (1981) examined metal uptake by a thiol-containing resin. The solubility of metal species in the lipid bilayer was simulated by using solvent extraction or a non-polar resin. Further research is in progress, using this more fundamental approach to modeling the biological transport of metal species, in an attempt to obtain measurements which more closely relate to the toxicity of species in a natural water system.

CONCLUSIONS

The immediate aim of metal speciation studies should be the defining of toxicity to aquatic biota in terms of metal species. There now exist a number of well-defined operational procedures for the quantification of selected groups of metal species, and considerable data for a range of natural water systems have now been accumulated (Florence and Batley, 1980). Significant among these findings is the existence of high proportions of the total metal concentrations in complexed or colloidally-associated forms. The ways in which sampling and measurement procedures perturb the natural species have not yet been fully investigated. Toxicity testing is also required to evaluate the biological significance of operational classifications and assist in the development of new or more meaningful schemes. Such experiments are also needed to establish "safe" metal concentration limits below which application of the currently lengthy procedures for speciation measurement serve little purpose.

REFERENCES

Anderson, D.M., and Morel, F.M.M., 1978, Copper sensitivity of Gonyaulax tamarensis, Limnol. Oceanogr., 23:283.
Andrew, R.W., Biesinger, K.E., and Glass, G.E., 1977, Effects of inorganic complexing on the toxicity of copper to Daphnia magna, Water Res., 11:309.
Anon., 1980, Metallothionein in trace metal metabolism, Nutr. Rev., 38:286.
Baier, R.W., 1977, Lead distribution in the Cape Fear River estuary, J. Environ. Qual., 6:205.
Batley, G.E., 1981, Electroanalytical techniques for the determination of heavy metals in seawater, Mar. Chem., in press.
Batley, G.E., and Farrar, Y.J., 1978, Irradiation techniques for the release of bound heavy metals in natural waters and blood, Anal. Chim. Acta., 99:288.
Batley, G.E., and Florence, T.M., 1974, An evaluation and comparison of some techniques of anodic stripping voltammetry, J. Electroanal. Chem., 55:23.
Batley, G.E., and Florence, T.M., 1976, Determination of the chemical forms of dissolved cadmium, lead and copper in seawater, Mar. Chem., 4:347.
Batley, G.E., and Gardner, D., 1978, A study of copper, lead and cadmium speciation in some estuarine and coastal marine waters, Estuarine Coastal Mar. Sci., 7:59.
Beneš, P., and Steinnes, E., 1974, In situ dialysis for the determination of the state of trace elements in natural waters, Water Res., 8:947.
Beneš, P., and Steinnes, E., 1975, Migration forms of trace elements in natural fresh waters and the effect of the water storage, Water Res., 9:741.
Betz, M., 1979, Separation of naturally occurring high molecular weight complexes from seawater, Mar. Chem., 7:165.
Brown, S.D., and Kowalski, B.R., 1974, Pseudopolarographic determination of metal complex stability constants in dilute solution by rapid scan anodic stripping voltammetry, Anal. Chem., 51:2133.
Cassidy, R.M., and Elchuck, S., 1980, Trace enrichment methods for the determination of metal ions by high performance liquid chromatography, J. Chromatogr. Sci., 18:217.
Chau, Y.K., and Wong, P.T.S., 1981, Some environmental aspects of organo-arsenic, lead and tin. Proceedings of a N.B.S. Workshop on Environmental Speciation and Monitoring Needs for Trace Metal Containing Substances from Energy-Related Processes, Washington, D.C., 1981, in press .
Davis, J.A., and Leckie, J.O., 1978, Effect of adsorbed complexing

liquids on trace metal uptake by hydrous oxides, Environ. Sci. Technol., 12:1309.

Figura, P., and McDuffie, B., 1980, Determination of the labilities of soluble trace metal species in aqueous environmental samples by anodic stripping voltammetry and Chelex column and batch methods, Anal. Chem., 52:1433.

Filby, R.H., Shah, K.R., and Funk, W.H., 1974, Role of neutron activation analysis in the study of heavy metal pollution of a lake-river system, in: "Proc. 2nd Int. Conf. Nuclear Methods in Environ. Res.", J.R. Vogt and W. Meyer, eds., NTIS, Springfield, Va.

Florence, T.M., 1977, Trace metal species in fresh waters, Water Res., 11:681.

Florence, T.M., and Batley, G.E., 1980, Chemical speciation in natural waters, CRC Crit. Rev. Anal. Chem., 9:219.

Florence, T.M., and Batley, G.E., 1981, A new scheme for chemical speciation of copper, lead, cadmium and zinc in seawater, in: "Proceedings of an International Conference on Heavy Metals in the Environment", Amsterdam, in press.

Foster, E.O., and Morris, A.W., 1971, The seasonal variations of dissolved ionic and organically associated copper in the Menni Straits, Deep-Sea Res., 18:231.

Giesy, J.P., and Briese, L.A., 1977, Trace metal transport by particulates and organic carbon in two South Carolina streams. Verh. Internat. Verein. Limnol., 20:1401.

Gjessing, E.T., 1965, Use of "Sephadex" gel for the estimation of molecular weight of humic substances in natural water, Nature, 208:1091.

Gnassia-Barelli, M., Romeo, M., Laumond, F., and Pesando, D., 1978, Experimental studies on the relationship between natural copper complexes and their toxicity to phytoplankton, Mar. Biol., 47:15.

Great Lakes Science Advisory Board, 1980, International Joint Commission Report of the Aquatic Ecosystem Objectives Committee, pp. 63.

Green, D.E., Fry, M., and Blondin, G.A., 1980, Phospholipids as the molecular instruments of ion and solute transport in biological membranes, Proc. Natl. Acad. Sci. USA, 77:257.

Harrison, R.M., and Laxen, D.P.H., 1980, Physicochemical speciation of lead in drinking water, Nature, 286:791.

Hart, B.T., and Davies, S.H., 1977a, A new dialysis-ion exchange technique for determining the forms of trace metals in water. Aust. J. Mar. Freshwater Res., 28:105.

Hart, B.T., and Davies, S.H., 1977b, A batch method for the determination of ion-exchangeable trace metals in natural waters. Aust. J. Mar. Freshwater Res., 28:397.

Hart, B.T., and Davies, S.H., 1981, Trace metal speciation in the freshwater and estuarine regions of the Yarra River, Victoria, Estuarine Coastal Mar. Sci., 12:353.

Harvey, G.R., Boren, D.A., and Tokar, J.M., 1981, Structures of seawater fulvic and humic acids derived from proton NMR studies and historical data, Mar. Chem., in press.

Hoffman, M.R., Yost, E.C., Eisenreich, S.J., and Maier, W.J., 1981, Characterization of soluble and colloid-phase metal complexes in river water by ultrafiltration. A mass-balance approach, Environ. Sci. Technol., 15:655.

Jackson, G.A., and Morgan, J.J., 1978, Trace metal-chelator interactions and phytoplankton growth in seawater media: Theoretical analysis and comparison with reported observations, Limnol. Oceanogr., 23:268.

Jenne, E.A., ed., 1979, Chemical modeling in aqueous systems-speciation, sorption, solubility and kinetics, A.C.S. Symposium Series 93, American Chemical Society, Washington, D.C.

Klapow, L.A., and Lewis, R.H., 1979, Analysis of toxicity data for California marine water quality standards, Jour. Water Poll. Control Fed., 51:2054.

Lee, J., 1979, A scheme for the separation and characterization of possible metal-organic species in natural waters: some preliminary data, Geol. Surv. Can., Paper 79:121.

Lee, J., 1981, The use of reverse phase liquid chromatography for studying trace metal-organic associations in natural waters, Water Res., 15:507.

Leonard, J.D., and Crewe, N., 1981, Study on the extraction of organic compounds from seawater with XAD-2 resin, in: Proceedings of a Marine Chemistry Symposium, Halifax, N.S., Canada, June, p.2.

Leppard, G.G., Massalski, A., and Lean, D.R.S., 1977, Electron-opaque microscopic fibrils in lakes: their demonstration, their biological derivation and their potential significance in the redistribution of cations, Protoplasma, 92:289.

McKnight, D.M., and Morel, F.M.M., 1980, Copper complexation by siderophores from filamentous blue-green algae, Limnol. Oceanogr., 25:62.

Mantoura, R.F.C., 1979, Organometallic interactions in natural waters: a review, in: "Organic Chemistry of Sea Water", E.K. Duursma and R. Dawson, eds., Elsevier Oceanography Series, Elsevier, Amsterdam.

Montgomery, J.R., and Santiago, R.J., 1978, Zinc and copper in 'particulate' forms and 'soluble' complexes with inorganic and organic ligands in the Guanajibo River and coastal zone, Puerto Rico, Estuarine Coastal Mar. Sci., 6:111.

Ovchinnikov, Y.A., 1979, Physico-chemical basis of ion transport through biological membranes: Ionophores and ion channels, Eur. J. Biochem., 94:321.

Pankow, J.F., Leta, D.P., Lin, J.W., Ohl, S.E., Shum, W.P., and Janauer, G.E., 1977, Analysis for chromium traces in the aquatic ecosystem, Sci. Total Environm., 7:17.

Piotrowicz, S.R., Harvey, G.R., Springer-Young, M., Courant, R.A., and Boren, D.A., 1981, Studies of cadmium, copper and zinc complexation by marine fulvic and humic materials in seawater using anodic stripping voltammetry, in: "Trace Metals in Seawater", C.S. Wong, J.D. Burton, E. Boyle, K. Bruland, and E.D. Goldberg, eds., Plenum, N.Y.

Raspor, B., Valenta, P., Nürnberg, H.W., and Branica, M., 1978, The chelation of cadmium with NTA in seawater as a model for the typical behaviour of trace metal chelates in natural waters, Sci. Total Environm., 9:87.

Selwyn, M.J., and Dawson, A.P., 1977, Model membranes and transport systems, Biochem. Soc. Trans., 5:628.

Sharp, J.H., 1973, Size classes of organic carbon in seawater, Limnol. Oceanogr., 17:494.

Shuman, M.S., and Michael, L.C., 1978, Application of the rotated disk electrode to measurement of copper complex dissociation rate constants in marine coastal samples, Environ. Sci. Technol., 12:1069.

Skogerboe, R.K., Wilson, S.A., and Osteryoung, J.G., 1980, Exchange of comments on scheme for classification of heavy metal species in natural waters, Anal. Chem., 52:1960.

Slowey, J.F., Jeffrey, L.M., and Hood, D.W., 1967, Evidence for organic complexed copper in sea water, Nature, 214:377.

Smith, R.G., 1976, Evaluation of combined applications of ultrafiltration and complexation capacity techniques to natural waters, Anal. Chem., 48:74.

Steinberg, C., 1980, Species of dissolved metals derived from oligotrophic hard water, Water Res., 14:1239.

Sugimura, Y., Suzuki, Y., and Miyake, Y., 1978, Chemical forms of minor metallic elements in the ocean, J. Oceanogr. Soc. Japan, 34:93.

Sylva, R.N., and Davidson, M.R., 1979, The hydrolysis of metal ions. Part I. Copper (II), J. Chem. Soc., Dalton Trans., 232.

Van den Berg, C.M.G., Wong, P.T.S., and Chau, Y.K., 1979, Measurement of complexing materials excreted from algae and their ability to ameliorate copper toxicity, J. Fish. Res. Board Can., 36:901.

Wershaw, R.L., and Pickney, D.J., 1977, Chemical structure of humic acids, Part 2. The molecular aggregation of some humic acid fractions in N, N-dimethylformamide, J. Res. U.S. Geol. Surv., 5:571.

Whitfield, M., and Turner, D.R., 1979, Critical assessment of the relationship between biological, thermodynamic and electrochemical availability, in: "Chemical Modeling in Aqueous Systems", E.A. Jenne, ed., ACS Symposium Series 93, American Chemical Society, Washington, D.C.

Young, J., Gurtisen, J.M., Apts, C.W., and Crecelius, E.A., 1979, The relationship between the copper complexing capacity of seawater and copper toxicity in shrimp zoeae, Mar. Environ. Res., 2:265.

DISCUSSION: G. BATLEY

M. BRANICA
I question your very fundamental approach and I propose that: (1) bioassays are strong and useful research tools for enlarging the knowledge base to interpret whole-organism phenomena; (2) when the dissolved metal concentration is very low, the speciation can be very important to studies necessary to evaluate biological concentration factors; (3) bioavailability is not one box, and interactions have to take into account the specific physiology of biological species and the state of the foodweb; (4) additional important methods are neutron activation analysis and isotope dilution approaches.

G. BATLEY
There are two approaches which should be taken in metal speciation studies:
(1) to establish operationally-defined chemical procedures, supported by biological data, to define the toxic fraction of the total metal;
(2) to more fully understand the physical chemistry of toxic metal species in water and their interactions with biota.

I support both approaches. However, I see the second as a longer-term research exercise, whereas the first is essential to meet the current needs of environmental authorities.

PHYSICAL SEPARATION TECHNIQUES

IN TRACE ELEMENT SPECIATION STUDIES

E. Steinnes

Department of Chemistry
University of Trondheim - NLHT
7055 Dragvoll, Norway

INTRODUCTION

A variety of physical separation techniques has been applied to the study of trace element speciation in natural waters (or has at least been reported to be potentially useful for this kind of work). The techniques of concern separate particulates and/or dissolved species according to size, specific gravity or charge. Since methods based on size separation are by far the most commonly used, this presentation concentrates on such methods. Those separations based on electric charge will be mentioned only to the extent that they do not involve chemical reactions.

The most simple approach to be taken involving size separation techniques would appear to be the separation of "dissolved" and "particulate" forms of the element concerned. However, this is not a straight forward issue because there is no clearly-defined borderline between the two groups (but rather there is a continuous scale) and, also, because the separation media available generally do not show a clear cut-off at a certain diameter or molecular weight. The situation becomes even more complex when it comes to more ambitious schemes where more than one physical separation step is involved.

Table 1 represents an attempt to classify, according to size, the compounds with which trace elements may be associated in natural waters. The figures given for diameter and molecular weight ranges, associated with each size group, are listed only to indicate the orders of magnitude involved and are not to be regarded as definitive in any way.

Below are described briefly some physical separation techniques

Table 1. Association of trace elements with compounds of different size ranges.

Category	Examples of compounds	Approximate molecular weight range	Approximate diameter in nm
Simple dissolved compounds	Free inorganic ions Inorganic ion pairs Inorganic complexes Small organic molecules	< 200	< 1
Larger dissolved compounds	Fulvic acids Fatty acids Polyhydroxo complexes Polysilicates	$200 - 10^4$	$1 - 10$
Colloidal material	Humic acids Proteins Inorganic colloids (metal hydroxides, clay minerals, etc.)	$10^4 - 10^6$	$10 - 100$
Suspended material	Inorganic particles Organic particles	$> 10^6$	> 100

that are currently employed or that may be considered useful for future application in trace element speciation studies in aquatic systems. The advantages and difficulties associated with each technique are discussed. A more extensive survey of these techniques is found in a recent textbook on trace chemistry of aqueous solutions (Beneš and Majer, 1980).

FILTRATION AND ULTRAFILTRATION

The most extensively used physical separation technique, in trace element speciation work, is filtration of the sample solution through a thin porous membrane having pores of molecular dimensions (or not much larger) in order to effect a separation away from the solubles of particles or colloids of greater dimensions than the pore diameter. Filters covering fairly-well the entire range of pore sizes of 1-1000 nm average diameter are commercially available. Depending on the pore size of the filter, two different experimental arrangements are commonly used. At pore sizes of about 100 nm and above, the conventional suction filtration, such as by means of a vacuum pump, is employed. When filters of smaller pore size are used, it may be necessary to introduce a high pressure condition in order to obtain a satisfactory filtration rate. This arrangement is called ultrafiltration and is normally based on the use of a compressed inert gas (such as nitrogen or argon). Alternatively the ultrafiltration can be carried out in a centrifuge. Ultrafiltration cells are often provided with a magnetic stirrer in order to prevent filter clogging during the ultrafiltration of solutions containing appreciable amounts of particles too large to pass through the membrane.

A special version of ultrafiltration is diafiltration which can be used to concentrate colloids from a large solution volume prior to further fractionation. In this case, new solution is added either discontinuously or continuously so that the solution volume remains nearly constant.

In a major number of speciation studies reported in the literature, the filtration of the sample through a 450 nm membrane filter, prior to further chemical separation and determination steps, is employed because of its rapidity and simplicity (see for example the review by Florence and Batley, 1977). As already indicated, this step does not yield a true separation of soluble and particulate material because a large part of the colloidal fraction is likely to pass through the filter along with the soluble species. On the other hand, clogging of the filter pores may in many cases cause an irreproducible fraction of the < 450 nm particulates to be retained. The 450 nm filtration is therefore not likely to give very conclusive results in general.

In order to obtain more information about the distribution of trace elements among different size fractions and to allow a more proper separation of the dissolved species, some investigators have proposed schemes involving several ultrafiltration steps using filters with different pore sizes on aliquots of a sample (Guy and Chakrabarti, 1976) or successive filtrations on the same sample (Hoffman et al., 1981).

In interpreting the results from filtration experiments it is necessary to have in mind that all the pores in a particular filter are not equally large. The pore diameter may show considerable variation, especially for ultrafilters having small pore sizes. Consequently, a precise determination of the size of colloidal particles, by means of ultrafiltration, is difficult.

Furthermore, rather serious errors may be associated with filtration procedures. Disturbance of the separation due to clogging of filter pores has already been mentioned. This effect appears to be less troublesome in ultrafiltration, when magnetic stirring is employed, than in conventional filtration. Contamination from the filtering apparatus and, in particular, from the filter itself is an obvious source of error in studies involving extremely low concentrations of the species of interest. In addition to releasing the trace elements themselves, the filter materials could possibly release organic compounds that could interfere with trace element species.

Perhaps the most serious source of error in filtration experiments, however, is likely to be losses due to adsorption, in particular with regard to fresh water samples having a low total content of dissolved species. Adsorption may occur on all surfaces to which the solution is exposed during the experiment, and in particular on the filter. This problem becomes especially important in ultrafiltration through filters of small pore size because of their relatively large surface area and relatively long separation time. In a study involving ultrafiltration of 100 ml of river water, through a Diaflo PM 10 filter, the loss due to adsorption was found to be 60% or more for the elements Fe, Al and Sc. By a filtration through a 450 nm membrane filter, using the same water, the corresponding figures were less than 10% (Beneš and Steinnes, 1974). On the other hand, in filtration experiments where samples of potable water were spiked with lead to a concentration of 20 µg l^{-1}, adsorption losses of 20 - 40 % to 450 nm filters were observed, the actual value depending somewhat on the pH-value of the sample (Gardner and Hunt, 1981). Apparently adsorption effects cannot be excluded even in cases of membrane filters with large pore sizes.

Adsorption losses are very disturbing in speciation studies, even if the total loss can be accounted for, because different soluble species of an element may be adsorbed to a different extent. For example, the species $PbCO_3$ is much more strongly sorbed on negatively charged silica than is Pb^{2+} (Stumm and Bilinski, 1973); this means that further speciation studies on the remaining solution may give false results.

In a recent study (Laxen and Harrison, 1981), fresh water samples were filtered through Nuclepore filters of various pore sizes in an ultrafiltration cell. These filters, consisting of a thin (5-10 µm) polycarbonate membrane with individual pores of uniform diameter (± 20%), present a much reduced area for adsorption and may yield considerable improvement in size separations. In accordance with this, the authors found a significantly better performance of the Nuclepore filters compared with ordinary ultrafilters. One disadvantage reported was a greater tendency of clogging with the Nuclepore filters.

DIALYSIS

The dialysis technique, which is based on the fact that colloidal particles cannot pass through membranes with pore sizes smaller than the particle diameter, is similar to ultrafiltration in many respects. The main difference is that no external pressure is applied to force a solution through the membrane in dialysis. The membranes used in speciation studies so far have average pore sizes of the order of 1-5 nm, corresponding to molecular weights of about $10^3 - 10^4$.

In the classical dialysis arrangement, the sample solution is contained in a bag made from the membrane, or in a vessel the bottom of which consists of the membrane. This bag or vessel is submerged in a solution which is in a larger vessel and which does not contain the species to be dialyzed. Rona et al. (1962) used this approach to remove ions from sea water and retain high-molecular-weight compounds of manganese and zinc. Guy and Chakrabarti (1976, 1977) used an opposite arrangement in studies of heavy metal complexation with naturally-occurring, high-molecular-weight compounds, where uncomplexed metal ions, or low-molecular-weight complexes of these metals, were dialyzed into the bag. A third type of arrangement consisting of a dialysis cell with two equal compartments, one containing the sample and the other one distilled water, was used by Beneš and Steinnes (1975, 1976) in speciation studies on river water and in investigations on the use of radionuclides in the study of trace element behaviour in natural waters. When operating this cell, vigorous shaking was employed all of the time.

Dialytic equilibrium is established when the concentration of a diffusible species is the same on each side of the membrane. The time necessary to approach this equilibrium depends on the charge and size of the species concerned, as well as the general conditions under which the experiment is carried out. Since the membranes are negatively charged, positively charged or neutral species are dialyzed more readily than negatively charged ones of similar size.

Dialysis experiments suffer from the same sources of error as previously described for ultrafiltration. Contamination from leaching of the membrane may affect the contents of the dialysis vessel. Adsorption to the membrane or to the walls of the dialysis cell may also be significant.

Another difficulty that may arise occurs when there exists a chemical equilibrium between a metal and a non-dialyzable complexing agent such as a humic acid. If the metal ion is dialyzed from the small vessel into a larger reservoir and the humic acid is retained in the vessel, additional metal ions may be released from the metal-humic acid complex, and thus the original speciation pattern is distorted. The same effect may occur during ultrafiltration of metal-humus complexes.

A method which is not appreciably affected by the sources of error discussed above is dialysis *in situ* (Beneš and Steinnes, 1974). It consists of immersing a dialysis cell containing pure water into the natural water to be sampled in the river, lake or sea. Only molecules or ions smaller than the pore diameter of the dialysis membrane are able to diffuse from the natural water into the dialysis cell. This diffusion proceeds until the same concentrations of individual species are attained outside and inside the cell, and until adsorption equilibrium is established between the species in the inside solution and the dialysis membrane (or dialysis cell). After the dialysis and adsorption equilibria have been established, the content of the dialysis cell can be analyzed for the total concentration of dissolved elements and for individual ionic and molecular species present. The sampling *in situ* also helps in avoiding errors that might have occurred if conventional sampling in bottles and subsequent transport to the laboratory for separations had been employed (such as errors resulting from adsorption to the walls of the bottle and certain speciation changes which may occur during transport and storage). Recently, a modification of the dialysis *in situ* technique has been proposed (Beneš, 1980). Dissolved species diffuse through the membrane into the cell where they are removed from the solution by adsorption, ion exchange, and etc. If the concentration of dissolved species in the solution inside the cell is maintained at a very low value as compared with the outside solution, then the diffusion flow is directly proportional to the concentration of the dissolved species in the water. The average water concentration during the sampling period can be determined then from the total amount of the species accumulated within the cell.

A somewhat similar set-up for the concentration of dissolved species from water samples was reported by Hart and Davies (1978). In that case, the water is pumped through a dialysis cell and the solution passing the membrane is circulated through a column with a chelating resin.

ELECTRODIALYSIS

This technique is based on the fact that the passage of ions through a dialysis membrane can be accelerated by applying a potential gradient across the membrane. An electrodialysis cell consisting of three compartments separated by membranes may be used for this purpose. If the sample solution is contained in the central compartment, colloids and neutral dissolved species will be retained there, while charged species will migrate into the respective eletrode compartments. Applications of this technique for trace element speciation studies in natural waters have not yet been reported but the method is potentially interesting for this purpose.

ELECTROPHORESIS

Electrophoretic separations are usually performed on supports such as cellulose, silica, or alumina. The use of these materials in speciation studies on water samples is not feasible because of the severe adsorption effects involved (Beneš and Glos, 1975). The so-called free-liquid electrophoresis, where the separation process takes place in aqueous solution, seems more suitable for the purpose (Beneš and Steinnes, 1975; Beneš and Glos, 1979).

CENTRIFUGATION

This technique is based on the use of centrifugal forces to separate colloidal and suspended particles from dissolved species. In the centrifugal field, a force acts upon the particles of a dispersed phase having a density different from that of the dispersing medium, an action which causes particle movement in the same or in the opposite direction to that of the action of the force, depending on particle density relative to that of the dispersing medium. The radius of the sedimented particles (r) can be calculated from the following formula (Beneš and Majer, 1980):

$$r^2 = \frac{9\eta \ln (x_2/x_1)}{2(\rho-\rho_o)\omega^2 t}$$

where t is the centrifugation time, x_1 and x_2 are the initial and final distances of the particle from the axis of rotation, ρ and ρ_o are the densities of respectively the particle and the dispersing medium, η is the viscosity of the dispersing medium, and ω is the angular rotation velocity.

The use of a normal laboratory centrifuge (having a rotation speed of 3000 - 6000 rpm) enables complete centrifugation of colloids with radii of 30 - 40 nm and greater in a few hours, assuming a density in the range of 2 - 3. For smaller particle sizes, a high-speed centrifuge (up to 15,000 rpm) or an ultracentrifuge is necessary.

The use of centrifugation for the separation of suspended and colloidal particles from water was discussed in detail by Lammers (1967). The technique has so far had few applications in speciation studies. Comparisons of centrifugation and filtration techniques, however, indicate that the former may in some cases be preferable (Beneš and Steinnes, 1974, 1975; Salbu, 1981).

GEL FILTRATION

The technique of gel filtration, or gel chromatography, is especially useful for separating large molecules such as proteins. Gel filtration has been used for the separation of humic compounds in water but ultrafiltration seems to be more appropriate for this purpose (Gjessing, 1976). Gel filtration does not appear to be very promising for trace element speciation studies in water.

COMBINATIONS OF TECHNIQUES

Several papers reporting combinations of different physical and chemical techniques for speciation studies in natural waters have been reported in recent years. Besides the numerous reports on chemical fractionations and determinations following a 450 nm membrane filtration, a variety of alternative schemes have been suggested. Some examples are listed in the following.

Dialysis *in situ* - neutron activation (Beneš and Steinnes, 1974).
Dialysis - centrifugation - ion exchange - electrophoresis-neutron activation (Beneš and Steinnes, 1975).
Ultrafiltration - dialysis - atomic absorption (Guy and Chakrabarti, 1976).
Centrifugation - ion exchange - ultrafiltration - neutron activation (Beneš et al., 1976).
Ultrafiltration - dialysis - chelating resin - anodic stripping voltammetry - atomic absorption (Hart and Davies, 1978).
Centrifugation - anodic stripping voltammetry (Salim and Cooksey, 1979).
Ultrafiltration - chelating resin - anodic stripping voltammetry (Laxen and Harrison, 1981).
Dialysis - chelating resin - filtration - centrifugation - ultracentrifugation (Salbu, 1981).
Ultrafiltration - anodic stripping voltammetry - atomic absorption (Hoffman et al., 1981).

The future development of trace element speciation studies in natural waters will depend on further progress in the application of individual physical and chemical separation techniques. Even more important, however, are the ways in which these techniques are combined in order to characterize the systems involved.

REFERENCES

Beneš, P., 1980, Semicontinuous monitoring of truly dissolved forms of trace elements in streams using dialysis in situ - I. Principle and conditions, Water Res., 14:511.

Beneš, P., Gjessing, E.T., and Steinnes, E., 1976, Interactions between humus and trace elements in fresh water, Water Res., 10:711.

Beneš, P., and Glos, J., 1979, Radiotracer analysis of the physicochemical state of trace elements in aqueous solutions - II. Electromigration method, J. Radioanal. Chem., 52:43.

Beneš, P., and Majer, V., 1980, "Trace Chemistry of Aqueous Solutions", Academia, Prague.

Beneš, P., and Steinnes, E., 1974, In situ dialysis for the determination of the state of trace elements in natural waters, Water Res., 8:947.

Beneš, P., and Steinnes, E., 1975, Migration forms of trace elements in natural fresh waters and the effect of the water storage, Water Res., 9:741.

Beneš, P., and Steinnes, E., 1976, On the use of radionuclides in the study of behaviour and physico-chemical state of trace elements in natural waters, Intern. J. Environ. Anal. Chem., 4:263.

Florence, T.M., and Batley, G.E., 1977, Determination of the chemical forms of trace metals in natural waters, with special reference to copper, lead, cadmium and zinc, Talanta, 24:151.

Gardner, M.J., and Hunt, D.T.E., 1981, Adsorption of trace metals during filtration of potable water samples with particular reference to the determination of filterable lead concentration, Analyst, Lond., 106:471.

Gjessing, E.T., 1976, "Physical and Chemical Characteristics of Aquatic Humus", Ann Arbor Science Publ., Ann Arbor, Michigan.

Guy, R.D., and Chakrabarti, C.L., 1976, Studies of metal-organic interactions in model systems pertaining to natural waters, Can. J. Chem., 54:2600.

Guy, R.D., and Chakrabarti, C.L., 1977, Graphite furnace atomic absorption spectrophotometer as a detector in speciation of trace metals, in: "Environmental Analysis", G.W. Ewing, ed., Academic Press, New York.

Hart, B.T., and Davies, S.H.R., 1978, A study of the physico-chemical forms of trace metals in natural waters and waste waters, Australian Water Resources Council, Technical Paper No. 35.

Hoffmann, M.R., Yost, E.C., Eisenreich, S.J., and Maier, W.J., 1981, Characterization of soluble and colloidal-phase metal complexes in river water by ultrafiltration. A mass-balance approach, Environ. Sci. Technol., 15:655.

Lammers, W.T., 1967, Separation of suspended and colloidal particles from water, Environ. Sci. Technol., 1:52.

Laxen, D.P., and Harrison, R.M., 1981, A scheme for the physicochemical speciation of trace metals in freshwater samples,

Sci. Total Environm., 19:59.

Rona, E., Hood, D.W., Muse, L., and Buglio, B., 1962, Activation analysis of manganese and zinc in sea water, Limnol. Oceanogr., 7:201.

Salbu, B., 1981, New trends in analysis of ground water, Mikrochim. Acta (Wien) II, p. 351.

Salim, R., and Cooksey, B.G., 1979, The analysis of river water for metal ions (lead, cadmium and copper) both in solution and adsorbed on suspended particles, J. Electroanal. Chem., 105:127.

Stumm, W., and Bilinsky, H., 1973, Trace metals in natural waters; difficulties of interpretation arising from our ignorance on their speciation, in: "Advances in Water Pollution Research - Jerusalem - 1972", S.H. Jenkins, ed., Pergamon Press, Oxford.

DISCUSSION: E. STEINNES

E.K. DUURSMA   Studies by Dawson in 1972 (see R. Dawson and E.K. Duursma, 1974, Neth. J. Sea Res., 8:339) showed that the time of equilibration with dialysis is proportional to the hydrated ionic radius (determined using more than 10 radionuclides). This fact could be useful with a technique that you mentioned (for example, having an adsorbent in one compartment and studying the diffusion through the dialysis membrane separating compartments). Thus an indication could be found relating to ionic or complex radii. I note that the explanation for why such a relation exists has still to be given.

E. STEINNES   This seems to be a possibility that should be tested further.

R.F. VACCARO   I wish to comment that experience in our laboratory suggests inequalities in Nuclepore filters with respect to their staining characteristics. One will have to be careful therefore before assigning clean separations of trace metal species to these filters.

E. STEINNES   On the other hand, these filters are known from air partition work to show low blank values for a number of trace elements. It would certainly be useful if someone would look more in detail on Nuclepore filters, with respect to their possible extended use in trace element speciation studies in water.

F.H. FRIMMEL   Have you looked at the concentrations adsorbed outside the cell in your in situ experiments?

E. STEINNES   We have not. Adsorption on the external surface of the dialysis bag is, however, not likely to affect the over-all result of the dialysis *in situ*.

M. BERNHARD   Do you have any problem with growth of micro-organisms (bacteria, microalgae) during your experiments both within the dialysis tubes and on their surface? A microbiological test could show if contamination existed.

E. STEINNES   The experiments that we have carried out so far on dialysis *in situ* were all performed during the winter season when no major growth problem would be expected. We are aware of the fact that this problem might become serious.

J.P. GIESY   Do you have a method to avoid sorption of humic-type compounds to gel permeation materials?

E. STEINNES   I do not have any personal experience in that area, but I think these effects can hardly be avoided.

VOLTAMMETRIC STUDIES ON TRACE METAL SPECIATION IN NATURAL WATERS

PART I: METHODS

Pavel Valenta

Chemistry Department, Institute of Applied Physical
Chemistry, Nuclear Research Center (KFA) Juelich,
Federal Republic of Germany

INTRODUCTION

Heavy and transition metals occur in natural waters both in the dissolved state and bound to the surface layers of suspended particulate matter and sediments. Although the concentrations of toxic trace metals in the dissolved state are rather low, the dissolved state remains important. In and from this state occurs the transfer of toxic metals to and from suspended matter, sediments and aquatic organisms. For these various interactions and, ultimately the toxic effects, the various chemical species in which the trace metals exist in natural waters have a key function.

Some trace metals (Cd, Co, Cu, Hg, Ni, Pb, Tl, Zn, etc.) are well accessible to polarography and voltammetry. With respect to the determination sensitivity, low accuracy risks and low cost requirements, voltammetry is at present one of the most powerful analytical methods for the determination of the dissolved overall levels of toxic metals in natural waters (Nürnberg et al., 1976; Nürnberg, 1979; Mart et al., 1982). Moreover, the substance specificity of polarography and voltammetry provides an informative and versatile approach to speciation studies of dissolved trace metals such as Cd, Cu, Pb and Zn in natural waters at or at least sufficiently close to their rather low, real concentration levels (Nürnberg et al., 1976; Nürnberg and Valenta, 1982).

The selection of the respective voltammetric or polarographic procedure depends on the species concentration level, the nature of the speciation type and its electrochemical properties. In general, two limiting cases can be distinguished; labile complexes and inert complexes. This paper will be devoted to the methodological aspects, while in part II (Nürnberg, this volume), applications and emerging conclusions of general significance for chemical oceanography and chemical limnology of trace metals will be treated.

SPECIATION BY LABILE COMPLEXES

Natural waters contain ligands of various kinds and origins. The inorganic ligands significant for trace metal speciation are Cl^-, OH^-, CO_3^{2-}, HCO_3^- and SO_4^{2-}. They form complexes with heavy metal ions and therefore, in many natural waters, metals are only present in a rather marginal amount as aquocomplexes. In general, the equilibrium between the metal aquocomplex, $M(H_2O)$, and complexes with the inorganic ligands, X, mentioned above, is attained rather rapidly and is shifted towards the complexed metal ion, MX_j, according to the value of the respective stability constant β_{MXj}. The complexation equilibrium and the overall stability constant β_{MXj} are expressed by equations (1) and (2).

$$M(H_2O) + j X \rightleftharpoons MX_j + j H_2O \quad (1)$$

$$\beta_{MXj} = \frac{(MX_j)}{(M)(X)^j} \quad (2)$$

The overall stability constants, β_{MXj} used here, are conditional constants, that is, they refer to the ionic strength of the considered medium and contain, therefore, the activity coefficients of the reactants due to the general salt effects. Thus the β's are related to the concentrations of the reactants of the complex equilibria and not to their activities.

In the following, the metal ions bound in aquocomplexes will be designated simply by M. In general, a sequence of complexes MX_j with different ligand numbers (j = 1,2,3 etc.) is formed and the consecutive stability constant, K_{MXj}, is defined by

$$K_{MXj} = \frac{(MX_j)}{(MX_{(j-1)})(X)} = \frac{k_f}{k_d} \quad (3)$$

which corresponds to the equilibrium between the consecutive complexes, and

$$MX_{(j-1)} + X \underset{k_d}{\overset{k_f}{\rightleftharpoons}} MX_j \quad (4)$$

Here the formation rate constant and the dissociation rate constant are designated by k_f and k_d respectively.

Rapid attainment of the equilibrium given by equation (4) depends both on the formation rate constant, k_f, and the dissociation rate constant, k_d. For most types of complexes, k_f remains large because, in a recombination mechanism, complex formation is predominantly or fully diffusion controlled. Then, for a series of similar complexes the dissociation rate constant, k_d, is more or less inversely proportional to the stability constant K_{MXj}.

Complex equilibrium and the shift of the half wave potential

If the conditions for a rapid attainment of all complex equilibria during the measuring time (about 3s in classical polarography) are fulfilled, and the electrode process itself is reversible, an approach based on the analysis of the reversible polarographic wave can be applied to determine the overall stability constant, β_{MXj}, of the metal complex, MXj, and its ligand number, j. In the case of the reversible reduction of the complex MXj according to

$$MX_j + ne^- \rightleftharpoons M(Hg) + jX \tag{5}$$

both β_{MXj} and j can be obtained from the difference of the half wave potentials of the uncomplexed metal ion $(E_{1/2})_f$ and the complexed metal ion $(E_{1/2})_c$, which depends on the ligand concentration (X) as follows.

$$E_{(1/2)c} - (E_{1/2})_f = \Delta E_{1/2} = -\frac{RT}{nF} \ln (\beta_{MXj}(X)^j) \tag{6}$$

Thus, from the slope of the dependence, $\Delta E_{1/2}$ vs. ln (X) (Figure 1), the ligand number j can be obtained, and from the inter-

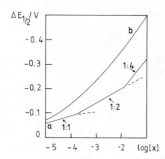

Figure 1. Dependence of the half wave potential on free ligand concentration (X).
    a. Predominant presence of complex species with indicated ligand number in certain ranges of (X)
    b. Simultaneous presence of all possible complexes due to consecutive series of complex equilibria

cept with the Y axis at (X) = 1, the overall complexity constant β_{MXj} can be evaluated.

In the case of consecutive complex equilibria, these quantities are evaluated according to the method of De Ford and Hume (Crow, 1969) using a generalized equation (7) for the shift of the half wave potential, $\Delta E_{\frac{1}{2}}$.

$$\Delta E_{\frac{1}{2}} = -\frac{RT}{nF} \ln \sum_{o}^{m} \beta_j (X)^j \tag{7}$$

From equation (7) follow the Leden-functions,

$$F_o(X) = \sum_{o}^{m} \beta_j (X)^j = \exp\left(-\left(\frac{nF}{RT}\Delta E_{\frac{1}{2}}\right) + \ln \frac{I_m}{I_c}\right) \tag{8}$$

and $$F_j(X) = \frac{F_{(j-1)}(X) - \beta_{(j-1)}}{(X)} \tag{9}$$

where R is the gas constant, T the temperature in °K, n the number of electrons transferred in the electrode process (usually for the trace metals considered, n equals 2), F is the Faraday (96500 C mol^{-1}), I_m is the diffusion current of the uncomplexed metal ion, I_c is the diffusion current of the complexed metal ion and $\beta_o = 1$. For the consecutive stability constant, K_{MXj}, defined by equation (3), it follows that

$$K_j = \frac{\beta_j}{\beta_{(j-1)}} \tag{10}$$

With the aid of these equations, the overall stability constants β_j, the coordination numbers j and the consecutive stability constants K_j of a series of consecutively formed complexes may be evaluated either from a graph or preferably with the aid of computer programs. Accurate data can be obtained for a number of complexes, provided the precision in the measurement of $E_{\frac{1}{2}}$ is ± 1 mV or better.

As the pH of natural waters is usually slightly alkaline (7.5-8), such as is the case for sea water, some mixed complexes containing the ligand X and the hydroxyl ion OH$^-$ may also occur, especially for Cu and Zn.

In the case of the reduction of a mixed complex $ML_x(OH)_y^{(n-y)+}$ according to

$$ML_x(OH)_y^{(n-y)+} + ne \longrightarrow M(Hg) + xL + y\, OH^- \tag{11}$$

the shift of the half wave potential $\Delta E_{\frac{1}{2}}$ is given by

$$\Delta E_{1/2} = -\frac{RT}{nF} \ln K + x \frac{RT}{nF} \ln(L) + 2.3y \frac{RT}{nF} (pH-14) \qquad (12)$$

Thus, at a constant pH, the slope of a plot of $E_{1/2}$ vs. ln (L) yields the value of x and analogously, at a constant ligand concentration (L), the slope of a plot of $E_{1/2}$ vs. pH yields the value of y. The overall stability constant β can be obtained from the intercept of these dependencies with the Y axis as in the previously mentioned case. In the more general but still realistic case of consecutive complex equilibria with more than one ligand, the effects of the various complexes add to the final shift of the half wave potential, $\Delta E_{1/2}$, as given by

$$\Delta E_{1/2} = -\frac{RT}{nF} \ln \left[1 + \sum_{i=1}^{N} \sum_{j=1}^{M} \beta_{M(L_i)_j} (L_i)^j \right] \qquad (13)$$

where L_i with i=1,2 ... N represents various ligands.

Determination of complexation at trace metal levels

The determination of the stability constant and the ligand number by classical polarography is possible at a metal concentration of at least 10^{-5}M which is many orders of magnitude higher than its concentration in natural waters. However, results rather close to reality are to be expected if all major parameters of the water sample, particularly pH and the concentration of major components, are held constant while only the concentration of the metal studied is raised during the determination.

To be sure that the increase in the metal ion concentration will not noticeably affect specific side reactions relevant at the trace level for the equilibrium parameters of the studied complex, it is preferable to work at the lowest possible concentration of the metal studied. Then, the determined conditional equilibrium constant is strictly valid for a given medium, and can be used in the evaluation of the distribution of a given metal among various species in the natural water system. There is another important reason to work at a concentration level close to the natural concentration of the given metal. If the conditional complexity constants of all species are determined, a proof of the validity of the distribution of the given metal among various species in natural waters can be performed in the following manner. Some quantity which depends on the distribution is measured and the result is compared with the computed value of the quantity. If all equilibria have been taken into account, an agreement between the two values is found; otherwise some unknown ligand taking part in the equilibrium was not taken into account. As will be shown later, the reversible half wave potential of the reduction of a given metal ion in a natural water system is such a quantity which can be used for the verification of the calculated distribution.

The need for the lowering of the metal level in speciation studies has led to the application of advanced polarographic and voltammetric methods, enabling one to work at metal concentration levels down to 10^{-9} M, such as at or near the natural concentration level. These methods are: phase sensitive A.C. polarography; differential pulse polarography; and an adapted version of anodic stripping voltammetry, (ASV). In all these cases, quantities equivalent to the half wave potential are measured and the treatment outlined above is applied to get the conditional overall stability constants and the ligand numbers of the labile complexes.

Thus, by employing phase sensitive A.C. polarography, the dominating Cu complexes in coastal water from Bretagne have been identified at their natural level (about 5×10^{-8}M), and their stability constants were determined via $E_{1/2}$-shifts and pH-dependence (Odier and Plichon, 1971) with the aid of equation (12). Cu is present mainly as $CuCl^+$ and $Cu(HCO_3)_2(OH)^-$ in addition to hydrated Cu^{2+}. In the same way, using differential pulse polarography, the composition and the stability constant of the copper morpholine complex $CuL_2(OH)(H_2O)^+$ has been determined in slightly alkaline solution (Narasimhan and Valenta, 1982). This complex is present in effluent waters from thermal power stations if morpholine is used as a pH-regulating, anticorrosion agent. A still more sensitive method, permitting work at the 10^{-9}M level, is based on conventional ASV applying a linear potential scan in the stripping stage (Sipos et al., 1980) (Figures 2-4). Here, the pseudo-polarogram (Bubić and Branica, 1973) is constructed from the heights of ASV-peaks corresponding to a series of cathodic deposition potentials, E_d, adjusted in the potential range of the hypothetical D.C.-polarogram (Figure 4). The measurements are performed over an extended range of the ligand concentration (X), and the relationship of the pseudo-half wave potential $E^*_{1/2}$ vs. ln (X) is evaluated for all consecutive MX_j which can be present in the natural water studied. Using this method, the complex equilibria of Pb(II) carbonates existing in sea water could be studied at a realistic total Pb(II) concentration in artifical sea water containing all significant components of genuine sea water. Two carbonato complexes of Pb(II), $(PbCO_3)^0$ and $Pb(CO_3)_2^{2-}$, have been identified and their conditional stability constants in sea water medium have been determined (see Figure 5). Then, in artifical sea water, the dependence of $\Delta E^*_{1/2}$ on pH has been measured by the outlined procedure to get pseudo-polarograms of the Pb(II) reduction. The action of the carbonate buffer system operative in the oceans has been simulated over an extended pH-range controlling the partial pressure of CO_2 in the N_2 used for deaeration. The measured values of $E^*_{1/2}$ fitted well to the theoretical curve obtained from the sea water model of Pytkowicz and Hawley (1974) with the aid of equation (13), and the validity of this chemical model of sea water, with the conditional stability constants used, was proved. Moreover, it was proved in this manner that, among the labile Pb(II) complexes present in sea water, the rather stable carbonato complexes obviously prevail.

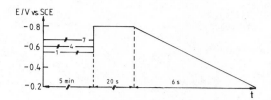

Figure 2. Course and timing of polarizing voltage to obtain a pseudo-polarogram.

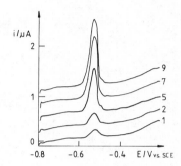

Figure 3. Family of ASV-curves obtained at different cathodic deposition potentials.

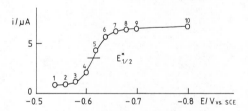

Figure 4. Resulting pseudo-polarogram. Numbers refer to the curves in Figure 3.

The labile complexes are not limited to inorganic ligands present in natural waters. Also, some components of dissolved organic matter (DOM), like certain amino acids, form, with some heavy metals, rather weak labile organic complexes which have been studied in natural water systems by voltammetric methods described above. Thus, the composition and the conditional stability constants of Cd(II)-glycine complexes in sea water have been studied applying differential pulse polarography (Simões Gonçalves et al., 1982). In the De Ford and Hume treatment, modified Leden-functions have been used to correct the competition of labile Cd chlorocomplexes.

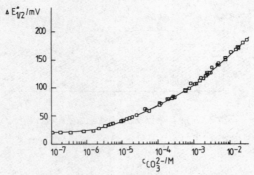

Figure 5. Dependence of the half wave potential shift of the pseudopolarogram for Pb(II) as a function of the carbonate concentration in artificial sea water (total Pb(II) concentration is 6×10^{-9}M).

Two Cd complexes with glycine have been identified, CdL and CdL_2, and their conditional stability constants, β_1 and β_2, were determined from the best fit of the experimental points of the dependence of the Leden-function, $F*_1(L)$, on the free ligand concentration, (L), expressed by

$$F*_1(L) = \frac{F*_o(L) - 1 - \beta'_1(Cl^-) - \beta'_2(Cl^-)^2 - \beta'_3(Cl^-)^3}{(L)}$$

$$= \beta_1 + \beta_2 (L) \qquad (14)$$

where β'_1, β'_2, and β'_3 are the stability constants of $CdCl^+$, $CdCl_2$, and $CdCl_3^-$, respectively.

SPECIATION BY INERT COMPLEXES

The second limiting case of complex equilibria studied by polarographic and voltammetric methods is that of inert metal complexes, formed mostly with chelating organic ligands, L. During the short measuring time, the concentrations of the metal, (M), and of the complex, (MLj), do not change noticeably. Thus, the degree of the complexation can be determined by measuring voltammetrically that concentration (M) which remains uncomplexed at an adjusted concentration (L). Usually the reduction of ML occurs at a sufficiently more negative potential than that of M and both concentrations (ML) and (MLj) can be determined by their respective polarographic or voltammetric responses. However, the reduction of MLj proceeds irreversibly and, therefore, just the measurement of (M) is generally preferred. Two types of investigations for those speciation studies can be performed. The uncomplexed amount of (M) may be actually present in the form of labile inorganic complexes, MeXj, treated in the foregoing section.

(a) In a natural water sample or in a model solution, an appropriate overall concentration of the trace metal to be studied is adjusted. Increasing concentrations of the studied ligand are added and, after the reattainment of the equilibrium, the reversible response due to the concentration of trace metal (MXj) remaining uncomplexed by the inert complex, or a chelate-forming ligand, L, is measured. The complexation degree and the stability constants of the inert complexes ML are determined from the resulting plot for the titration of the metal by the ligand L.

(b) In a natural water sample or in a model solution, an appropriate concentration of the chelator L, to be studied, is adjusted. Then the solution is titrated by a standard metal solution and the response due to the amount of trace metal not complexed in inert species is measured. The amount of the complexed metal and the stability constant of the complex MLj are determined from the resulting plot for the titration of the ligand by the metal ion M.

Both procedures (the titration of the metal by the ligand and the titration of the ligand by the metal) approach the end point of titration from one or from the other side and lead therefore principally to the same result. The advantage of the second approach, (b), is that the equilibrium of the ligand with other components of the DOM is substantially free of perturbation during the titration, whereas, in the first case, the equilibrium has to be attained after every change of the ligand concentration. The advantage of the first procedure, (a), is that the ligand is in a large excess and, thus, side reactions with the ligand will not influence substantially its total concentration.

Titration of the metal by the ligand

The titration procedure starts usually with the adjustment of an appropriate concentration of the trace metal in the natural water sample or the model solution. The metal concentration is always kept rather small and the required level depends on the applied voltammetric method. If differential pulse polarography (DPP) at the dropping mercury electrode (DME) is used, a typical total level of c_M of $10^{-7}M$ or more is required, whereas the usage of differential pulse anodic stripping voltammetry (DPASV) permits one to lower c_M to the $10^{-9}M$ level. The measured voltammetric or polarographic response is due to aquocomplexes, M, and labile complexes, ΣMXj, of the metal studied, and it decreases with increasing concentration of the inert, complex species-forming ligand, L. As the concentration of the ligand (L) changes in general by several orders of magnitude, it is convenient to record the semilogarithmic relationship between the decrease of unchelated (MXj) (in percent) and log (L) (Figure 6).

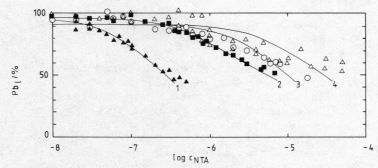

Figure 6. Sequence of titration curves of 4 µg/l total Pb(II) with NTA in (1) 0.55 M NaCl, (2) 0.01 M CaCl$_2$, (3) artificial sea water, (4) Ligurian sea water, pH 7.8.

As an example, the chelation of Pb(II) with NTA$^-$ in sea water, and in model solutions having the ionic strength of sea water (I = 0.7), will be discussed (Raspor et al., 1978; Raspor et al., 1980a).

All relevant equilibria interacting with the metal and the ligand, in chelation of Pb(II)-traces with NTA^{3-} in sea water, are shown in Figure 7. These are the equilibria with the major salinity components of sea water, for example Cl$^-$, CO$_3^{2+}$, OH$^-$, Ca^{2+}, Mg^{2+} and Na$^+$. Also, the protonation equilibrium of NTA^{3-} has to be taken into account, as at pH 8, HNTA^{2-} is the prevailing form. From the trace metals only, Zn(II) has to be taken into account, as it is present at a similar level as Pb(II) and both metals form NTA-chelates of about the same stability. The mass balance equations for the total metal concentration, c_M, and the total ligand concentration, c_L, are expressed by

$$c_{Pb} = (Pb^{2+})+(PbCl^+)+(PbCl_2)+(PbCl_3^-)+(PbOH^+)+(Pb(OH)_2)$$

$$+(PbCO_3)+(Pb(CO_3)_2^{2-})+(PbNTA^-) = (Pb_1)+(PbNTA^-) \quad (15)$$

$$c_L = (NTA^{3-})+(HNTA^{2-})+(NaNTA^{2-})+(CaNTA^-)+(ZnNTA^-)$$

$$+(PbNTA^-) \quad (16)$$

where (Pb$_1$) represents the sum of the concentrations of all labile Pb complexes and of Pb^{2+}, yielding a common reversible voltammetric Pb(II)-response (Figure 6). From the titration plot, the apparent stability constant, β_{app}, can be determined if the total concentration of ligand added for a 50% decrease of the (Pb$_1$) is evaluated. Then according to equation (17) for (Pb$_1$) = (PbNTA$^-$), at this point $\beta_{app} = 1/c_L$ is obtained.

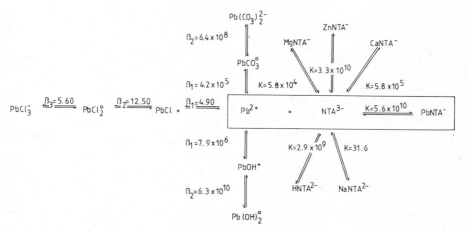

Figure 7. Complex equilibria interacting in chelation of Pb(II) with NTA^{3-} in sea water.

$$\beta_{app} = \frac{(PbNTA^-)}{(Pb_1)c_L} \qquad (17)$$

The real conditional overall stability constant β_{PbNTA^-} for the complex $PbNTA^-$ can be then determined from β_{app} and equations (15) and (16) if all other terms are known (such as the concentrations of these species in sea water and their appropriate stability constants). For model solutions, the equations (15) and (16) reduce to corresponding simpler relations, as less terms have to be taken into account. In Figure 6 the titration has been performed not only for natural and artificial water but also for some model solutions to follow the individual effects of major salinity components on the complexation equilibrium with NTA. Compared with the alkaline earth model solutions, the demand in NTA-concentration for 50% Pb-chelation has, in artificial sea water, further increased about 2 times towards 10^{-5}M NTA. This is due to the accumulative action of all competitive effects of the salinity components on Pb-chelation. For natural seawater the demand of NTA for 50% Pb-chelation has again increased compared with artificial sea water. This increased NTA demand is partially caused by the larger Zn-concentration in the natural water sample and by the competition for the ligand by various other trace metals dissolved in natural sea water.

A similar treatment is applicable for the investigation of the speciation of certain heavy trace metals with amino acids. In this manner, the conditional stability constants of the weaker but still rather inert complexes of Zn, of the type ML and ML_2, with 1-aspartic acid (Sugawara and Valenta, 1981) and with glycine (Simões Gonçalves and Valenta, 1982) were determined using differential pulse polarography.

In principle, the voltammetric measurement always perturbs somewhat the complex equilibrium in the vicinity of the electrode, as the electrode process consumes a minute amount of metal ions. This will induce dissociation of ML_j. This effect can be, however, minimized by having a sufficient excess of (L) over c_M and by selecting an appropriate voltammetric method with small measuring time, for example, preferentially a pulse technique or, at very low c_M-levels, DPASV with small cathodic deposition times. A practical approach is to prove experimentally the absence of noticeable perturbations of the complex equilibria by showing that the reversible response, due to the labile complexed metal ion concentration, is diffusion controlled (Raspor et al., 1978; Raspor et al., 1980a). A particularly good proof for the negligible perturbation of the equilibrium of the studied inert complex is to evaluate its common conditional stability constant, β_{ML_j}, from every point of the relationship in Figure 6. Contrary to the apparent stability constant β_{app}, the constant β_{ML_j} is no more affected by any specific side effects operative in the respective medium but only by the general salt effect due to the existing ionic strength of the medium, for example, 0.7 M for sea water. If the obtained values of β_{ML_j} do not exhibit a systematic trend, but agree within a reasonable random scatter around a mean value, the validity of the outlined procedure is ensured.

It should be stressed that from the viewpoint of aquatic chemistry, the major aim of the outlined procedure is not to determine exactly the overall stability constants of the inert complexes ML_j, but to evaluate directly in the natural medium, via the dependence shown in Figure 6, the respective concentrations of the inert complexes forming ligand, L, which are required to achieve, with a certain heavy metal, a certain chelation degree in the studied water type. The stability constants β_{ML_j} calculated as shown above serve only as an inherent control of the validity and consistency of the procedure.

Titration of the ligand by the metal ion

In this type of investigation, an appropriate concentration of the given inert complex or chelate species, forming ligand L, exists or is adjusted using a natural water sample or a model solution. Increasing concentrations of the studied metal (M) are added and, after the reattainment of equilibrium, the response due to the unchelated remaining trace metal is measured. This response increases with increasing metal concentration to give a line with two or more branches of different slopes (Figure 8) (Shuman and Woodward, 1977).

Supposing that only one complex ML is formed, the following equilibria exist in the studied water sample. In part A of the titration plot, the concentration of the labile, bound, metal ions (M_1) and that of the inert species forming free ligand (L) are

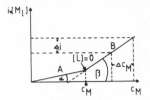

Figure 8. Schematic plot for the titration of ligand forming inert complexes with a heavy metal.

given by equations (18) and (19):

$$(M_1) = (M) - (ML) \qquad (18)$$

$$(L) = c_L - (ML) \qquad (19)$$

Once the intersection point is reached at the total metal concentration c_M, the free ligand concentration (L) decreases to zero. Then in the steeper branch B of the plot the concentration M_1 is given by equation (20)

$$(M_1) = (M) - c_L \qquad (20)$$

and (M_1) increases linearly with the free metal concentration (M). c_L corresponds to the constant total concentration of the inert complexes forming ligand L. If the successive formation of complex species with the ligand L occurs in concentration ranges of the ligand well-separated from each other, their conditional stability constant can be evaluated from the slopes of the respective branches. In this way, complexity constants of the humic acid complexes and the EDTA-complex with Cd(II), in model solutions and in sea water, have been determined (Sugawara et al., 1982)

Determination of the complexation capacity of natural waters

Another interesting application of this type of titration plot is the determination of the complexation capacity of natural waters (Duinker and Kramer, 1977). This is an empirical diagnostic parameter corresponding to the sum of the concentrations of all material which forms inert complexes with the titrant heavy metal. As the concentrations of those components of dissolved organic matter (DOM) able to bind metal ions strongly are frequently rather low, particularly in sea water, the complexation capacity is conveniently determined by voltammetry with its superior sensitivity. Among the heavy metals, Cu and Pb are the preferential choice for complexation capacity determinations as they form inert complexes with many DOM- components and are readily-determined by polarography and voltammetry.

The titration plot in complexation capacity measurements has two
branches as in the foregoing case and its interpretation has been
given already (Figure 9). In the present case, however, the intersection point is of focal interest and not the slopes of the two
branches. The concentration of the titrant heavy metal corresponding
to the intersection point is the complexation capacity of the tested
water type with respect to the titrant heavy metal, such as Pb. In
practice, one observes frequently a rather gradual change of the
slope in the vicinity of the endpoint of the voltammetric titration.
This affects the accuracy of the determined complexation capacity.
In this case, the initial and the final parts of the titration plots
have to be extrapolated to find the right intersection point of the
plot. As the tendency to form inert complexes with the DOM-components
depends on the nature of the heavy metal, the respective complexation capacity will be different for various heavy metals. Generally
the tendency for forming these complexes follows the sequence
Cu>Pb>Cd.

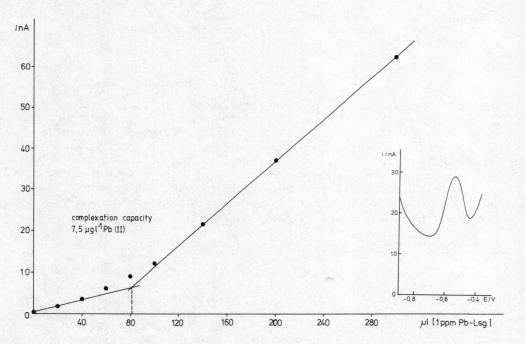

Figure 9. Determination of the complexation capacity of surface
water from the Lake of Constance. On the right: DPASV
voltammogram of Pb. Hanging mercury drop electrode, deposition time 2 min., deposition potential −0.9 V, scan
rate 2 mV s^{-1}, clock time 0.5 s, pulse height 50 mV,
equilibration time for each measurement 10 min.

Knowing the complexation capacity, one can evaluate also the average apparent stability constant K' of all the inert complexes formed with the reference metal, by using the titration curve under the presumption that only complexes of the type ML are formed. K' is defined by equation (21)

$$K' = \frac{\Sigma(ML)}{(M_1) \ \Sigma(L)} \tag{21}$$

where $\Sigma(ML)$ is the sum of the concentrations of all inert complexes and $\Sigma(L)$ is the sum of the free concentrations of all respective ligands. The slope, $tg\alpha_B$, of part B of the dependence, $i=f(c_M)$ (Figures 8 and 9), is defined by equation (22)

$$tg\alpha_B = \frac{\Delta i}{\Delta c_M} = m \tag{22}$$

where Δi is the increment of the current, corresponding to the increment Δc_M, determined from the experimental points of the dependence, being sufficiently distant from the intersection point. On the other side, the slope $tg\alpha_A$, of part A is defined by equation (23)

$$tg\alpha_A = \frac{m(M_1)}{c_M} = \frac{m}{1 + K'(L)} \tag{23}$$

which can be obtained readily from equation (18) and the apparent conditional stability constant K' defined by equation (21).

If the reference metal is already present in the water sample its concentration has to be included in the term c_M. It can be seen from equation (23) that the slope $tg\alpha_A$ changes during the titration. However, at the beginning of the titration, the free ligand concentration $\Sigma(L)$ can be taken as equal to the total ligand concentration c_L, $(\Sigma(L) = c_L)$, and for the slope $tg\alpha_A$ at the beginning of the titration plot, equation (24) holds as a consequence.

$$(tg\alpha_A)_{c_M \to 0} = \frac{m}{1 + K' c_L} \tag{24}$$

Knowing the constant m from the part B of the titration plot, and the complexing capacity c_L from the intersection point of parts A and B, the apparent stability constant K' can be evaluated from the limiting value of the slope $(tg\alpha_A)_{c_M \to 0}$.

Another procedure for the determination of the complexation capacity c_L and the apparent stability constant K', for complexes of the type ML with the reference metal, has been proposed by Ruzić as is reported in Plavšić et al. (1982). He has shown that, for a 1:1 complex, equation (25) is valid.

$$\frac{(M_1)}{c_M - (M_1)} = \frac{(M_1) + 1/K'}{c_L} \qquad (25)$$

By plotting the left side of equation (25) vs. M_1, a straight line is obtained with the slope $1/c_L$. The intersection of this line with the y-axis at $(M_1) \to 0$ gives the value of $1/K'c_L$ according to equation (26).

$$\left(\frac{(M_1)}{c_M - (M_1)}\right)_{(M_1) \to 0} = \frac{1}{K'c_L} \qquad (26)$$

Thus, by knowing the complexation capacity c_L from the slope of equation (25), one can obtain also the apparent stability constant K'.

Kinetics and mechanism of the chelation of heavy metals in trace amounts

The voltammetric procedure outlined can also be applied to kinetic studies of chelation in natural waters. In this case, the time dependence of the concentration of unchelated heavy metal is followed by DPASV or ASV after adding the organic ligand L. As equal concentrations of the ligand and of the trace metal are adjusted, second order kinetics are operative. The formation rate constant, k_f, is evaluated from analysis of the time dependence of the concentration of unchelated heavy metal. As an example, the determination of the formation rate constant of the Cd(II)-EDTA chelate in North Sea water (Heligoland) by ASV is shown (Raspor et al., 1977) (Figure 10). For comparison, the time dependence in model solutions, 0.59 M NaCl, 0.01 M $CaCl_2$, 0.0536 M $MgCl_2$, and in artificial sea water, have been determined also. The model solutions correspond to the major salinity constituents of sea water. The experimental results were evaluated according to the expression for a homogeneous, second order reaction,

$$k_f t = \frac{c_i - c_t}{c_i c_t} \qquad (27)$$

where c_i is the initial concentration (the same for both reactants) and c_t is the concentration of the reactants at time t. The rate constant k_f is determined from the slope of the best least-square-fit of equation (27). From the values of the formation rate constants it could be shown (Raspor et al., 1977; Raspor et al., 1980b) that the chelation mechanism of heavy metal traces, such as Cd, Pb

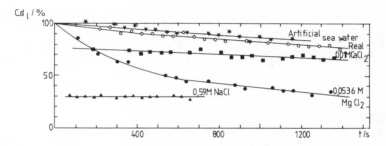

Figure 10. Time dependence of unchelated Cd(II) concentration at a 1:1 concentration ratio of Cd(II) and EDTA and at an initial concentration of 5×10^{-7} M in different model solutions of salinity components and in North Sea water (Heligoland).

and Zn in the sea, is not a simple recombination mechanism between, for example, hydrated Cd^{2+} and ligand L^-, but rather a ligand exchange mechanism with the existing chelates of alkaline earth metals.

The initial concentrations c_i of the heavy metal, and of the chelating ligand L, and the measuring time t of the voltammetric response have to be selected to fit to the kinetic window of the applied method (Raspor et al., 1980b). For a range of formation rate constants k_f, the reaction half-times $t_{1/2}$ resulting for the corresponding initial reactant concentration c_i, presuming equal values for both reactants and a consequent second order kinetics, can be calculated according to

$$t_{1/2} = 1/c_i k_f \tag{28}$$

Depending on the parameters used in DPASV or in DPP, the values of k_f which can be determined lie in the range of 10^2 to 10^6 1 $mole^{-1}s^{-1}$, if the initial concentrations are adjusted in the range 10^{-6} to 10^{-8}M.

CONCLUDING REMARKS

The potentialities of polarography and voltammetry in studies of trace metal speciation in natural waters have not yet been fully exploited. Thus, the use of solid electrodes, such as the rotating gold electrode, will extend the studies to metals not amenable to determination by the mercury electrode, metals such as Hg (Sipos et al., 1979). A further promising aspect is the combination of voltammetry with a separation technique, such as high performance liquid chromatography (HPLC) using a voltammetric detector, to determine

the stability constants of chelates formed with given fractions of
the DOM. In this way, the final task of the elucidation of the
speciation of trace metals dissolved in natural waters, by well-
defined organic ligands expected in natural waters, could be pur-
sued further.

REFERENCES

Bubić, S., and Branica, M., 1973, Voltammetric characterization of
    ionic state of cadmium present in sea water, Thalassia Jugosl.,
    9:47.
Crow, D.R., 1969,"Polarography of Metal Complexes",Academic Press,
    London.
Duinker, J.C., and Kramer, C.J.M., 1977, An experimental study on
    the speciation of dissolved zinc, cadmium, lead and copper in
    River Rhine and North Sea water, by differential pulsed anodic
    stripping voltammetry, Mar. Chem., 5:207.
Mart, L., Rützel, H., Klahre, P., Sipos, L., Platzek, U., Valenta,
    P., and Nürnberg, H.W., Comparative studies on the distribution
    of heavy metals in the oceans and in coastal waters, in: "Trace
    Metals in Sea Water", C.S. Wong, ed., Plenum Press, New York,
    in press.
Narasimhan, S.V., and Valenta, P., Speciation of copper in the pre-
    sence of morpholine in alkaline solution studied by d.c. and
    differential pulse polarography, J. Electroanal. Chem., in
    press.
Nürnberg, H.W., 1979, Polarography and voltammetry in studies of
    toxic trace metals in man and his environment, Sci. Total En-
    vironm., 12:35.
Nürnberg, H.W., in: this volume.
Nürnberg, H.W., and Valenta, P., Potentialities and applications
    of voltammetry in chemical speciation of trace metals in the
    sea, in: "Trace Metals in Sea Water", C.S. Wong, ed., Plenum
    Press, New York, in press.
Nürnberg, H.W., Valenta, P., Mart, L., Raspor, B., and Sipos, L.,
    1976, Applications of polarography and voltammetry to marine
    and aquatic chemistry. II. The polarographic approach to the
    determination and speciation of toxic metals in the marine
    environment, Fresenius' Z. Analyt. Chem., 282:357.
Odier, M., and Plichon, V., 1971, Le cuivre en solution dans l'eau
    de mer: forme chimique et dosage. Etude par polarographie
    à tension sinusoïdale surimposée, Anal. Chim. Acta, 55:209.
Plavšić, M., Krznarić,D., and Branica, M., Determination of the
    apparent copper complexing capacity of seawater by anodic
    stripping voltammetry, Mar. Chem., in press.
Pytkowicz, R.M., and Hawley, J.E., 1974, Bicarbonate and carbonate
    ion-pairs and a model of sea water at $25^\circ C$, Limnol. Oceanogr.,
    19:223.
Raspor, B., Nürnberg, H.W., Valenta, P., and Branica, M., 1980a,

The chelation of lead by organic ligands in sea water, in: "Lead in the Marine Environment", M. Branica and Z. Konrad, eds., Pergamon Press, London.

Raspor, B., Nürnberg, H.W., Valenta, P., and Branica, M., 1980b, Kinetics and mechanism of trace metal chelation in sea water, J. Electroanal. Chem., 115:293.

Raspor, B., Valenta, P., Nürnberg, H.W., and Branica, M., 1977, Application of polarography and voltammetry to speciation of trace metals in natural waters. II. Polarographic studies on the kinetics and mechanism of Cd(II)-chelate formation with EDTA in sea water, Thalassia Jugosl., 13:79.

Raspor, B., Valenta, P., Nürnberg, H.W., and Branica, M., 1978, The chelation of cadmium with NTA in sea water as a model for the typical behaviour of trace heavy metal chelates in natural waters, Sci. Total Environm., 9:87.

Shuman, M.S., and Woodward, G.P. Jr., 1977, Stability constants of copper-organic chelates in aquatic samples, Environ. Sci. Technol., 11:809.

Simões Gonçalves, M.L.S., and Valenta, P., Determination of the stability constants of zinc(II)-glycine complexes in sea water by differential pulse polarography and by potentiometry, J. Electroanal. Chem., in press.

Simões Gonçalves, M.L.S., Valenta, P., and Nürnberg, H.W., 1982, Voltammetric and potentiometric investigations on the complexation of Cd(II) by glycine in sea water, J. Electroanal. Chem., 132:357.

Sipos, L., Golimowski, J., Valenta, P., and Nürnberg, H.W., 1979, New voltammetric procedure for the simultaneous determination of copper and mercury in environmental samples, Fresenius' Z. Analyt. Chem., 298:1.

Sipos, L., Valenta, P., Nürnberg, H.W., and Branica, M., 1980, Voltammetric determination of the stability constants of the predominant labile lead complexes in sea water, in: "Lead in the Marine Environment", M. Branica and Z. Konrad, eds., Pergamon Press, London.

Sugawara, M., and Valenta, P., 1981, Voltammetric studies on the speciation of trace metals by amino acids in sea water, Rapp. Comm. Int. Mer Médit., 27:9.

Sugawara, M., Valenta, P., and Nürnberg, H.W., 1982, Voltammetric studies on the complexation of Cd(II) by humic acid in sea water, J. Electroanal. Chem., in press.

DISCUSSION: P. VALENTA

G. BATLEY   The measurement of complexing capacity in natural systems is not always straightforward. In a recent study of the copper complexing capacity of fibrillar colloids by ASV, we observed, with added copper, the appearance of a new adsorption wave on the anodic side of the copper reduction wave. This phenomenon was due to the copper(I)-fibril complex formed by the reduction of Cu(II) through mixing.

P. VALENTA   It is true that such complexation can arise but, as voltammetry offers a signal which is species sensitive, such distinctions can be more readily detected by it than by other methods.

F. FRIMMEL   Do you see any experimental possibility of getting information about the complexation competition of the heavy metals originally present in natural samples as compared to the reference metal used for the determination of complexation capacity?

P. VALENTA   It depends on the type of water. In sea water, due to a large excess of alkaline earths, such as calcium and magnesium, all ligands forming inert complexes will be bound presumably to calcium and magnesium, with the traces of heavy metals present having no influence on the complexation capacity. In fresh waters, however, such information is possible in principle. In simple cases, the equilibrium constant of the ligand exchange reaction can be obtained from the titration curves as in my Figure 8 and equation (24).

Y.K. CHAU   Do you have any thermodynamic definition (such as in terms of conditional stability constants) to show what type of ligands your complexation capacity measurement can measure?

P. VALENTA   Yes. From the slope of the part A of the titration curve (see my Figure 8), the average conditional complexity constant can be evaluated. Check my reference to Sugawara et al., 1982.

M. BRANICA   By a fast mixing procedure (or a short drop time) can one exclude a kinetic addition to determinations of complexation capacity? For a practical approach, it seems very important to avoid any addition of Hg^{++} to titrated samples for determination of complexation capacity.

P. VALENTA   In the case of the rotating electrode, the increase of the rotation speed can sometimes (but not generally) help to eliminate the kinetic component. The use of short drop time has no influence when the pulse technique is used, such as in the differential pulse mode. The addition of Hg salts is of course not allowed.

COMPLEX FORMATION IN SOLUTION AND IN HETEROGENEOUS SYSTEMS

F. H. Frimmel

Institut für Wasserchemie und Chemische Balneologie
Technische Universität
München, Federal Republic of Germany

INTRODUCTION

Tragedies like the Minamata disease show that trace metals, according to their speciation, can have great effects on human life. However, despite its impact on man and nature, the importance in aquatic ecosystems of complex formation is incompletely understood. Therefore, detailed information is needed on the occurrence and on the properties of metal species in the aquatic environment, with surface waters being of particular interest.

THE GENERAL SITUATION IN SURFACE WATERS

Analytical methods for trace element speciation in surface waters have to be considered in connection with the natural concentration ranges. Heavy metals, and most of the organic and inorganic ions and molecules which can act as ligands, have been measured in aqueous systems for several years. Table 1 shows typical concentration values which are found in the River Rhein (stream km 915; Waal/ Netherlands; Rijncommissie, 1981). They are compared with recommended standards and maximum concentrations for drinking water (Europäische Gemeinschaften, der Rat, 1980). In addition there are the values for tap water of München as a typical pre-alpine ground water with calcium, magnesium and bicarbonate as dominating ions.

In fresh water the alkaline ions are as high as 7 mmol/l, the alkaline earth ions are up to 3.2 mmol/l and dissolved heavy metals add up to concentrations of some µmol/l. Additionally there is a significant amount of heavy metals bound in particulate form in the

Table 1. Chemical parameters of the River Rhein (R) and tap water of München (MUC) and, also, the recommended drinking water standards of the European Community (EC).

	R mean	R max.	MUC	EC (max)
Temperature (°C)	12.0	21.0	9.8	12 (25)
pH value	7.6	8.0	7.5	6.5-8.5
electrical conductivity (µS/cm)	838	1210	482	400
Na^+ (mg/l)	82	156	2,8	20
Ca^{2+} (mg/l)	79	95	76,8	100
Mg^{2+} (mg/l)	11.7	14.3	23.0	30
Cl^- (mg/l)	160	272	5.0	25
SO_4^{2-} (mg/l)	70	92	21.0	25
HCO_3^- (mg/l)	157	186	316	
HPO_4^{2-} (mg/l)	1.21	2.06	<0.04	0.27
Fe (mg/l)			0.02	0.05 (0.2)
Fe* (mg/l)	1.05	2.60		
Cu (µg/l)	7	9	7.6	100
Cu* (µg/l)	11	24		
Zn (µg/l)	37	65	219	100
Zn* (µg/l)	84	130		
Cd (µg/l)	0.4	1.1	0.2	(5)
Cd* (µg/l)	1.0	3.1		
Hg (µg/l)	0.1	0.3	<0.01	(1)
Hg* (µg/l)	0.2	1.0		
Pb (µg/l)	3	8	3.5	(50)
Pb* (µg/l)	15	34		
DOC (mg/l)	5.0	7.5	0.3	

* dissolved plus particulate

River Rhein (twice to five times the dissolved amount).

As ligands, one must consider inorganic ions in a concentration range of 10 mmol/l. Organic ligands tend towards about 50% of the total dissolved organic carbon (DOC) in the River Rhein. This portion is due mainly to humic substances (Sontheimer and Gimbel, 1977). The amount of DOC decreases as the run-off goes down, and, the less polluted a surface water system is, the higher the percentage of humics within the total amount of DOC.

Synthetic ligands have to be taken into account depending on special situations in surface waters which get polluting effluents (such as from sewage plants,etc.). There is some experience relating to the use of nitrilotriacetate (NTA) in detergents in Canada and Sweden (Woodiwiss et al., 1979). The situation which might be expected in central Europe points to a maximum concentration of a few μmol/l for that ligand in river water (Golterman et al., 1976).

Of course, there are great differences to be expected for the concentrations of metals and ligands in different surface waters, depending on natural and industrial influences like seasons, meteorological conditions, waste water effluents, etc.. However, experience shows that the concentration range which is important for coordination chemistry in fresh waters is fairly well reflected by the numbers above. They can, therefore, be used for equilibrium calculations and for the validation of special experimental results.

EQUILIBRIUM CONCEPT

The interaction of metals and ligands leads only in a few cases to complexes which are so stable that they can be isolated and determined as such (e.g.,alkyl lead, alkyl mercury and mercury-thio compounds). For the most part, the coordination compounds can be quantitatively characterized only by their equilibrium reactions.

The general situation for a specific metal ion (Me) in an aqueous system is described in Figure 1. The hydrated metal ion (Me^{2+} aq.) reacts with the ligand L to form MeL. The equilibrium constant K_1 is defined according to Sillén and Martell (1964, 1971). Side reactions are the protonation of the ligand L which leads to HL, and the complexation of Me with other ligands L', characterized by the equilibrium constants K_2 and K_1' respectively. Me forms, with the ligands X and X', compounds MeX and MeX' which precipitate and are characterized by the solubility products (K_{so}). Analogous reactions have to be considered for all other metals present in natural aqueous systems.

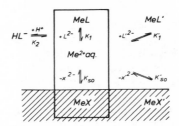

Figure 1. General model for complex formation in aqueous systems.

This will lead to an additional competition for the ligands L and L'. The simplified model is based only on the formation of 1:1 complexes or compounds. Polynuclear species can be taken into account by their gross constants (β) but are mostly omitted.

This scheme is used in most thermodynamic equilibrium concepts (Vuceta and Morgan, 1978; Stumm and Morgan, 1981). The most important parameters for a natural water system are used together with stability constants and solubility products to calculate an equilibrium model. This approach to describing waters is certainly a most useful one and has been extended to kinetic and extrathermodynamic considerations (Hoffmann, 1981). However, the quality of the model depends strongly on the available basic information about the system to be described; very often, analytical and physico-chemical data are insufficient. Kinetic correlations may be poor as well, especially for heterogeneous systems. This leads to a marked uncertainty of calculated values of metal species even for fairly simple systems (Stolzberg, 1981). Therefore, the analytical approach to metal speciation in surface waters with the aim of making the species quantitatively measurable is of pressing importance at the present time.

CLASSIFICATION

Metal ion species can be classified into three main groups relevant to surface waters: dissolved; particulate; and biologically-bound species (Table 2).

The separation of dissolved and particulate substances can be

Table 2. Classification of metal species in surface waters.

    I. Dissolved species
          Hydrates (ionic)
          Small complexes
          Chelates
          Metal organic compounds

    II. Particulate species
          Sparingly soluble compounds
          Bound by adsorption
          Bound by ion exchange
          Bound by complexation (chelation)

    III. Species bound to biological systems
          Bound by adsorption
          Incorporated

done by membrane filtration (0.45 μm). Though not free of random effects, the method is generally accepted in water analysis. However, problems arising from colloidal substances (size range of 1 to 100 nm) have to be kept in mind. Moreover, many species may belong to more than one group of the classification scheme.

DIRECT EXPERIMENTAL APPROACH

There are very few analytical methods in routine analysis which are suited for direct metal speciation. Therefore, physico-chemical methods and chemical reactions are often used to differentiate between the various forms of metals. Analytical methods with low detection then can be applied successfully. In general, one does not get information about the molecular structure of the coordination compound in this manner; rather, one gets information about the reaction ability under defined conditions. One has to bear in mind, however, that the preparatory operation may include significant influences on the natural system being probed and on its equilibria. The decision on whether the results are still useful or not depends on the possibility of getting quantitative information about the disturbance. For comparative results it may be sufficient to make sure that the reactions are performed under the same conditions.

The determination of the chemical forms of trace metals in natural waters has been reviewed by several authors (Florence and Batley, 1977; Schwedt, 1981). Some typical examples, with the methods used, are listed in Table 3. The main reaction and the procedure employed lead, together with the applied analytical method, mostly to operationally-defined results (Skogerboe et al., 1980). The cited results reflect the major aims of the work and give information about general speciation, the distribution of metals according to ligands (inert, labile, ionic, organically-bound), mobilization and the complexation capacities of ligands. Some well-defined conditions even lead to stability and rate constants. In addition, a simple characterization is given for the system under investigation (see also the list of abbreviations at the end of Table 3).

Titration Methods

Most of the determinations on solutions are based on titrations. The unknown sample is titrated with a defined solution of a special metal or vice versa. The reaction can be controlled by electrochemical methods. Polarographic and voltammetric determinations lead to a speciation of labile and electrochemically inert forms. However, the results can include some disturbances caused by irreversible reactions and sorption of natural polyelectrolytes on the electrode surfaces (Hanck and Dillard, 1977b; van Leeuwen, 1979). Detailed information about the electrochemical properties of the system can be derived from the shape of the voltammogram and from the application

Table 3. Methods for the determination of chemical forms of trace metals.

Reaction	Metal	Analyt. Method	Result	Ref.
Titration	Cu	ASV, DPP	K, rate const. (FA, HUS, L)	/14-16/
	Pb	ISE	K (FA)	/17/
	Fe, Pb	Polarogr. ASV	Speciation (FA, HUS)	/18/
	Mn, Fe, Cu, Zn, etc.	ASV, DPP, ISE	K, interaction (carbonate, FA, HUS, Gly)	/19-22/
Titration Computation	Cu	ISE, VIS	Equilibrium function, CC, speciation (FA, NAT)	/23-25/
Complexation	Co	DPP	CC (NAT)	/26/
Ultrafiltration	Mn, Fe, Cu, Cd, Pb	ASV, AAS	Distribution, CC, K (NAT)	/27/
Ultrafiltration Titration	Cu	ASV, BIO	Complexation, bio effect (NAT)	/28/
Ultrafiltration Dialysis	Mn, Fe, Cu, Cd, Pb	ASV, ISE, AAS, FLU	Speciation: labile/inert, molecular size, K, CC, distribution, complexation (FA, HUS, NAT)	/20,27, 29,30/
Ultrafiltration Gel chromatog.	Al, Fe, Cu, Zn, Pb	UV, VIS, AAS	MW distribution (FA, HUS)	/31/
Ion exchange	Co, Ni, Cu, Zn, Cd, Pb	ASV, AAS	Speciation: labile/inert, classification (HUS, NAT, L)	/32-35/
Adsorption (Polystyrene)	Fe	AAS, DPP	Speciation: non labile (FA)	/36-38/
Adsorption (Bentonite, MnO_2, $Fe(OH)_3$, humus)	Fe, Cu, Zn, Cd, Pb	AAS, ASV	Distribution, K (carbonate, HUS, tannic acid)	/39,40/

Table 3. (continued)

Reaction	Metal	Analyt. Method	Result	Ref.
Solubilization (Hydroxide, carbonate, phosphate)	Cu	AAS	Mobilization, CC (L)	/41-43/
Solubilization (Oxides, NTA, hydroxides, ion exchanger)	Fe, Ni, Cu, Cd, Zn, Pb	AAS	Mobilization (sediment, L)	/44-46/
Centrifugation Ultrafiltration Ion exchange	Mn, Fe, Cu, Cd, etc.	UV, VIS, NAA	Association (HUS)	/33/
Extraction	Cr, Mn, Co, Ni, Cu, etc.	AAS	Me-content (FA, HUS, sediment)	/47/
Digestion	Cu, Zn, Cd, Pb	ASV	Classification: ionic/organically bound	/34,48/
Enrichment Reduction Hydrolysis Complexation	Fe	Mößbauer, EPR	Coordination sphere (HUS)	/49,50/

List of abbreviations:

AAS	atomic absorption spectroscopy
ASV	anodic stripping voltammetry
BIO	bioassay
CC	complexation capacity
DPP	differential pulse polarography
EPR	electron paramagnetic resonance
FA	fulvic acid
FLU	fluorescence spectroscopy
Gly	glycine
HUS	humic substances
ISE	ion selective electrode
K	stability constant
L	model ligand
MW	molecular weight
NAA	neutron activation analysis
NAT	natural water sample
UV	ultraviolet range spectroscopy
VIS	visible range spectroscopy

of different independent methods. Variation of measuring parameters (for example, drop size, pulse amplitude) or the use of cyclic voltammetry are well suited to prove the validity of the results.

The calculation of stability constants cannot always rely on the assumption that polarographic or voltammetric peak currents are only due to "free" metals. However, measurements under identical conditions can lead to a useful comparison of different samples. The detection limits of the voltammetric determination of complexation capacities are in the range of $\mu mol/l$.

The use of ion selective electrodes seems to be applicable even at lower concentrations. A lack of identity with results gained by other methods suggests that there are also some problems here which are not well understood.

Separation Methods

The group of determinations based on separation techniques includes ultrafiltration, dialysis, ion-exchange and adsorption. Ultrafiltration gives a realistic distribution pattern of the trace metals in natural complexes of different molecular sizes. Problems arise from disturbance of equilibria and from precipitation, both of which occur within ultrafiltration cells during increases in concentration. Changes of the pH value and of the total ion content may lead to a number of additional influences which cannot be quantified. There is no possibility to get information on the coordination sphere of the trace metals. Determining the metal concentrations with the most sensitive methods like NAA and AAS leads to detection limits of less than a $\mu g/l$.

Ion-exchange experiments and ASV allow a distinction of very-to-slowly labile and inert species, with a number of intermediate steps depending on the technique used. Again, influences on the metal/ligand equilibria may become important for the ion-exchanger acts as a competitive ligand. The results depend strongly on the chemical form of the resin, its capacity and the composition of the aqueous sample. The behaviour of complex species with different gross charges, and the interaction of ligands with metals strongly bound to the resin, have to be controlled carefully. The observable effects reach down to metal concentrations smaller than ppb.

Adsorption reactions as indications of different metal species lead mostly to rather complicated heterogeneous systems. They can only be interpreted clearly in the cases of relatively simple model experiments.

Solubilization Methods

Solubilization experiments rely mainly on classical heterogeneous reaction systems. The dissolved ligands react with a scarcely soluble compound of the metal of concern. All reaction parameters have to be kept defined and constant. Therefore the method is only useful for model experiments. Natural ligands have to be isolated and pretreated by standardized procedures. The irreproducible quality of the solid compounds, uncontrolled side reactions which influence the solubility, and equilibria with complicated kinetic behaviour are the main reasons for the difficulties in interpreting results.

The reaction of some dissolved model ligands and solid Cu-phosphate (Figure 2) under standardized conditions show clearly that the yield is not always a linear function of the ligand concentration. Sites with different reactivity can explain that effect.

Blank values of a few µmol/l are typical for these methods and indicate a relatively high detection limit. No reasonable results can be obtained for raw water samples and any comparison of solid compounds with natural sediments has to have highly random features.

Chemical Reaction Methods

Special chemical reactions, as pretreatment for a determination, lead to a rough speciation of trace metals in natural samples. Extractions, competitive complexation (such as by EDTA or dithizone), and different digestion methods (UV irradiation, acid treatment, oxidation, etc.) are the most common principles. The results can only be interpreted as operational values and do not include any information about the molecular structure. Detection limits are in general as good as the applied analytical method. However, a chemical reaction can lead to an increasing blank for the overall determination.

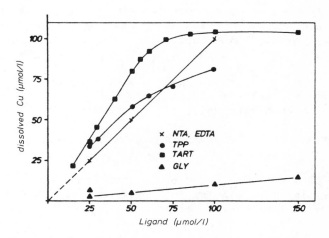

Figure 2. Solubilization of Cu-phosphate in solutions of different ligands (Frimmel et al., 1980).

Spectroscopic Methods

Information about the coordination sphere (such as that of Fe and Cu in aqueous humus) can be gained by advanced structural determination techniques like Mössbauer spectroscopy, NMR and EPR. The interpretation of natural samples suffers from a lack of sufficient comparable substances (Goodman, 1980). Furthermore, a fundamental problem is that spectroscopic methods cannot be applied to naturally-occurring concentration levels; the necessity to concentrate the sample poses the general question of whether the structural information gained about it is valid for the original dilute solution.

REFERENCE METALS

The applied methods (Table 3) could have been listed also according to the metals of concern. Two groups would be dominant:

a. methods which are suitable for determining species of <u>all</u> relevant metals present in the sample. These methods give the real natural concentrations but have to cope with the low values for the individual species.
b. methods which can only use one or two metals as reference. This implies a simplification of the system, for all metal/ligand interactions are represented by the equilibria of the reference metals. Cu forms complexes with relatively high stability constants, when compared to most of the other trace metals. Therefore Cu is very often used as a reference element.

The methods of group "b" include a marked chemical interference which is caused by the reaction of the reference metal with the original sample. The complexation capacity as a final result can only be seen in connection with the specific reference metal, and should more precisely be called "<u>free</u> complexation capacity". All other present metal/ligand equilibria, which are of course in competition reactions with the reference metal, are mostly unknown and leave the final interpretation uncertain.

A pretreatment of natural samples with the aim of isolating all trace and earth alkaline metals from the ligands, and applying the ligands only in the alkaline (or anionic) form to the reference metals, may lead to a "<u>total</u> complexation capacity". In this case, not the determination reaction but, instead, the pretreatment step includes most of the possible errors.

CONCLUSION

A quantitative description of all or at least some of the most important metal species in aquatic systems needs a strong basis of theoretical understanding and experimental determinations. The lack of reliable data on the concentrations, on the identities of organic ligands, and on the metal/ligand and metal/surface interactions points out the importance of chemical analysis and the investigation of chemical reactions.

Different analytical approaches to trace metal speciation in aquatic systems show that there is no method for general application at the moment. As is often the case in the study of intricate natural systems, one has partly to abandon the need to determine well-defined chemical structures. Therefore, most of the determinations lead to results which are operationally-defined and have to be interpreted carefully. There have to be applied different independent methods to the same sample to reach a certain generalization and a theoretical explanation. In practice, complicated situations in nature must be simplified by trying to keep special parameters constant or by splitting the system into smaller and better-defined units. Model systems and model reactions are of great importance for the validation of results.

Some applied methods need a rather complicated pretreatment of the samples. This situation has to be characterized quantitatively to lead correctly from experimental results to conclusions on the natural system. One may also approach the natural conditions experimentally, step by step, in an effort to improve a model.

REFERENCES

Allen, H.E., and Boonlayangoor, C., 1978, Mobilization of metals from sediment by NTA, Verh. Internat. Verein. Limnol., 20:1956. /44/

Batley, G.E., and Florence, T.M., 1976, A novel scheme for the classification of heavy metal species in natural waters, Anal. Lett., 9:379. /34/

Beneš, P., Gjessing, E.T., and Steinnes, E., 1976, Interactions between humus and trace elements in fresh water, Water Res., 10:711. /33/

Blutstein, H., and Smith, J.D., 1978, Distribution of species of Cu, Pb, Zn and Cd in a water profile of the Yarra River estuary, Water Res., 12:119. /48/

Buffle, J., Deladoey, P., and Haerdi, W., 1978, The use of ultrafiltration for the separation and fractionation of organic ligands in fresh waters, Anal. Chim. Acta, 101:339. /31/

Buffle, J., Greter, F.L., and Haerdi, W., 1977, Measurement of complexation properties of humic and fulvic acids in natural waters with lead and copper ion-selective electrodes, Anal. Chem., 49:216. /21/

Buffle, J., Greter, F.L., Nembrini, G., Paul, J., and Haerdi, W., 1976, Capabilities of voltammetric techniques for water quality control problems, Z. Analyt. Chem., 282:339. /18/

Campbell, P.G.C., Bisson, M., Gagné, R., and Tessier, A., 1977, Critical evaluation of the copper (II) solubilization method for the determination of the complexation capacity of natural waters, Anal. Chem., 49:2358. /41/

Dietz, F., and Frank, H.-D., 1977, Anwendung anorganischer und organischer Austauschermassen als Bodenkörper für Remobilisierungsversuche, Z. Wasser Abwasser Forsch., 10:109. /46/

Dietz, F., Frank, H.-D., and Koppe, P., 1975, Bestimmung des Remobilisierungspotentials von Wässern gegenüber Schwermetallverbindungen, Z. Wasser Abwasser Forsch., 8:104. /45/

Ernst, R., Allen, H.E., and Mancy, K.H., 1975, Characterization of trace metal species and measurement of trace metal stability constants by electrochemical techniques, Water Res., 9:969. /22/

Europäische Gemeinschaften, der Rat: Richtlinie des Rates, vom 15. Juli 1980, über die Qualität von Wasser für den menschlichen Gebrauch. Amtsblatt der Europäischen Gemeinschaften Nr. L 229/11 (30.8.80). /2/

Figura, P., and McDuffie, B., 1977, Characterization of the calcium form of Chelex-100 for trace metal studies, Anal. Chem., 49:1950. /35/

Figura, P., and McDuffie, B., 1980, Determination of labilities of soluble trace metal species in aqueous environmental samples by anodic stripping voltammetry and chelex column and batch methods, Anal. Chem., 52:1433. /32/

Florence, T.M., and Batley, G.E., 1977, Determination of the chemical forms of trace metals in natural waters, with special reference to copper, lead, cadmium and zinc, Talanta, 24:151. /11/

Frimmel, F.H., 1979, Polarographische Untersuchungen von Kupferkomplexen, Z. Wasser Abwasser Forsch., 12:206. /16/

Frimmel, F.H., 1980, Komplexierung von Metallionen durch Gewässerhuminstoffe. III. Modellentwicklung und Gewässerbezug, Z. Wasser Abwasser Forsch., 13:217. /38/

Frimmel, F.H., and Dietz, F., 1980, Zur summarischen Bestimmung der Komplexbildungsfähigkeit. II. Umsetzungen des Kupferphosphatbodenkörpers mit definierten Liganden, Z. Wasser Abwasser Forsch., 13:12. /43/

Frimmel, F.H., Geywitz, J., and Quentin, K.-E., 1981, Modelle für Eisen-Huminstoffkomplexe, Vom Wasser, 57:185. /37/

Frimmel, F.H., and Niedermann, H., 1980, Komplexierung von Metallionen durch Gewässerhuminstoffe. I. Ein Braunwassersee als Huminstofflieferant, Z. Wasser Abwasser Forsch., 13:119. /36/

Gamble, D.G., Underdown, A.W., and Langford, C.H., 1980, Copper (II) titration of fulvic acid ligand sites with theoretical, potentiometric, and spectrophotometric analysis, Anal. Chem., 52:1901. /23/

Golterman, H.L., and van Weelden, R.H., 1976, Schatting von NTA-concentratie in Nederlands oppervlaktewater bij vervanging van fosfaten uit wasmiddelen. H_2O, 9:57. /5/

Goodman, B.A., 1980, Exchange of comments on the simulation of electron paramagnetic resonance spectra of a copper – fulvic acid complex, Anal. Chem., 52:1770. /53/

Guy, R.D., and Chakrabarti, C.L., 1975, Analytical techniques for speciation of trace metals, in: "Proceedings of Inter. Conf. on Heavy Metals in the Environment", Toronto, Canada. /20/

Guy, R.D., Chakrabarti, C.L., and Schramm, L.L., 1975, The application of a simple chemical model of natural waters to metal fixation in particulate matter, Can. J. Chem., 53:661. /39/

Hanck, K.W., and Dillard, J.W., 1977a, Determination of the complexation capacity of natural water by cobalt (III) complexation, Anal. Chem., 49:404. /26/

Hanck, K.W., and Dillard, J.W., 1977b, Evaluation of micromolar complexometric titrations for the determination of the complexing capacity of natural waters, Anal. Chim. Acta, 89:329. /52/

Hoffmann, M.R., 1981, Thermodynamic, kinetic, and extrathermodynamic considerations in the development of equilibrium models for aquatic systems, Environ. Sci. Technol., 15:345. /9/

Hoffmann, M.R., Yost, E.C., Eisenreich, S.J., and Maier, W.J., 1981, Characterization of soluble and colloidal-phase metal complexes in river water by ultrafiltration. A mass-balance approach, Environ. Sci. Technol., 15:655. /27/

Jones IV, D.R., and Manahan, S.E., 1977, Elimination of copper-carbonate complex interference in chelating agent analysis by copper solubilization, Anal. Chem., 49:10. /42/

Lakatos, B., Korecz, L., and Meisel, J., 1977, Comparative study on the Mössbauer parameters of iron humates and polyuronates, Geoderma, 19:149. /50/

McCrady, J.K., and Chapman, G.A., 1979, Determination of copper complexing capacity of natural river water, well water and artificially reconstituted water, Water Res., 13:143. /24/

Nriagu, J.O., and Coker, R.D., 1980, Trace metals in humic and fulvic acids from Lake Ontario sediments, Environ. Sci. Technol., 14:443. /47/

Rijncommissie Waterleidinbedrijven: Jahresbericht '80; Teil A: Der Rhein, 1981, Amsterdam. /1/

Saar, R.A., and Weber, J.H., 1980a, Comparison of spectrofluorometry and ion-selective electrode potentiometry for determination of complexes between fulvic acid and heavy-metal ions, Anal. Chem., 52:2095. /30/

Saar, R.A., and Weber, J.H., 1980b, Lead(II)-fulvic acid complexes. Conditional stability constants, solubility, and implication for lead(II) mobility, Environ. Sci. Technol., 14:877. /17/

Schwedt, G., 1981, Methoden zur Bestimmung von Element-Spezies in natürlichen Wässern, in: "Analytiker Taschenbuch Vol. 2", R. Bock, ed., Springer-Verlag, Berlin. /12/

Senesi, N., Griffith, S.M., and Schnitzer, M., 1977, Binding of Fe^{3+} by humic materials, Geochim. Cosmochim. Acta, 41:969. /49/

Shuman, M.S., and Cromer, J.L., 1979, Copper association with aquatic fulvic and humic acids. Estimation of conditional formation constants with a titrimetric anodic stripping voltammetry procedure, Environ. Sci. Technol., 13:543. /14/

Shuman, M.S., and Woodward, G.P., Jr., 1977, Stability constants of copper-organic chelates in aquatic samples, Environ. Sci. Technol., 11:809. /15/

Sillén, L.G., and Martell, A.E., 1964, Stability Constants of Metal-Ion Complexes, Second Edition, Special Publication No. 17, The Chemical Society, London. /6a/

Sillén, L.G., and Martell, A.E., 1971, Stability Constants of Metal-Ion Complexes, Supplement No. 1, Special Publication No. 25, The Chemical Society, London. /6b/

Skogerboe, R.K., Wilson, S.A., and Osteryoung, J.G., 1980, Exchange of comments on scheme for classification of heavy metal species in natural waters, Anal. Chem., 52:1960. /13/

Sontheimer, H., and Gimbel, R., 1977, Untersuchungen zur Veränderung der Fracht an organischen Wasserinhaltsstoffen mit der Wasserführung am Beispiel des Rheins, gwf-wasser/abwasser, 118:165. /3/

Srna, R.F., Garrett, K.S., Miller, S.M., and Thum, A.B., 1980, Copper complexation capacity of marine water samples from southern California, Environ. Sci. Technol., 14:1482. /28/

Stella, R., and Ganzerli-Valentini, M.T., 1979, Copper ion-selective electrode for determination of inorganic copper species in fresh water, Anal. Chem., 51:2148. /25/

Stolzberg, R.J., 1981, Uncertainty in calculated values of uncomplexed metal ion concentration, Anal. Chem., 53:1286. /10/

Stumm, W., and Morgan, J.J., 1981, "Aquatic Chemistry", John Wiley & Sons, New York. /8/

Swallow, K.C., Hume, D.N., and Morel, F.M.M., 1980, Sorption of copper and lead by hydrous ferric oxide, Environ. Sci. Technol., 14:1326. /40/

Truitt, R.E., and Weber, J.H., 1981, Determination of complexing capacity of fulvic acid for copper(II) and cadmium(II) by dialysis titration, Anal. Chem., 53:337. /29/

van Leeuwen, H.P., 1979, Complications in the interpretation of pulse polarographic data on complexation of heavy metals with natural polyelectrolytes, Anal. Chem., 51:1322. /51/

Vuceta, J., and Morgan, J.J., 1978, Chemical modeling of trace metals in fresh waters: Role of complexation and adsorption, Environ. Sci. Technol., 12:1302. /7/

Wilson, S.A., Huth, T.C., Arndt, R.E., and Skogerboe, R.K., 1980, Voltammetric methods for determination of metal binding by fulvic acid, Anal. Chem., 52:1515. /19/

Woodiwiss, C.R., Walker, R.D., and Brownridge, F.A., 1979, Concentrations of Nitrilotriacetate and certain metals in Canadian wastewaters and streams: 1971 - 1975, Water Res., 13:599. /4/

COMPLEX FORMATION IN SOLUTION AND HETEROGENEOUS SYSTEMS

DISCUSSION: F.H. FRIMMEL

H.W. NURNBERG   There are also water types (oceans) with infinite amounts of particulates where no separation problems exist. However, ion-selective electrodes are usually not sensitive enough to be useful tools. In ASV, as a rule, the use of the differential pulse mode is encouraged with care advised to avoid problems due to adsorption. However, this possible interference, which is readily detectable, has been substantially exaggerated in some recent papers on applications of voltammetry to speciation studies.

F.H. FRIMMEL   Yes, I think this is true. Anyway, one can readily detect interferences by the form of the peaks or by changing the parameters of the method.

G.G. LEPPARD   Would you care to comment on model reactions and their usefulness?

F.H. FRIMMEL   To learn about metal complexation in complicated natural systems and to overcome problems arising from insufficient basic data, it is useful to investigate model reactions which are derived from our knowledge of the average concentrations of the various components in water.

The reactions of mercury are among the most highly interesting ones because of the exceptional chronic toxicity of this metal and its property to form very stable and specific complexes. The coordination of mercury to thio-groups has to be expected in surface waters; the Hg-complexes of the thio-ligands can be treated by UV-irradiation, a procedure which leads to distinctive yields of HgS (Frimmel, F.H., Sattler, D., and Quentin, K.-E., 1980, Vom Wasser, 55:111).

$$HgO\ aq. + HS-R \rightleftharpoons [Hg-SR]OH + H_2O$$
$$\downarrow h\nu$$
$$\{[HgS\cdots R]OH\}$$
$$\downarrow$$
$$HgS\downarrow + R'$$

The reaction of dissolved mercury goes on with increasing irradiation time, and comes finally to a ligand specific yield. The results prove that, even under aerobic conditions, thio-ligands act as scavengers for mercury. The photochemical reaction

F.H. FRIMMEL
(continued)

can be seen as a model for the degradation of the complexes and for the formation of highly insoluble sulfide. The latter feature explains the rapid decrease of dissolved mercury concentrations in surface waters to levels which could not be expected if Hg-species other than the sulfide were dominating. A further realistic model is found in the interaction of iron and ligands of the hydroxamate type (Abu-Dari, K., Cooper, S.R., and Raymond, K.N., 1978, Inorg. Chem., 17:3394). The reactions occurring under conditions which are similar to those of surface waters lead to the conclusion that such a type of reaction can play a significant role in preventing the formation of scarcely-soluble ferric compounds (Frimmel, F.H., Geywitz, J., and Quentin, K.-E., 1981, Vom Wasser, 57:185); this has to be considered in a discussion of the role of iron with regard to aquatic humus.

The importance of model reactions has to be seen in fairly-simplified but well-defined initial conditions, as well as in the light of possible approaches to reality, by taking into account more and more parameters which are typical for natural waters.

DIRECT SPECIATION ANALYSIS OF MOLECULAR AND IONIC ORGANOMETALS

Y.K. Chau

National Water Research Institute
Environment Canada
Canada Centre for Inland Waters
Burlington, Ontario, Canada. L7R 4A6

and

P.T.S. Wong

Great Lakes Biolimnology Laboratory
Fisheries and Oceans Canada
Canada Centre for Inland Waters
Burlington, Ontario, Canada. L7R 4A6

INTRODUCTION

   The presence of organometals in the environment was thought to be derived from anthropogenic sources until recently when it was recognized that some biotic and abiotic processes in the environment could produce organometals. Thus a new horizon of knowledge has been opened up; one which focusses on the formation, pathways and fate of organometals and organometalloids in ecosystems, phenomena which are difficult to predict by conventional chemistry theories. Some of these compounds are volatile and are important transport species in biological and geochemical cycles (Brinckman et al., 1982). These types of compounds not only have direct effects on metal mobilization, but also have serious impacts on the environment because of their modified toxicological properties.

   Direct analysis of the chemical forms of an element at their environmental concentration levels is perhaps one of the most diffi-cult tasks confronting analytical chemists today. Metals in ionic form in an ecosystem are speciated with reference to their oxidation numbers, free forms (aquo-complexed) and complexed forms which include both organic and inorganic ligands. Most research in this area has been done using electrochemical techniques. Organometals and organometalloids, however, exist in the environment in both molecular and ionic forms, and also incorporated into large molecules of biological origin. Techniques of analysis first require a separation method, followed by a determination technique specific to either one part of or the whole molecule. The most favorable methodology, therefore, is the use of an element-selective detector such as AAS, AFS, or AES coupled to a separation technique, such as GC or HPLC, to achieve separation and element specific analysis.

Some speciation techniques involve a separation step, after which the species of interest is decomposed and analyzed in its inorganic forms. Serious contamination may result at the final step and the analyst can be totally unaware of it. Other methods determine only one form and depend on calculation by difference to estimate the concentration of the other species from the total analysis. Such practice can magnify any analytical errors arising from several single analyses, resulting in a wide range of variation. The most ideal speciation technique should be one that analyzes directly the molecular or ionic forms of interest in their authentic forms as they are present in the original samples. For example, in the analysis of tetraalkyllead species, the chance of contamination is extremely rare, and there is no calculation by difference.

The present paper discusses the capabilities and limitations of the two most basic and popular systems, GC-AAS and LC-AAS, and reviews recent developments. A comprehensive literature review on these topics up to 1977 is already available (Fernandez, 1977).

THE GC-AAS SYSTEM

The combination of GC and AAS has proven successful in the direct speciation analysis of several molecular forms of metal alkyls (Chau and Wong, 1977). The choice of stationary phase in the GC is important for clean separation of components without tailing of peaks. For organometals in general, non-polar silicone phases such as the OV-1, OV-101 types are suitable. Light loadings (3-5%) of stationary phase are generally preferred for well defined peaks. The oven and transfer line temperatures are chosen according to the boiling points of the compounds. Oven temperatures which are too high sometimes cause decomposition of the organometals in the column; the situation is likewise for the transfer line.

The interfacing of the two instruments is relatively simple. A transfer line is used to connect the column outlet of the GC to the furnace of the AAS. Several types of tubings (2mm o.d.) can be used as a transfer line, depending on the chemical properties of the analytes. For example, methylmercury compounds are decomposed in a heated stainless steel line and, therefore, teflon tubing should be used instead. Other materials, such as glass-lined stainless steel tubing and all glass tubing are satisfactory. The temperature of the transfer line can be controlled by heating tapes.

Other element-specific detectors, such as atomic fluorescence (AFS) (van Loon et al., 1977), atomic emission (AES) (Braman and Tompkins, 1979), and a microwave plasma detector (MPD) based on the emission of the element excited by microwave plasma (Uden, 1981), have been used successfully. Figure 1 illustrates the interfacing of a GC to an AAS.

THE AA FURNACE

Because AAS is used as a dedicated detector for the GC, its furnace or burner has to be heated during operation to the atomization temperatures of the organometals in order to receive properly the analytes emerging from the GC column. When the flame mode AAS is used, the interfacing can be achieved simply by connecting the transfer line directly to the nebulizer of the burner, and the flame can be continuously operated during the analysis. Such a setup, however, is limited by the sensitivity of the flame AAS, which is generally at the µg range. For higher sensitivity, the flameless furnace must be used. Commercially-available graphite furnaces operate on

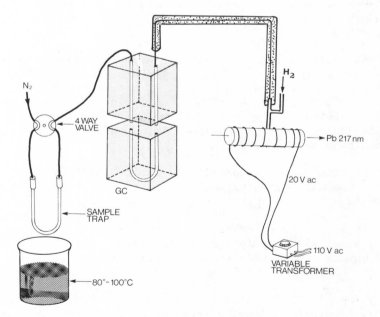

Figure 1. Schematic diagram showing the interfacing of GC and AAS.

cycles of drying, ashing and atomizing. Although it is feasible to maintain the furnace at its atomizing temperature continuously, over a period of analysis of 10 minutes or more, such practice would shorten the life of the graphite tubes. Prolonged heating would also deteriorate the graphite, causing changes in sensitivity. To avoid all these problems, an electrothermal furnace is constructed from silica tubing (7 mm i.d., 4 cm long) with open ends. It is wrapped with 26 gauge Chromel wire to give a resistance of about 5 ohms. The furnace tube is wrapped with several layers of wet asbestos tape and dried slowly in an oven before use. Hydrogen and air can be introduced to the furnace to enhance atomization and to elevate the furnace temperature. With an applied voltage of about 20 V, a.c., the furnace temperature, with hydrogen burning, can readily reach about $900^\circ C$ which is sufficient for most of the organometals studied. The electrothermally-heated furnace can be operated continuously at its maximum temperature for the whole operation day without deterioration or damage. It is conveniently mounted on top of the burner head and aligned to the optical path of the instrument.

THE SAMPLING TECHNIQUES

The GC-AAS system can handle gaseous and liquid samples. For volatile organometals, the sample is trapped cryogenically in a U-tube packed with ordinary chromatographic materials (e.g., 3% OV-1). Then the trapped sample is injected into the GC through a 4-way valve installed between the injection port and the carrier gas inlet. Liquid samples can be directly injected into the GC through an ordinary injection port.

The volatile molecular forms of tetraalkyllead of R_4Pb type (such as Me_4Pb, $Me_nEt_{4-n}Pb$, Et_4Pb), Me_4Sn and mixed alkyltins (such as $Me_nBu_{4-n}Sn$, etc.), have a relatively high vapor pressure and can be cryogenically trapped at $-150^\circ C$. A typical example is the determination of tetraalkyllead in which all five mixed methylethyl lead compounds can be separated and identified in gaseous or liquid samples with detection limits of 0.1 ng as Pb (Figure 2).

In sampling air for volatile organolead compounds using cryogenic trapping, the main difficulties are the condensation of moisture in air, which often blocks the trap, and water interferences in the subsequent GC separation. Such difficulties have been overcome by using a large pre-trap, containing glass beads and immersed in a cryogenic bath at $-130^\circ C$ (De Jonghe et al., 1980). After sampling, the trapped sample is desorbed at $60^\circ C$ into a U-adsorption tube held at $-190^\circ C$ for the subsequent GC-AAS analysis. Large volumes of air (400 litres) can be sampled in a short time without the problems caused by water. By such a device, tetraalkyllead species in ambient air can be determined readily without the usual lengthy sampling procedure. This sampling technique can be applied to other volatile

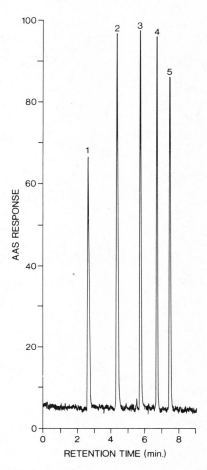

Figure 2. GC-AAS chromatogram showing separation of five tetra-alkyllead compounds. Each peak represents ca. 5 ng of Pb. 1-Me_4Pb, 2-Me_3EtPb, 3-Me_2Et_2Pb, 4-$MeEt_3Pb$, 5-Et_4Pb.

metal alkyls such as tetramethyltin, methyl selenides, methylarsines, and etc. A special gas sampler has been designed to collect evolved gases from sediments in the lake bottom for studying the volatile forms of some elements evolved as a result of methylation processes (Chau et al., 1977).

For biological samples such as fish, aquatic weeds and sediments, molecular organometals such as tetraalkyllead readily can be extracted into hexane because of their lipophilicity. The extract then can be directly injected to the GC-AAS system for analysis (Chau et al., 1980).

Ionic organometal compounds such as Me_3Pb^+ and Et_3Pb^+ are highly soluble in water and are not extractable into any non-polar solvents. They can be extracted only by polar solvents such as iso-amyl alcohol in the presence of saturated sodium chloride. These two compounds have adequate boiling points for GC separation in a polar column (Hayakawa, 1971). We have evaluated this technique using GC-AAS analysis and found that the recovery was about 85-90% and the sensitivity was much less than that for tetraalkyllead (Chau and Wong, unpublished data). A GC-AAS chromatogram for the separation of Me_3Pb^+ and Et_3Pb^+ is illustrated in Figure 3. Other ionic organometals with high boiling points can be determined after derivatization described in the following sections.

DERIVATIZATION TECHNIQUES

For the highly polar and solvated organometalloids and organometals, such as the methylarsenic acids, and alkyltin ions, $Me_nSn^{(4-n)+}$, a chemical derivatization is required to convert the

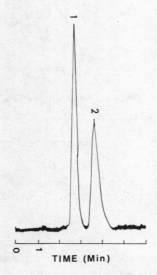

Figure 3. GC-AAS analysis of $Me_3PbCl(1)$ and $Et_3PbCl(2)$. Each peak represents 20 ng of Pb.
Column: 1.2m glass, 5% DEGS on Chromosorb W, 80-100 mesh, pre-coated with 5% NaCl.
Temperature Programs: 140°C isothermal 3 min; then 15° min^{-1} to 170°; injection port, 165°; transfer line, 165°. Carrier gas N_2, 85 mL min^{-1}.
AAS: Pb (217nm line); D_2 background corrector; furnace, 900°, H_2, 130 mL min^{-1}.

DIRECT SPECIATION ANALYSIS OF ORGANOMETALS 93

analytes to more volatile forms for efficient GC separation. The most common derivatization used is the hydride reduction by reacting the polar alkyl metal species in water with a hydride reagent (sodium borohydride) to change them to their corresponding hydrides. For example, different alkyl derivatives of arsenic acids have been reduced to their arsines, which are trapped and distilled to the detector with and without the use of a GC (Wong et al., 1977; Braman et al., 1977).

$$CH_3AsO(OH)_2 + BH_4^- \longrightarrow CH_3AsH_2 \quad (bp\ 2°C)$$
$$(CH_3)_2AsO(OH) + BH_4^- \longrightarrow (CH_3)_2AsH \quad (35.6°C)$$
$$(CH_3)_3AsO + BH_4^- \longrightarrow (CH_3)_3As \quad (70°C)$$

Similarly, alkyltin (IV) compounds can also be converted to their respective hydrides, separated and determined by emission (Braman and Tompkins, 1979) and by AAS (Hodge et al., 1979).

$$CH_3SnCl_3 + BH_4^- \longrightarrow CH_3SnH_3 \quad (bp\ 0°C)$$
$$(CH_3)_2SnCl_2 + BH_4^- \longrightarrow (CH_3)_2SnH_2 \quad (35°C)$$
$$(CH_3)_3SnCl + BH_4^- \longrightarrow (CH_3)_3SnH \quad (59°C)$$
$$Sn(IV) + BH_4^- \longrightarrow (CH_3)_4Sn \quad (76.8°C)$$

This reaction can be applied to other alkyl metals of $R_nM^{(4-n)+}$ type, such as Pb(IV), Ge(IV), Sb(IV), and etc.

The second technique is peralkylation which involves the use of a selected R' Grignard reagent to convert the alkylated metal, $R_nM^{(4-n)+}$ to $R_nMR'_{(4-n)}$ in solvent medium. The tetraalkylated metal derivatives are now volatile and readily separated by GC. For example, butyltin species have been determined, after extraction into tropolone, and methylated to the tetraalkylated forms by GC-MS (Meinema et al., 1978). This technique was later modified by using a pentyl Grignard reagent to make the butylpentyl derivatives which have higher boiling points, thus facilitating solvent evaporation (Maguire and Huneault, 1981). Further improvements were made by using a "salting out" technique to extract the highly polar and solvated methyltin species from natural waters into tropolone-benzene solution, followed by butylation of the methyltins to the low-boiling methylbutyltin derivatives suitable for GC-AAS analysis (Chau et al., 1982). The reactions involve:

$$MeSn^{3+} + BuMgCl \longrightarrow MeSnBu_3 \quad (bp\ 122-124°C)$$
$$Me_2Sn^{2+} + BuMgCl \longrightarrow Me_2SnBu_2 \quad (70°C)$$
$$Me_3Sn^+ + BuMgCl \longrightarrow Me_3SnBu \quad (41-42°C)$$
$$Sn(IV) + BuMgCl \longrightarrow Me_4Sn \quad (145°C)$$

The four butyl derivatized methyltin compounds separate well in the GC and give excellent AA signals (Figure 4). The absolute detection limit is 0.1 ng Sn. Thus for a 5 litre water sample used for extraction, a detection limit of 0.04 ppb Sn can be achieved.

Using the above two derivatization techniques, the occurrence of certain alkyltin compounds has been detected for the first time in environmental samples. Such information is vitally essential, not only to follow the fate of industrial organotin compounds in the environment, but also to understand the transformation of these compounds in ecosystems. A summary of the occurrence of alkylmetals in the environment is given in Table 1.

The derivatization methods modify a part of a molecule to render it more volatile without changing the original structure of the alkyl groups bonded to the metal nucleus. Thus the authenticity of the original form is not altered. The hydride generation method is more sensitive because it is a total sample consumption technique. We have, however, observed rearrangement reactions of the alkyl groups during the hydride reduction of methylarsenic compounds, and this may cause confusion in identification. The peralkylation reactions are instantaneous and quantitative, but slightly inferior in sensitivity because only an aliquot of the final solution containing the derivatives is used for analysis.

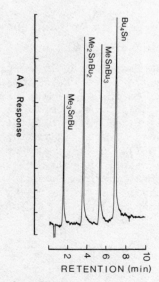

Figure 4. GC-AAS chromatogram of butyl derivatized methyltins and Sn(IV) compounds. Each peak represents ca. 8 ng Sn.

THE LC-AAS SYSTEM

High pressure liquid chromatography has been used successfully in separating high-molecular-weight and large organic molecules. HPLC used in combination with an element-selective detector would, therefore, provide a powerful system for speciating the polar and solvated organometal compounds that are not readily analyzed by the GC-AAS. There have been many designs to interface the LC and AAS; however, the lack of a "perfect" interface is still the main road block of this technique. Attempts to solve these problems have not been completely satisfactory.

The inherent difficulties of interfacing LC and AA lie in the limitations of the furnace to handle the continuous flow of the LC effluent. With flame AAS as the detector, the interface can be achieved simply by directly feeding the LC effluent into the burner nebulizer (Figure 5a). If the LC flow is too slow compared to that of the AA nebulizer, an auxiliary solvent flow is required to balance the flow (Yoza and Ohashi, 1973). Copper chelates in several amino carboxylic acids have been determined by this technique (Jones and Manahan, 1976). Unfortunately, the flame AAS detector does not provide sufficient sensitivity for metal speciation work in environmental studies. Electrothermal furnace atomizers can generally provide detection limits three orders of magnitude lower than those

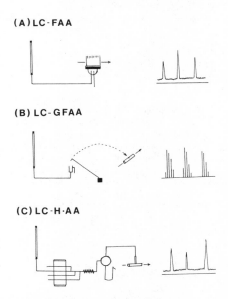

Figure 5. Schematic diagrams showing modes of LC-AAS interfacing. (A) Direct LC to flame AAS. (B) via an autosampler to a graphite furnace. (C) through an automatic hydride generation system to an electrothermal furnace.

Table 1. Occurrence of alkylated metals and metalloids in environmental samples.

Arsenic

Location	sample	s(III),As(V)	MMA	DMA	Technique	Ref
Moira R., Ont.	water(ng/mL)	146	4	52	GC-AAS	11
Bend Bay, Ont.	water	453	9	93	GC-AAS	11
Scripps, Calif.	seawater	0.02, 1.75	0.02	0.12	Trap-AAS	18
L. Carrol, Fla.	water	50	4.39	6.31	Trap-AES	13
Moira R., Ont.	sediment[a]	1.92	2.14	3.25	GC-AAS	12
	sediment	1.16	3.75	2.29	GC-AAS	12
Bay Quinte, Ont.	yellow perch (ng/g)	68.0	0.74	2.45	GC-AAS	11
	white sucker	0.67	0.60	0.24	GC-AAS	11
	white perch	16.90	0.84	nd	GC-AAS	11
	Alewife	1.33	nd	nd	GC-AAS	11

MMA - monomethylarsenic acid; DMA - dimethylarsinic acid; a - 50 g sediment and 140 mL lake water incubated for 7 days, the solution was analyzed for different As species. Results in µg/L as As. nd - not detected.

Lead

Location	sample	I	II	III	IV	V	Technique	Ref
College St. Toronto	air (ng/m^3)	nd	1	2	1	8	GC-AAS	19
London, UK	urban air	total tetraalkyllead 40-110					Adsorption AAS	20
Morecambe Bay, U.K	rural air	total tetraalkyllead 0.5-230					Trap-AAS	21
Baltimore H. tunnel, USA	air	21-66	0.5-9	11-16	1-6	11-42	GC-MPD	6
Antwerp, Belgium	street air	23	4.5	1.6	0.8	3.0	GC-AAS	7
Vineland Ck, Ont.	Coho salmon (ng/g)	nd	4.3	2.3	2.8	nd	GC-AAS	9
Stoco L., Ont.	yellow perch	1.7	nd	7.1	4.4	2.7	GC-AAS	9
	rock bass	0.6	nd	nd	nd	2.8	GC-AAS	9

I - Me$_4$Pb, II - Me$_3$EtPb, III - Me$_2$Et$_2$Pb, IV - MeEt$_3$Pb, V - Et$_4$Pb

Table 1. (continued)

Location	sample	Me$_3$Pb$^+$	Et$_3$Pb$^+$	Me$_4$Pb	Et$_4$Pb	Technique	Ref
St.Lawrence R.	sediment (μg/g)	nd	3.80	nd	nd	GC-AAS	11
		nd	0.76	nd	nd	GC-AAS	11

<u>Tin</u> (water, ng/L)

Location	sample	MeSn^{3+}	Me$_2$Sn^{2+}	BuSn^{3+}	Bu$_2$Sn^{2+}	Bu$_3$Sn$^+$	Sn(IV)	Technique	Ref
San Diego Bay Calif.	water	2-8	15-45	nd	nd		6-38	Trap-AAS	14
Grand Haven, Mich.	water	13	nd	1220	1600		490	Trap-AAS	14
Hamilton H. Ont.	water		10	10	80		40	GC-FPD	22

Location	sample	MeSn^{3+}	Me$_2$Sn^{2+}	Me$_3$Sn$^+$	Sn(IV)	Technique	Ref
Florida	rain (10)	5.9	7.4	0.2	11	Trap-AES	5
	tapwater(2)	4.3	1.3	1.5	2.2	Trap-AES	5
	freshwater(15)	2.0	1.4	1.5	4.2	Trap-AES	5
	saline	0.6	1.4	0.5	1.7	Trap-AES	5
Baltimore H.	water	nd	5-20	5-10	nd-400	GC-FPD	1
Toronto H., Ont.	water	960	290	nd	540	GC-AAS	17
Port Dover, Ont.	water	610	160	nd	140	GC-AAS	17

GC - gas chromatography; AAS - atomic absorption spectrometry; AES - atomic emission spectrometry; FPD - flame photometric detector; Trap - either cryogenic or chemical, e.g., ICl; MPD - microwave plasma detector; nd - not detected. Bracketed numbers indicate no. of samples taken for average.

of flame atomizers. There are two basic approaches to solve these problems. The first is to install a sampling device between the LC and the AA furnace to sample the effluent and to inject it into the furnace (Figure 5b). A multiport sampling and injection valve, controlled by a sequencer, was used (Cantillo and Segar, 1975) to take an aliquot of the effluent and inject it into the furnace, while the main LC effluent stream was stopped or bypassed during the heating cycles of the atomizer. The major drawback of this technique is the lengthy analysis time because of the cycles required in the furnace operation. Commercial auto-samplers have been used to sample the LC effluent from an overflowing well (Brinckman et al., 1977) or on line (Stockton and Irgolic, 1979). All these techniques result in pulse signals the sum of which can be taken as the peak height of an analyte. Because the sampling rate is governed by the rate of graphite furnace operation, which consists of three cycles plus cooling time (about 40-50 seconds per sample), only broad chromatographic peaks can be analyzed. Thus, if the chromatographic peak is narrow and only lasts for 2-3 minutes, the AA analysis can vary considerably depending on how well the auto sampler operation is synchronized to the emergence of the peak. By this technique, As(III), As(V), methylarsonic acid, dimethylarsinic acid (Woolson and Aharonson, 1980), arsenobetaine, arsenocholine and inorganic arsenic (Stockton and Irgolic, 1979), as well as di- and trialkyltin compounds (Jewett and Brinckman, 1981), have been speciated. Current improvements include storage of the chromatographic peaks for detailed analysis of the peak-containing aliquot off-line. Thus the number of AA analyses is no longer limited by the flow rate nor restricted by the furnace cycles (Vickrey et al., 1979). Greater numbers of analyses can be performed on an analyte peak to achieve accurate calculation of its area. The method is lengthy, but it certainly improves the description of a chromatographic peak.

The second approach is to convert the analytes on line to volatile derivatives for continuous feeding to the furnace (Figure 5c). For elements that readily form covalent hydrides (such as As(III) and As(V), Ge(IV), Sn(IV), Pb(IV), Sb(V), etc.), the on line hydride generation has shown some degree of success. In this regard, a procedure has been developed for the determination of As(III), As(V), monoarsonic acid, dimethylarsinic acid and p-amino phenyl arsonate by hydride atomization after ion chromatographic separation (Ricci et al., 1981). At the same time, we also developed independently a procedure for the analysis of As(III), As(V), monoarsonic acid, and dimethylarsinic acid by coupling a conventional ion exchange column, containing a strongly-acidic cation exchange resin, to an automatic hydride generation system (Figure 6). Because of the mild separation conditions with ion exchange chromatography and the individual hydride reaction, the possible rearrangement of alkyl groups was avoided. The elution was simple. A chromatogram is given in Figure 7 to illustrate the separation. These two examples fully demonstrate that other forms of chromatography can be used to achieve separation.

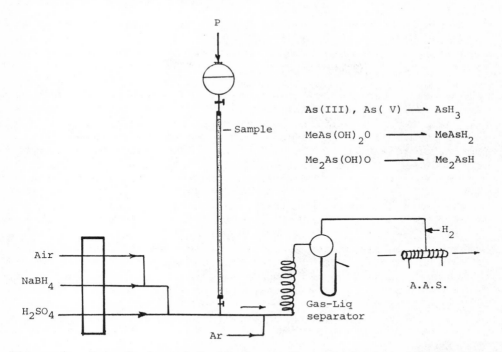

Figure 6. Interfacing of Ion Exchange Chromatography and AAS through a hydride generation system for the determination of organoarsenics.
Column: 0.5cm x 25cm AG50Wx8, 100-200 mesh; eluants: 20ml 0.2M trichloroacetic acid and 20ml 1M ammonium acetate; flow rate, 2.3ml min^{-1}; Hydride system: 1% $NaBH_4$, 2ml min^{-1}; air, 4ml min^{-1}; 2.4N H_2SO_4, 4ml min^{-1}; Ar, 500ml min^{-1}.

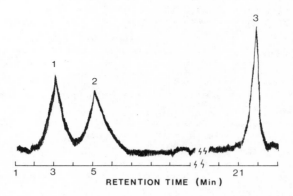

Figure 7. Chromatogram showing separation of (1) As(III) (2) $MeAs(OH)_2O$ (3) $Me_2As(OH)O$ by Ion Exchange Chromatography-Hydride AAS. Each peak represents 50 ng As.

Recently, an HPLC interface with AAS has been developed through an automatic hydride generator for the analysis of alkyltin compounds with excellent sensitivity (2-20 pg); this system is currently the state-of-the-art in LC-AAS interfacing (Burns et al., 1981).

Other spectrometric detector systems coupled to LC, such as flame emission photometric detectors (Freed, 1975), have also been used but they generally lack the required sensitivity at the low ng level. Electrochemical detectors based on the detection of the oxidation or reduction of the analyte have long been used for organic compounds. They have good sensitivities but do not have adequate element specificity and versatility compared to the spectrophotometric detectors. Recently, alkyl- and phenyl-derivatives of Hg, Sn, Pb and Sb have been determined by cyclic voltammetry with ng sensitivity after HPLC separation (MacCrehan et al., 1977). Further improvements in selectivity were achieved by using the differential pulse mode of amperometry to discriminate against the interferences of other reducible species such as Cd^{2+}, Pb^{2+} and Cu^{2+} (MacCrehan and Durst, 1978). Not enough research and developmental work has been done in this area. It is predicted that an increasing use of electrochemical detectors will appear in the future.

REFERENCES

Andreae, M.O., 1977, Determination of arsenic species in natural waters, Anal. Chem., 49:820. /18/
Braman, R.S., Johnson, D.L., Foreback, C.C., Ammons, J.M., and Bricker, J.L., 1977, Separation and determination of nanogram amounts of inorganic arsenic and methylarsenic compounds, Anal. Chem., 49:621. /13/
Braman, R.S., and Tompkins, M.A., 1979, Separation and determination of nanogram amounts of inorganic tin and methyltin compounds in the environment, Anal. Chem., 51:12. /5/
Brinckman, F.E., Blair, W.R., Jewett, K.L., Iverson, W.P., 1977, Application of a liquid chromatograph coupled with a flameless atomic absorption detector for speciation of trace organometallic compounds, J. Chromatogr. Sci., 15:493. /26/
Brinckman, F.E., Jackson, J.A., Blair, W.R., Olson, G.J., and Iverson, W.P., Ultratrace speciation and biogenesis of methyltin transport species in estuarine waters, in: NATO Adv. Res. Inst. "Trace metals in sea water", Erice, Italy, 1981, Plenum Publ. Corp., in press. /1/
Burns, D.T., Glockling F., and Harriott, M., 1981, Investigation of the determination of tin tetraalkyls and alkyltin chloride by atomic absorption spectrometry after separation by gas-liquid or high performance liquid-liquid chromatography, Analyst, Lond. , 106:921. /32/
Cantillo, A.Y., and Segar, D.A., 1975, Metal species identification in the environment - A major challenge for the analyst, in:

"Proc. Int. Conf. Heavy Metals in the Environment", Toronto, Canada, I-183-204. /25/

Chau, Y.K., Snodgrass, W.J., and Wong, P.T.S., 1977, A sampler for collecting evolved gases from sediment, Water Res., 11:807. /8/

Chau, Y.K., and Wong, P.T.S., 1977, An element- and speciation-specific technique for the determination of organometallic compounds, in: "Environmental Analysis", G.W. Ewing, ed., Academic Press, New York. /3/

Chau, Y.K., and Wong, P.T.S., unpublished data. /11/

Chau, Y.K., Wong, P.T.S., and Bengert, G.A., 1982, Determination of methyltin and Sn(IV) species in water, Anal. Chem., 54:246. /17/

Chau, Y.K., Wong, P.T.S., Kramar, O., Bengert, G.A., Cruz, R.B., Kinrade, J.O., Lye, J., and van Loon, J.C., 1980, Occurrence of tetraalkyllead compounds in the aquatic environment, Bull. Environm. Contam. Toxicol., 24:265. /9/

De Jonghe, W.R.A., Chakraborti, D., and Adams, F.C., 1980, Sampling of tetraalkyllead compounds in air for determination by gas chromatography-atomic absorption spectrometry, Anal. Chem., 52:1974. /7/

Fernandez, F.J., 1977, Metal speciation using atomic absorption as a chromatography detector, At. Absorpt. Newsl., 16:33. /2/

Freed, D.J., 1975, Flame photometric detector for liquid chromatography, Anal. Chem., 47:186. /33/

Harrison, R.M., and Laxen, D.P.H., 1978, Natural source of tetraalkyllead in air, Nature, 275:738. /21/

Harrison, R.M., and Perry, R., 1977, The analysis of tetraalkyllead compounds and their significance as urban air pollutants, Atmos. Environ., 11:847. /20/

Hayakawa, K., 1971, Microdetermination and dynamic aspects of in vivo alkyllead compounds, Jap. J. Hyg., 26:377. /10/

Hodge, V.F., Seidel, S.L., and Goldberg, E.D., 1979, Determination of tin(IV) and organotin compounds in natural waters, coastal sediments and macroalgae by atomic absorption spectrometry, Anal. Chem., 51:1256. /14/

Jewett, K.L., and Brinckman, F.E., 1981, Speciation of trace di- and triorganotins in water by ion-exchange HPLC-GFAA, J. Chromatogr. Sci., 19:583. /29/

Jones IV, D.R., and Manahan, S.E., 1976, Aqueous phase high speed liquid chromatographic separation and atomic absorption detection of amino carboxylic acid-copper chelates, Anal. Chem., 48:502. /24/

MacCrehan, W.A., and Durst, R.A., 1978, Measurement of organomercury species in biological samples by liquid chromatography with differential pulse electrochemical detection, Anal. Chem., 50:2108. /35/

MacCrehan, W.A., Durst, R.A., and Bellama, J.M., 1977, Electrochemical detection in liquid chromatography: Application to organometallic speciation, Anal. Lett., 10:1175. /34/

Maguire, R.J., personal communication. /22/

Maguire, R.J., and Huneault, H., 1981, Determination of butyltin species in water by gas chromatography with flame photometric detection, J. Chromat., 209:458. /16/

Meinema, H.A., Burger-Wiersma, T., Versluis-de Haan, G., and Gevers, E.Ch., 1978, Determination of trace amounts of butyltin compounds in aqueous systems by gas chromatography/mass spectrometry, Environ. Sci. Technol., 12:288. /15/

Radziuk, B., Thomassen, Y., van Loon, J.C., and Chau, Y.K., 1979, Determination of alkylated lead compounds in air, Anal. Chim. Acta, 105:255. /19/

Ricci, G.R., Shepard, L.S., Colovos, G., and Hester, N.E., 1981, Ion chromatography with atomic absorption spectrometric detection for determination of organic and inorganic arsenic species, Anal. Chem., 53:610. /31/

Stockton, R.A., and Irgolic, K.J., 1979, The Hitachi graphite furnace-Zeeman atomic absorption spectrometer as an automated, element-specific detector for high pressure liquid chromatography: the separation of arsenobetaine, arsenocholine and arsenite/arsenate, Intern. J. Environ. Anal. Chem., 6:313. /27/

Uden, P.C., 1981, Specific element detection in chromatography by plasma emission spectroscopy, in: NBS Special Publication 618 "Environmental speciation and monitoring needs for trace metal-containing substances from energy-related processes", F.E. Brinckman and R.H. Fish, eds. /6/

van Loon, J.C., Lichwa, J., and Radziuk, B., 1977, Non-dispersive atomic fluorescence spectroscopy, a new detector for chromatography, J. Chromat., 136:301. /4/

Vickrey, T.M., Howell, H.E., and Paradise, M.T., 1979, Liquid chromatogram peak storage and analysis by atomic absorption spectrometry, Anal. Chem., 51:1880. /30/

Wong, P.T.S., Chau, Y.K., Luxon, L., and Bengert, G.A., 1977, Methylation of arsenic in the aquatic environment, in: "Trace Substances in Environmental Health - XI", D.D. Hemphill, ed., Univ. of Missouri Press, Columbia, Missouri. /12/

Woolson, L.A., and Aharonson, N., 1980, Separation and detection of arsenical pesticide residues and some of their metabolites by high pressure liquid chromatography-graphite furnace atomic absorption spectrometry, J. Ass. Off. Anal. Chem., 63:523. /28/

Yoza, N., and Ohashi, S., 1973, The application of atomic absorption method as a flow detector to gel chromatography, Anal. Lett., 6:595. /23/

DISCUSSION: Y.K. CHAU AND P.T.S. WONG

R. BONIFORTI   Did you establish any relations between the biological activity of sediments and the alkylation of metals?

Y.K. CHAU   No.

H. MUNTAU   What is your estimate of the total uncertainty associated with the determination (sampling on the lake bottom) of gaseous metal organics formed in sediments?

Y.K. CHAU   The collection of volatile organic metal compounds of sediments is done for studies on the occurrence of these compounds in the environment; we did not produce statistics with regard to the collection efficiency.

H.W. NURNBERG   With the organometallics the contamination problems (such as contributions by the laboratory atmosphere and release of materials by container walls) are much less severe than one would guess. There remains, of course, the problem of losses incurred during sampling, clean up and derivatization.

Y.K. CHAU   We did not investigate the losses during sampling. However, we know that once the volatile metal alkyls are in the cryogenic trap and stored in the cold, there is no loss of the compounds. Derivatization reactions at the ultratrace level are astonishingly quantitative.

R.F. VACCARO   Can you differentiate between natural background concentrations and anthropogenic sources of organometallic compounds? Are natural organometallics measurable in clean waters such as those of alpine lakes and the Sargasso Sea?

Y.K. CHAU   Our present state is one of studying the occurrence of organometallics in the environment.

BIOAVAILABILITY, TRACE ELEMENT ASSOCIATIONS WITH COLLOIDS AND AN EMERGING INTEREST IN COLLOIDAL ORGANIC FIBRILS

Gary G. Leppard and B. Kent Burnison

National Water Research Institute
Environment Canada
Canada Centre for Inland Waters
Burlington, Ontario, Canada. L7R 4A6

INTRODUCTION

There is an urgent need for environmental managers to understand better how to assess bioavailability as it relates to the physicochemical status of biologically-important trace elements in surface waters. As several of the chapters in this volume will show, it should be particularly revealing to improve our knowledge of the natural aquatic associations between trace elements and both true particles and colloids. More and more it is realized that such associations determine much of the specific nature of trace element speciation in surface waters.

In addition to discussing bioavailability in relation to trace element speciation, this chapter will (1) outline some recent ideas on the roles of particulates in the respeciation of trace elements and (2) focus on an emerging interest in the activities of colloids. In particular, we will feature the recently-discovered colloidal organic "fibrils".

BIOAVAILABILITY AND RELATED CONCEPTS

Trace elements exist in several different physicochemical forms in surface waters, with the assemblage of the various forms of a given element being referred to as its speciation. If a given species of an element can be taken up readily by living cells, it is referred to as being a biologically available form of the element or, simply, as being "bioavailable". Bioavailability of lacustrine micronutrients is discussed concisely in Wetzel (1975) and related concepts are defined in this volume in the chapter by Baudo. Many factors can change the speciation of a given element in a particular aquatic situation and this topic is treated in some detail later in relation to bioavailability. Not the least among these factors is the ability of some organisms to exert an influence over trace element speciation in their immediate vicinity (see chapter by Giesy) by means of a secretion of organics specific to facilitating a control. Such a control mechanism can lead to adjustments in the speciation of a critical trace element so as to assist the secreting organism in an ecological competition (Murphy et al., 1976).

The importance of dissolved organic substances in regulating trace element speciation and, consequently, biological phenomena, is increasingly appreciated. Organic complexes (see Stumm and Morgan, 1981, for definitions and descriptions) of metals, both nutrient and toxic, have come under intensive study. The result of such complexation in the external milieu of common plankton tends to be beneficial in two ways. Essential nutrient metals present in insoluble form can interact with soluble organics to produce a soluble complex which is bioavailable. Concomitantly, a toxic metal present in free ionic form will have, after complexation, a greatly reduced toxicity. Biological detoxification mechanisms tend to be very general (see chapter by Albergoni and Piccinni). Conversely, the mechanism by which a nutrient metal is transported into cells by a natural complexing agent, such as an organic chelator, is very specific. For example, a cell can have a specific trans-membrane transport system for iron in which the iron is carried in by a specific, secreted, organic chelator (Emery, 1971).

Despite considerable strides in producing information about the nature of chemical species present in natural waters and in developing increasingly-sophisticated techniques to quantify them, there are major gaps in our knowledge. Little experimental research is done on aquatic biota to define, in detail, the dynamic interactions through time between an applied, given, trace element species and the physiology of the manipulated organism (see chapter by Barghigiani et al.). In particular, the response of aquatic microorganisms to heavy metal stress lacks a thorough documentation. Recently, Babich and Stotzky (1979) have illustrated the biological complexity of such a response by presenting data showing how the toxicity of heavy metals to aquatic microbes is dependent on the physicochemical characteristics of the environment. Depending on which abiotic factors dominate, heavy metal toxicity may be attenuated or potentiated. With regard to microbe response, there is too little data available on the metal desorption capability of cells and on the biodegradation of natural metal chelators.

Other major gaps in our knowledge hinder our understanding of how particulate material acts in the redistribution of trace elements within a given water body. Particulate materials can influence directly the speciation of trace metals (see chapter by Förstner and Salomons) as well as influencing their displacement by sedimentation and currents. The roles of particulates in the aquatic environment certainly must be important and increasing attention is being focussed on defining these roles in detail. Organic particulates definitely merit more attention. For example, in certain rivers, the particulate and dissolved phases are apparently not in equilibrium. In such cases, impinging organic matter can have a strong influence on trace metal associations (Eisenreich et al., 1980). With regard to particle-mediated transport, Jenne(1977) has proposed that the most significant role of clay-sized aluminosilicate minerals (in trace element sorption by soils and sediments and their transport in rivers) is as a mechanical substrate for the precipitation and flocculation of secondary minerals and dissolved-plus-colloidal organics. In all these studies, chemical fractionation schemes have been used to estimate the trace element speciation in the particulates. Future research must address itself to equating these fractions to truly available metals for aqueous and biotic interactions. Our own research has moved towards a greater use of physical fractionation schemes and some advantages of this attitude will become clear in later sections of this chapter.

TRACE ELEMENTS: SOME GENERAL ASPECTS OF THEIR INTERACTIONS WITH AQUATIC LIFE AND ORGANIC MATERIALS

Many scientific disciplines currently contribute to advances in the aquatic sciences and, as might be expected, some of them use a given scientific term in a manner specific to their particular needs and reflecting their particular history. Consequently, an element of confusion is currently being injected into limnological research by technology transfer, a confusion well-illustrated by the diverse usage of the term "trace element". To some, a trace element is an element required as an essential nutrient by organisms in extremely small quantities. Essential nutrients differ from one class of organism to another (Wetzel, 1975) and, consequently, lists of such differ according to the biological field of the author of the list. A useful list for limnologists interested in plant and algal life can be found in Vallentyne (1974). To others, a trace element is any element (nutrient or toxic) whose concentration in surface waters is usually below the detection limit of a standard chemical method, with the term "standard" being used loosely. Sometimes the term is arbitrarily defined with respect to a measured quantity, such as being less than 0.05mg/l in a fresh water sample (see McNeely et al., 1979, p.64).

With regard to the situation in a given lake, there are many factors (geological, physico-chemical, climatic and demophoric) which determine whether a given element is present in a trace amount or in a greater amount (Vallentyne, 1974; Wetzel, 1975). Sometimes the biota can act on the aquatic milieu so as to adjust concentration levels upwards or downwards with respect to bioavailable chemical species (see chapter by Giesy). Regarding this control phenomenon, research on iron speciation has been of particular interest (Simpson and Neilands, 1976; Bailey and Taub, 1980), with Murphy et al., (1976) suggesting that the excretion of iron-selective chelators can enable blue-green algae to dominate other algae.

One approach to selecting trace elements for intensive study is to examine journals from a variety of disciplines which impinge on the aquatic sciences and then to make lists of elements which are present in water at low levels and which are acceptable to many or most scientists as being important elements with respect to the health of biota, regardless of whether they are a micronutrient or a toxic substance. Our list is as follows: antimony, arsenic, cadmium, chromium, cobalt, copper, iron, lead, manganese, mercury, molybdenum, nickel, phosphorus, selenium, thallium, uranium and zinc. This list could be extended but speciation studies on other candidates are thus far too limited in scope. Our list is, of course, defective in that we have left aside for now a precise limnological definition of trace element. Interestingly, as a result of some recent findings on acidified lakes and streams, some investigators might now like to include the crustally-abundant element, aluminum, in a list of aquatic trace elements (Driscoll et al., 1980). Below is a brief guide to the literature on trace element species (and also to their relationships with organic components of natural waters). Some relationships between the bulk water and the bottom sediments can be found in the chapter by Förstner and Salomons.

A collection of pertinent facts which relates well to the elements in our list, at least in the context of Canadian surface waters, is outlined by McNeely et al. (1979). Major treatments of individual elements have been published recently as follows: cadmium (Förstner, 1980; Nriagu, 1980a); copper (Nriagu, 1979b); lead (Nriagu, 1978); mercury (Nriagu, 1979a); molybdenum (Chappell and Petersen, 1977; Jarrel et al., 1980); nickel (Nriagu, 1980b); zinc (Nriagu, 1980c). Extensive coverage of most of our list in an aquatic context can be found using Leland et al. (1978), MELIMEX (1979), Florence and Batley (1980), Hart (1981) and the chapter by Smies. Major works which relate well to the list in general include the books by Kavanaugh and Leckie (1980), Förstner and Wittmann (1981), Stumm and Morgan (1981) and the volume of which this chapter is a part.

Some general principles are emerging with respect to the interactions of trace elements and aquatic life, with much of the interaction being mediated by non-living organic materials. It is increasingly clear that various organic substances can play a major role in changing the speciation of trace elements and, consequently, subsequent effects on biota. Depending on the exact nature of an organic-metal association, soluble organics can serve to either facilitate binding of metals to sedimenting particles or compete with particles for soluble metal species so as to maintain a pool of metals in soluble form (see chapter by Förstner and Salomons). Organic-mediated speciation changes may be, in turn, mediated by changing physical conditions, such as changes in pH or redox potential, as well as by changes in the levels of inorganic substances, such as bicarbonate or chloride ion. The overall stress (or benefit) imposed by any given mix of bioavailable elements (both toxic and nutrient) on the biota can change drastically, of course, if any event occurs to change the ratios of the bioavailable elements within the mixture (see chapter by Albergoni and Piccinni). With regard to individual trace elements complexed by organics, chelated toxic metals tend to be much less toxic than the corresponding free metal ions (Hart, 1981), whereas chelation of many micronutrients tends to make them more bioavailable (Wetzel, 1975). These two latter diverse tendencies indicate that many metal speciation studies relating an applied complexing agent to a biological effect may be more complicated than their proponents realize (Jackson and Morgan, 1978). In addition to sending complexing agents out into the aquatic milieu to adjust trace metal speciation, some organisms can convert metals to organo-metallics having carbon to metal covalent bonds (Summers and Silver, 1978). This is apparently done to rid them of undesirable metals while inflicting their neighbors with an environment of greater toxic stress. The most interesting examples of this phenomenon are found in studies of the biomethylation of heavy metals. This major topic is treated in the chapters by Chau and Wong and by Smies.

THE AQUATIC MILIEU AS ENCOUNTERED BY THE LIVING CELL

Living cells in direct contact with surface waters (whether they be algal, bacterial, protozoan or fungal and whether they be organisms unto themselves or components of tissues such as the epithelia of animals) all tend to encounter trace elements in forms other than the simplest free ionic forms. Increasingly, one is impressed by the probable important roles of solids in binding and redistributing trace elements. For metals, the range of natural species can include many diverse physico-chemical forms (Florence and Batley, 1980) such as: the hydrated ion; simple organic and inorganic complexes; stable organic and inorganic complexes; adsorbed on inorganic colloids; adsorbed on organic colloids; bound in various ways to large particles which include solid organics, minerals and mixtures of various kinds of these. In terms of transport, the chemical species can be: truly dissolved; insoluble but suspended; or a component of colloids having intermediate characteristics between those of dissolved substances and sedimenting particles. In terms of stability, they can exist as extremely labile components of living metabolizing cells, or as virtually unavailable entities locked into the crystal structure of minerals or in any number of intermediate states. For non-metals the situation is similar, with phosphorous being a particularly interesting example (Wetzel, 1975; Stumm and Morgan, 1981).

The physiological barrier between the aquatic milieu and the contents of an immersed cell is, of course, the selectively-permeable cell membrane. In the case of many organisms, however, the outer surface of the cell membrane is not directly exposed to the bulk water (Marshall, 1976). Instead, there may be several layers of macromolecular materials external to the cell proper. In the case of macrophytes and many phytoplankton, the innermost of the layers is a rigid wall with all the layers being rich in polysaccharide (Preston, 1974). All external layers may have some capacity to impose a microenvironment on the cell membrane and thus modify its interactions with the bulk water. To express the situation anthropomorphically, the selective membrane "sees" what the external layers "show" to it. One kind of layer, the fibrillar slime layer of some organisms (Leppard et al., 1977; Massalski and Leppard, 1979b), will receive particular attention below.

The chemical speciation of an element determines the access of the biota to it (Bowen, 1979). Thus, it is obvious that measurements of micronutrient bioavailability and microelement toxicity cannot be simple measures. The capacity to make even a precise estimate from chemical and toxicological data is, of course, made even more complicated by the fact that the biota can adjust its own capacity to accept or reject a given trace element (see chapter by Albergoni and Piccinni) and can even adjust the chemical speciation in its immediate environment to suit its own particular needs (see chapters by Giesy and by Smies).

Many of the simpler chemical species of environmentally-significant metals are coming under increasingly more detailed scrutiny, both with regard to the quantitation of chemical species (see chapters by Batley, by Valenta, by Frimmel, and by Chau and Wong) and to the effects of a given chemical form on biota (see chapters by Barghigiani et al., by Albergoni and Piccinni, and by Smies). Among the non-metals, phosphorus presents the greatest challenge (Lean, 1973; Paerl and Downes, 1978; Eisenreich and Armstrong, 1980).

Through increasing successes at transferring previously underutilized technologies into limnology, more complex associations of trace elements and aquatic materials are becoming amenable to sophisticated analysis. In particular, the colloids are receiving a new burst of attention befitting their quantitative significance and their great capacity to bind other materials. Their chemical diversity is enormous but they all have the property of being physical units whose smallest dimensions fall approximately between 1 μm and 1 nm (Vold and Vold, 1966), with the different size classes forming a continuum between truly dissolved polymers and true particles which respond to the forces of sedimentation.

COLLOIDS AND THE INCREASING REALIZATION OF THEIR SIGNIFICANCE AS BINDERS OF TRACE ELEMENTS

Florence and Batley (1980) conclude in their review of chemical speciation in natural waters that colloids can account for much of all metal binding, with organic colloids often playing important roles. Recent studies have shown this to be true of lakes (Beneš and Steinnes, 1975), rivers (Florence, 1977), and estuarine waters (Batley and Gardner, 1978). In some cases, such as for phosphorus, complex transitions between colloidal, dissolved and particulate forms can occur and these presumably may be mediated directly by the biota (Lean, 1973). In a preliminary work by Allen et al. (1978), organic colloids in lake water appeared as strong binders of cadmium.

As a class of physical units, however, colloids have tended to be ignored by aquatic scientists because of technical difficulties in sizing them and in characterizing their native chemical structure. This attitude is changing as it becomes increasingly clear that colloids and their behaviour must be studied more profoundly to understand better the binding and transport of trace element species in natural waters (Breger, 1973; Florence and Batley, 1980). Inorganic colloids have been studied intensively within several scientific disciplines to considerable profit with respect to a general understanding of their properties. However, the inorganic colloids of greatest interest to us are those which form a part of complex larger particles, such as organically-coated clay particles. Such complex particles have proven difficult to analyze in detail in the surface water context (Jenne, 1977; Förstner and Patchineelam, 1980) even though the agricultural sciences have provided sound guidelines for such studies (Mortland, 1970). Organic colloids as entities are often removed from consideration completely as witnessed by the standard limnological practice of using a 0.45 µm filter to divide organic compounds into particulate (POC) and dissolved (DOC) fractions prior to further analyses (Wetzel, 1975). This practice utterly disregards the fact that many natural waters are rich in organic colloids which should be treated as a separate, intermediate class of carbon-rich materials for which the classical cutoff filter is inappropriate (Sharp, 1973).

A new appreciation of the roles of organic coatings on aquatic particles, both inorganic (Jenne, 1977; Davis and Leckie, 1978; Hunter, 1980; and the chapter by Förstner and Salomons) and organic (Massalski and Leppard, 1979b), is focussing attention anew on the contribution of various organic materials to trace element binding by larger sedimenting units. This appreciation is being coupled to the surprising observation that much of the colloidal organic material of the water of some lakes exists in the form of adhesive ribbon-like units called fibrils (Leppard et al., 1977; Massalski and Leppard, 1977a). New findings in both areas are contributing to increased interest by limnologists in organic-rich colloids and their natural associations. Consequently, we will focus the next section on a promising new line of research on this topic and refer the reader for a more global treatment to recent review works on inorganic colloids (Yariv and Cross, 1979) and aquatic particulates in general (Kavanaugh and Leckie, 1980). The latter work addresses many pertinent general questions such as sampling and measurement problems associated with characterization of particulates, the relationship of other things in the aquatic milieu to particulates, and the usefulness of certain measures for water quality management.

COLLOIDAL ORGANIC FIBRILS IN LAKE WATER: CAN WE ASSESS THEIR ROLE IN THE REDISTRIBUTION OF TRACE ELEMENT SPECIES?

Limnologists have employed electron microscopy (EM) for many years to look at the shapes and sizes of organic particles taken directly from lake waters. Particular attention has been given to morphological features relating to microbiological processing of non-living particles (e.g. aggregation, degradation) which transform them physically and chemically during sedimentation (Paerl, 1973: Paerl et al., 1975). This kind of research tended to use the scanning electron microscope (SEM) almost exclusively and, as a consequence of the relatively low resolving power of this kind of EM, colloids in the lower size range received little attention. The use of the transmission electron microscope (TEM) to examine lacustrine colloids (and those of other surface waters) in detail was initially discouraged by the bewildering variety of structures revealed by its much greater power of resolution.

In 1977, this latter situation changed with the observation that one kind of organic colloid (as defined by a combination of shape, dimensions and TEM staining characteristics) appeared to be dominant in water samples from two Canadian lakes (Leppard et al., 1977). This colloid has the appearance of a twisted ribbon with the diameter tending towards 5 nm (0.005 µm) and the length being much greater than the diameter (sometimes more than 100-fold greater). The discovery of this ribbon-like structure, or "fibril" (Figure 1), is providing a focus for new research. As will be shown below, its existence suggests experiments which could yield a conceptual jump forward over past descriptive studies (which struggled with the task of interpreting natural phenomena in terms of a great variety of structures which happened to be taken together as the result of falling into the same size class). Research on lacustrine colloids in our laboratories has focused on these colloidal fibrils since that time.

In 1979, a semi-quantitative microscopical technique revealed fibrils to be present in many Canadian lakes of different sizes and trophic states (Massalski and Leppard, 1979a). The one lake studied intensively had fibrils as a major colloidal component throughout the year at all depths sampled. Since this finding, a technology has been devised to isolate fibrils from lakes in high yield for purposes of

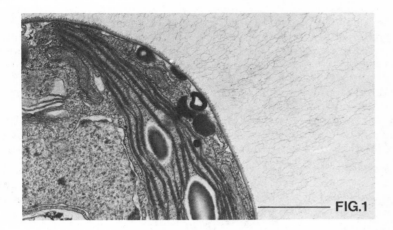

Figure 1. An eukaryotic alga with a dense coating of fibrils projecting from its outer surface into the surrounding lake water. The preparation of the samples for investigation by transmission electron microscopy is described in Burnison and Leppard (in preparation, 1981), modified after Leppard et al. (1977). The bar in Figure 1 and Figure 3 represents one µm.

(1) precise quantitation, (2) detailed chemical analyses and (3) assays for binding capacities with regard to trace substances (Burnison and Leppard, in preparation).

The following five facts suggest that the fibrils may play some important roles in lakes:

(1) Since 1976, 97% of lake water samples analyzed by us have revealed fibrils to be a readily-detected colloidal component;
(2) since 1976, 55% of lake water samples analyzed by us have revealed fibrils to be abundant relative to particulates in general;
(3) our work of the past year on fibril isolation and quantitation reveals that fibrils can comprise up to 35% of the mass of "dissolved" organic carbon in a water sample having a DOC-POC ratio of 10:1 (Burnison and Leppard, in preparation);
(4) our current research on colloidal aggregates continually supplements earlier published observations that fibrils form coatings on a great variety of particles in the water column (Massalski and Leppard, 1979a), including biota (Massalski and Leppard, 1979b);
(5) work with axenic cultures has shown that fibrils can be synthesized by several kinds of aquatic organisms, including

a green alga (Brown et al., 1976), a blue-green alga
(Leppard et al., 1977), and a floating macrophyte (Leppard,
unpublished), and also by the roots of a terrestrial plant
of major agricultural significance (Leppard and Ramamoorthy,
1975).

In addition to their widespread occurrence, their apparent quantitative importance and the immense surface area that small masses of them can provide for interaction with other things in the water, the following four facts suggest that they may be worthy subjects for studies on the binding and redistribution of trace elements:

(1) their major component is polysaccharide (Burnison, 1978), some of it acid polysaccharide, and such a molecular constitution is appropriate to the formation of associations with many diverse substances (Kertesz, 1951);

(2) fibrils will bind, from solution, conventional heavy metal stains used in electron microscopy (lead citrate, uranyl salts) and metal stains employed in light microscopy, such as ruthenium red (Leppard et al., 1977);

(3) fibrils will adhere to many kinds of particles and even cause aggregation of particles (Massalski and Leppard, 1979a), presumably affecting their sedimentation (Hutchinson, 1967) and, consequently, the sedimentation of bound trace elements (Ogura, 1977);

(4) fibrils have been shown to be present in masses of up to nearly 7 mg/l in a water sample from a mesotrophic lake (Burnison and Leppard, in preparation) and this concentration is great in comparison to a commonly accepted upper limit for trace elements in surface waters set at less than 0.05 mg/l.

The nine facts above provide a rationale for a program to assay fibrils for binding activity. Subsequently, this information is to be used to discover the impact of fibrils on the redistribution of trace elements within an aquatic ecosystem. No such assay work has ever been done on lake water fibrils but the procedure of Burnison and Leppard to isolate fibrils in high yields, using physical separation techniques, is currently being refined for trace element binding assays.

One might question why we focus at present on complex lake water samples after only a minimal effort at exploiting laboratory cultures which permit the manipulation of fibril production (Leppard et al., 1977). The reasoning is as follows. It has been suspected for many years (Haug et al., 1969) that both the detailed chemistry and the binding properties of some secreted polysaccharides may be a function of the water quality in which the secreting organism is immersed. Our laboratory work and some fortunate morphological observations from the field indicate that fibril biogenesis and fibril release

from cells are not simple processes. Stated more explicitly, the
sets of facts above suggest that fibrils generated biologically in
the laboratory from a monoculture may have atypical properties.
Also, in natural waters, organisms are exposed simultaneously to
many stresses of differing intensities. If secretion of fibrils
can occur in response to several different stresses, as we suspect,
then a natural mixture of fibril types is more likely to provide
information on the impact of fibrils on lacustrine processes than
is a sample of fibrils generated biologically as the result of a
single laboratory stress. The proper thrust for research on the
fibrils of surface waters does appear to be, then, the analysis of
natural fibrils, with laboratory situations exploited secondarily
to develop concepts as needed.

The basic isolation procedure is outlined schematically in
Figure 2 and our basic attitude is shown in Table 1. The fibril-
enriched product (see Figure 3) can be further enriched by an ultracentrifugation technique to remove non-fibrillar colloids and/or a
selective fractionation technique using ethanol; we have recently
succeeded in producing highly "purified" fibril preparations using
the latter approach. Assaying has begun and factors which complicate interpretation are being assessed. An experiment done recently
in Germany (Ghiorse and Hirsch, 1979), on fibrils which form a
coating around certain bacteria, gives us a preview of what we might
hope to demonstrate for lacustrine fibrils. These attached bacterial
fibrils were able to catalyze a reaction which produced manganese
oxides from manganese ions. When iron in various forms was added to
the water of the test system, the fibrils became associated with
microscopic deposits of iron oxides. Do these phenomena represent

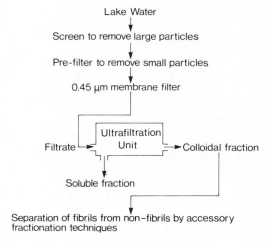

Figure 2. A generalized scheme for isolating fibrils from lakewater in high yield.

Table 1. An outline of the basic attitude adopted for our research on aquatic organic substances of very high molecular weight.

A COMMON APPROACH	OUR APPROACH
APPLY CHEMICAL TECHNIQUES DIRECTLY TO A COMPLEX NATURAL MIXTURE	ISOLATE A SPECIFIC PHYSICAL UNIT OF INTEREST, THEN DO ANALYSIS BY CHEMISTRY
DEGRADE SAMPLE AND DESCRIBE IT IN TERMS OF PRINCIPAL DEGRADATION PRODUCTS	APPLY QUANTITATIVE CHEMISTRY TO SPECIFIC PHYSICAL UNITS WHICH HAD BEEN STABILIZED
OBTAIN CHEMICAL UNITS DESCRIBING STARTING SAMPLE AS A WHOLE	OBTAIN CHEMICAL UNITS WHICH DESCRIBE A SPECIFIC IMPORTANT COMPONENT OF STARTING SAMPLE

Figure 3. The isolated colloidal fraction as seen in section view. Note the dominance of fibrils in this isolate.

part of a mechanism for the genesis of ferromanganese nodules in nature? Only more research can answer this question.

Fibrils were discovered as a result of technology transfer from the biomedical sciences into limnology. This transfer occurred late because it is costly, sophisticated and time-consuming. For these reasons, one can analyze few water samples for fibrils during the Canadian field season. With this constraint in mind we propose the following strategy to ascertain the impact of fibrils on lake metabolism:

(1) discover which trace elements are bound to a significant extent by fibrils;
(2) discover which of the major trace element-fibril associations from the list above lead to important phenomena such as bioavailability changes, detoxification, increased sedimentation of trace elements, etc.;
(3) study the most important fibril-mediated phenomena through time in a very small experimental lake which simultaneously is under intensive investigation with regard to its limnological parameters.

REFERENCES

Albergoni, V., and Piccinni, E., in: this volume.
Allen, H.E., Noll, K.E., Jamjun, O., and Boonlayangoor, C., 1978, Reactions of cadmium in Lake Michigan: kinetics and equilibria, Proc. Am. Chem. Soc. 176th National Meeting, Florida, Abstract 14.
Babich, H., and Stotzky, G., 1979, Physicochemical factors that affect the toxicity of heavy metals to microbes in aquatic habitats, in: "Aquatic Microbial Ecology", R.R. Colwell and J. Foster, eds., A Maryland Sea Grant Publication, University of Maryland, College Park.
Bailey, K.M., and Taub, F.B., 1980, Effects of hydroxamate siderophores (strong Fe(III) chelators) on the growth of algae, J. Phycol., 16:334.
Barghigiani, C., Ferrara, R., Ravera, O., and Seritti, A., in: this volume.
Batley, G.E., and Gardner, D., 1978, A study of copper, lead and cadmium speciation in some estuarine and coastal marine waters, Estuarine Coastal Mar. Sci., 7:59.
Batley, G.E., in: this volume.
Baudo, R., in: this volume.
Beneš, P., and Steinnes, E., 1975, Migration forms of trace elements in natural fresh waters and the effect of the water storage, Water Res., 9:741.
Bowen, H.J.M., 1979, "Environmental Chemistry of the Elements", Academic, London.

Breger, I.A., 1970, What you don't know can hurt you: organic colloids and natural waters, in: "Organic Matter in Natural Waters", D.W. Hood, ed., Institute of Marine Science, Occasional Publ. No. 1, College, Alaska.

Brown, D.L., Massalski, A., and Leppard, G.G., 1976, Fine structure of excystment of the quadriflagellate alga Polytomella agilis, Protoplasma, 90:155.

Burnison, B.K., 1978, High molecular weight polysaccharides isolated from lake water, Verh. Internat. Verein. Limnol., 20:353.

Burnison, B.K., and Leppard, G.G., 1981, Isolation of colloidal fibrils from lake water by physical separation techniques, in preparation.

Chappell, W.R., and Petersen, K.K., 1977, "Molybdenum in the Environment. Volume 2. The Geochemistry, Cycling, and Industrial Uses of Molybdenum", Marcel Dekker, New York.

Chau, Y.K., and Wong, P.T.S., in: this volume.

Davis, J.A., and Leckie, J.O., 1978, Effect of adsorbed complexing ligands on trace metal uptake by hydrous oxides, Environ. Sci. Technol., 12:1309.

Driscoll, C.T., Jr., Baker, J.P., Bisogni, J.J., Jr., and Schofield, C.L., 1980, Effect of aluminium speciation on fish in dilute acidified waters, Nature, 284:161.

Eisenreich, S.J., and Armstrong, D.E., 1980, Association of organic matter, iron and inorganic phosphorus in lake waters, Env. Int., 3:485.

Eisenreich, S.J., Hoffmann, M.R., Rastetter, D., Yost, E., and Maier, W.J., 1980, Metal transport phases in the upper Mississippi River, in: "Particulates in Water-Characterization, Fate, Effects, and Removal", M.C. Kavanaugh and J.O. Leckie, eds., Am. Chem. Soc., Washington, D.C.

Emery, T., 1971, Role of ferrichrome as a ferric ionophore in Ustilago sphaerogena, Biochemistry, 10:1483.

Florence, T.M., 1977, Trace metal species in fresh waters, Water Res., 11:681.

Florence, T.M., and Batley, G.E., 1980, Chemical speciation in natural waters, CRC Crit. Rev. Anal. Chem., 9:219.

Förstner, U., 1980, Cadmium, in: "The Handbook of Environmental Chemistry, Volume 3, Part A, Anthropogenic Compounds", O. Hutzinger, ed., Springer-Verlag, Berlin.

Förstner, U., and Patchineelam, S.R., 1980, Chemical associations of heavy metals in polluted sediments from the lower Rhine River, in: "Particulates in Water-Characterization, Fate, Effects, and Removal", M.C. Kavanaugh and J.O. Leckie, eds., Am. Chem. Soc., Washington, D.C.

Förstner, U., and Wittmann, G., 1981, "Metal Pollution in the Aquatic Environment, Second Edition", Springer-Verlag, Berlin.

Förstner, U., and Salomons, W., in: this volume.

Frimmel, F.H., in: this volume.

Gächter, R., and, Urech, J., in: this volume.

Ghiorse, W.C., and Hirsch, P., 1979, An ultrastructural study of iron and manganese deposition associated with extracellular polymers of *Pedomicrobium*-like budding bacteria, Arch. Microbiol., 123: 213.

Giesy, J.P., in: this volume.

Hart, B.T., 1981, Trace metal complexing capacity of natural waters: a review, Environ. Technol. Lett., 2:95.

Hunter, K.A., 1980, Microelectrophoretic properties of natural surface-active organic matter in coastal seawater, Limnol. Oceanogr., 25:807.

Hutchinson, G.E., 1967, The hydromechanics of the plankton, in: "A Treatise on Limnology, Volume 2", G.E. Hutchinson, John Wiley and Sons, New York.

Haug, A., Larsen, B., and Baardseth, E., 1969, Comparison of the constitution of alginates from different sources, in: "Proceedings of the Sixth International Seaweed Symposium", R. Margalef, ed., Subsecretaria de la Marina Mercante, Madrid.

Jackson, G.A., and Morgan, J.J., 1978, Trace metal-chelator interactions and phytoplankton growth in seawater media: theoretical analysis and comparison with reported observations, Limnol. Oceanogr., 23:268.

Jarrel, W.M., Page, A.L., and Elseewi, A.A., 1980, Molybdenum in the environment, Residue Rev., 74:1.

Jenne, E.A., 1977, Trace element sorption by sediments and soils - sites and processes, in: "Molybdenum in the Environment, Volume 2", W.R. Chappell and K.K. Petersen, eds., Marcel Dekker, New York.

Kavanaugh, M.C., and Leckie, J.O., eds., 1980, "Particulates in Water- Characterization, Fate, Effects, and Removal", Am. Chem. Soc., Washington, D.C.

Kertesz, Z.I., 1951, "The Pectic Substances", Interscience Publishers, New York.

Lean, D.R.S., 1973, Movements of phosphorous between its biologically important forms in lake water, J. Fish. Res. Board Can., 30:1525.

Leland, H.V., Luoma, S.N., Elder, J.F., and Wilkes, D.J., 1978, Heavy metals and related trace elements, Journal WPCF, 50:1469.

Leppard, G.G., and Ramamoorthy, S., 1975, The aggregation of wheat rhizoplane fibrils and the accumulation of soil-bound cations, Can. J. Bot., 53:1729.

Leppard, G.G., Massalski, A., and Lean, D.R.S., 1977, Electron-opaque microscopic fibrils in lakes: their demonstration, their biological derivation and their potential significance in the redistribution of cations, Protoplasma, 92:289.

Marshall, K.C., 1976, "Interfaces in Microbial Ecology", Harvard University Press, Cambridge, Mass.

Massalski, A., and Leppard, G.G., 1979a, Survey of some Canadian lakes for the presence of ultrastructurally discrete particles in the colloidal size range, J. Fish. Res. Board Can., 36:906.

Massalski, A., and Leppard, G.G., 1979b, Morphological examination of fibrillar colloids associated with algae and bacteria in lakes, J. Fish. Res. Board Can., 36:922.

McNeely, R.N., Neimanis, V.P., and Dwyer, L., 1979, "Water Quality Sourcebook, a Guide to Water Quality Parameters", Environment Canada, Inland Waters Directorate, Water Quality Branch, Ottawa.

Mortland, M.M., 1970, Clay-organic complexes and interactions, Adv. Agron., 22:75

MELIMEX, 1979, An experimental heavy metal pollution study. (Various titles and authors), Schweiz. Z. Hydrol., 41:pp. 165-314. Separatum Nr. 770.

Murphy, T.P., Lean, D.R.S., and Nalewajko, C., 1976, Blue-green algae: their excretion of iron-selective chelators enables them to dominate other algae, Science, 192:900.

Nriagu, J.O., editor, 1978,"The Biogeochemistry of Lead in the Environment, Part A. Ecological Cycles", Elsevier/North-Holland Biomedical Press, Amsterdam.

Nriagu, J.O., editor, 1979a, "The Biogeochemistry of Mercury in the Environment", Elsevier/North-Holland Biomedical Press, Amsterdam.

Nriagu, J.O., editor, 1979b, "Copper in the Environment. Part one: Ecological Cycling", John Wiley and Sons, New York.

Nriagu, J.O., editor, 1980a, "Cadmium in the Environment. Part one: Ecological Cycling", John Wiley and Sons, New York.

Nriagu, J.O., editor, 1980b, "Nickel in the Environment", John Wiley and Sons, New York.

Nriagu, J.O., editor, 1980c, "Zinc in the Environment. Part one: Ecological Cycling", John Wiley and Sons, New York.

Ogura, N., 1977, High molecular weight organic matter in seawater, Mar. Chem., 5:535.

Paerl, H.W., 1973, Detritus in Lake Tahoe: structural modification by attached microflora, Science, 180:496.

Paerl, H.W., and Downes, M.T., 1978, Biological availability of low versus high molecular weight reactive phosphorous, J. Fish. Res. Board Can., 35:1639.

Paerl, H.W., Thomson, R.D., and Goldman, C.R., 1975, The ecological significance of detritus formation during a diatom bloom in Lake Tahoe, California-Nevada, Verh. Internat. Verein. Limnol., 19:826.

Preston, R.D., 1974,"The Physical Biology of Plant Cell Walls", Chapman and Hall, London.

Sharp, J.H., 1973, Size classes of organic carbon in seawater, Limnol. Oceanogr., 18:441.

Simpson, F.B., and Neilands, J.B., 1976, Siderochromes in Cyanophyceae: isolation and characterization of schizokinen from Anabaena sp., J. Phycol., 12:44.

Smies, M., in: this volume.

Stumm, W., and Morgan, J.J., 1981, "Aquatic Chemistry, Second Edition", Wiley, New York.

Summers, A.O., and Silver, S., 1978, Microbial transformations of metals, Ann. Rev. Microbiol., 32:637.

Valenta, P., in: this volume.

Vallentyne, J.R., 1974,"The Algal Bowl - Lakes and Man", Environment Canada, Fisheries and Marine Service, Ottawa, Misc. Spec. Publ. 22:pp. 1-185.

Vold, M.J., and Vold, R.D., 1964, "Colloid Chemistry", Reinhold Publ.
   Corp., New York.
Wetzel, R.G., 1975, "Limnology", W.B. Saunders Co., Philadelphia.
Yariv, S., and Cross, H., 1979, "Geochemistry of Colloid Systems for
   Earth Scientists", Springer-Verlag, Berlin.

DISCUSSION: G.G. LEPPARD AND B.K. BURNISON

R.F. VACCARO   Would a lake water sample containing no micro-
                     bial cells remain stable with regard to its number
                     of fibrils or do numerical changes ensue with time?

G.G. LEPPARD   We have not done this experiment. We have noted
                     that fibrils can aggregate to form small flocs or
                     even macroscopic large flocs, but we do not under-
                     stand the mechanics of floc genesis. Also, fibril
                     degradation rate is not known and any starting sample
                     from lake water probably would contain a mixture of
                     fibril populations in different stages of degrada-
                     tion. These complications, coupled with the lack of
                     a standardized technique for counting individual
                     fibrils, have convinced us that experiments based
                     on counts of individual fibrils are not feasible at
                     present.

M. BRANICA   Did you attempt any specific physico-chemical
                     characterization of the colloidal organic fibrils?
                     I suggest looking at fibril effects on the stability
                     of known inorganic colloids?

G.G. LEPPARD   We delayed any detailed characterization be-
                     cause we had been making continual improvements in
                     the purification of isolated fibril fractions and
                     we wished to concentrate on specifics only when we
                     had achieved maximum purity. Recently, we attained
                     this goal and it is now appropriate to ask specific
                     questions related to fibril chemistry and function.
                     Your suggestion regarding a fibril behaviour study
                     is an excellent one.

H.W. NURNBERG   Since the fibrils contain polysaccharides, one
                     would expect, for heavy metals such as lead and
                     copper, a certain range of binding strengths. If
                     experimentally-measured binding were found to be
                     stronger, this could be attributed to other sub-
                     stances on the surface of the fibrils. Have you
                     done the binding measurements to permit inferences
                     about the fibril surface? Also, if you have looked
                     at binding in general, what heavy metals do you find
                     on fibrils?

G.G. LEPPARD Only recently have we been able to isolate fibrils in a quantity and purity sufficient to permit the experimental binding studies you suggest. We have at this moment only preliminary results. Work on the binding, in nature, of heavy metals by fibrils is still only preliminary.

R. BREDER What is the pressure that you use for membrane-filtration? Also, at what pressure do you think cells are broken or small organisms are destroyed?

G.G. LEPPARD We used 0.7atm. of positive pressure. There is no absolute answer to your second question; the pressure which causes breakage is a function of the kinds of organisms present. We did not have a breakage problem.

H.W. NURNBERG What was the typical concentration range for fibrils in the inland waters studied by your group? Also, do they occur in seawater?

G.G. LEPPARD Not enough samples were assayed to permit us to establish a "typical" range. The gross analyses ran from near 0 mg/litre up to 7 mg/litre. I have examined seawater once and have seen them in seawater once. No general survey of seawater has yet been made for fibrils.

E.K. DUURSMA Are the fibrils reacting as dissolved or as particulate material at a pH of 1? I note that this pH is often used to preserve samples and to allow particulates to settle.

G.G. LEPPARD We have not investigated them at a pH of 1. My preservation technique permits short term storage of them at pH values near 7, and my experience with complex biological materials in general has shown me that highly acidic preservation conditions can produce a great variety of artifacts.

G.E. BATLEY You mentioned that you chemically stimulate some organisms to produce fibrils. Specifically how do you do this?

G.G. LEPPARD We have had success with starvation stress. In particular, providing insufficient phosphate ion in a growth medium can induce fibril biogenesis (see Leppard et al., 1977).

BIOLOGICAL EFFECTS UNDER LABORATORY CONDITIONS

C. Barghigiani[+], R. Ferrara[+], O. Ravera[*] and A. Seritti[+]

[+]CNR - Istituto per lo Studio delle Proprietà
Fisiche di Biomolecole e Cellule, I-56100 Pisa, Italy

[*]Commission of the European Communities, Ispra
Establishment, Dept. of Physical and Natural Sciences
I-21020 Ispra (Va), Italy

INTRODUCTION

In the natural environment, organisms are exposed simultaneously to more than one stress and, therefore, they react to the entire environmental situation and not to individual stresses. As a consequence, it is difficult in the natural environment to establish relationships between the intensity of a single stress and its biological effects. The problem becomes more complex if the environment is polluted. For this reason, information on the effects of single pollutants on natural populations is very scarce.

In the case of the biological effects of metal speciation, we know that these are strictly related to various important environmental characteristics, such as temperature, pH and concentration of suspended particles. Consequently, to obtain information on the effects produced by a single stress (e.g., a given form of a heavy metal), it is necessary to carry out laboratory experiments in which each environmental factor can be maintained constant. These experiments are, generally, carried out on single species. If more than one species is used in the same container (such as one does with an aquarium), interspecific relationships (e.g., predation, parasitism) must be prevented and competition between individuals of the same species must be abolished or attenuated. Also, because experimental conditions are very artificial, it is difficult to predict the biological effects of any stress (e.g., heavy metal pollution) in the natural environment using results obtained from laboratory studies.

A useful compromise between laboratory experimentation and field observation is the "field microecosystem" ("enclosure") method which maintains a representative aliquot of the natural community in semi-natural conditions (see chapter by Gächter and Urech, this volume).

CHOICE OF CULTURE SYSTEM

The choice of experimental conditions and analytical methods presents several problems. In Table 1, the most commonly used experimental systems, with their advantages and disadvantages, are delineated. We note that batch experiments are very simple and may produce useful information such as determinations of lethal concentration, intracellular penetration and adsorption of a toxicant onto microorganism cell walls. On the other hand, in batch cultures, the influence of the pollutant is masked and modified with time by various factors, particularly by variations in those physical and chemical characteristics of the medium which result from biological activity. These variations are particularly detrimental in studies on the effects of different heavy metal species. In addition, the initial metal concentration in the medium may be reduced during the experiment by precipitation as well as by adsorption onto surfaces such as the walls of the container, suspended particles, cell surface structures of microorganisms and the epithelia of metazoa. Moreover, for mercury toxicity experiments, a loss by evaporation has to be considered (Huisman and ten Hoopen, 1978).

In experiments using microorganisms, in order to maintain metal concentration, cell concentration and medium composition constant with time, continuous flow culture systems of two types have been used (Ferrara et al., 1975; Premazzi et al., 1978; Tan, 1980). In one system, the flow rate of the medium (plus the pollutant) controls the population growth rate (chemostat system); in the other, the growth rate of the population controls the flow rate (turbidostat system). Open flow systems are also used in experiments on metazoa.

To test the effects of a given heavy metal under more natural conditions, experiments may be carried out on artificial communities maintained under constant conditions. This type of experimentation, when using metazoa, employs aquaria ("balanced aquaria", a laboratory microecosystem) containing different plant and animal species usually collected from one environment. The most important disadvantage of these experiments is that numerical proportions among the different species are rarely similar to those occurring in the natural environment.

Large, enclosed, water column systems represent a compromise between laboratory and field experiments (Takahashi et al., 1975; Kerrison et al., 1980). With this method, most of the drawbacks reported in Table 1 for laboratory experiments are overcome because a complete

Table 1. Advantages and disadvantages of different experimental systems

Experimental Systems		Advantages	Disadvantages
Small Volume Cultures	Batch Cultures	- Easy to carry out and maintain - Low cost	- Decrease of heavy metal with time - Increase of cell concentration with time - Accumulation of metabolites with time - Decrease of nutrients with time
	Continuous Flow Systems	- Constant heavy metal concentration with time - Constant cell concentration with time - No metabolite accumulation - Fresh medium with time	- High cost - Complexity of the system
Large Volume	Large enclosed water column	- Semi-natural experimental conditions	- High cost - Complexity of the experiment - Difficult to carry out and maintain

sample of the natural ecosystem can be studied. The limitation of
this method is the low reproducibility of the measured parameters
resulting from the seasonal variations of a highly complex system.
In addition, the horizontal water transport unfortunately is abolished
and the "wall effect" influences the results.

CHOICE OF MEDIUM

The choice between the use of natural or artificial culture
media has been discussed in the literature (Provasoli et al., 1957;
see also Ravera, 1977). Natural media exhibit a variable composition
with season and collection site, and they may themselves complex
heavy metals (Batley and Gardner, 1978; Hart, 1981). The presence of
chelators has a fundamental influence on metal speciation in culture
media. In Figure 1, the reduction of radioisotope uptake (Zn-65 and
Co-58) produced by synthetic chelators (SNTA and EDTA) in a freshwater copepod (Eudiaptomus padanus) is evident. Very similar results
have been obtained with the same chelators using a freshwater mussel

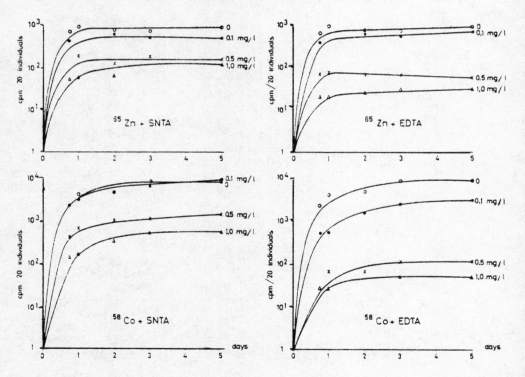

Figure 1. Effect of different concentrations of SNTA and EDTA on the
uptake of Zn-65 and Co-58 by Eudiaptomus padanus (in
Ravera et al., 1973).

(Unio mancus). In addition, the toxic effects of Cu, Cd, Ni, Hg, Zn and Pb were reduced in the unicellular green alga, Selenastrum minutum, in the presence of EDTA, SNTA and humic acids (Premazzi et al., 1977). In natural media, the influence of chelators may be abolished by destroying the organic molecules with UV irradiation. Artificial media have a defined composition, but for algal culture chelators are needed in order to complex some trace metal nutrients, especially iron (Gates and Wilson, 1960). For experiments on organisms which need an organic medium (e.g., protozoa, bacteria) it is obvious that chelators (e.g., aminoacids) must be present.

When artificial media are used, it must be remembered that commercially-available chemicals contain, as impurities, quantities of metals which are often of the same order of magnitude as those chosen for spiked-metal tests. This problem can be resolved by using chelating resins, such as Chelex 100 (Davey et al., 1970). It is to be remembered that most organisms would be expected to liberate products capable of forming complexes with heavy metals (Sharp, 1977; Fogg, 1977) irrespective of the type of medium used. This situation is discussed in other chapters of this volume.

SOME IMPORTANT PARAMETERS

Certain fundamental parameters (for example, light, oxygen concentration, concentration of suspended particles) must always be recorded during an experiment in order to evaluate the variations of the above mentioned parameters with time. For example, metal speciation is strongly affected by pH and temperature (Duinker and Kramer, 1977); as a consequence, these factors must be monitored in the medium for the duration of the experiment. Because aquatic organisms normally produce metabolites capable of producing changes in pH (Sunda, 1975), the measurement of this parameter is particularly important in batch experiments.

Metal speciation plays an important role because certain physico-chemical forms of a metal are more available to some organisms than to others. For this reason, the concentrations of ionic and complexed forms of the metal must be followed with time. Anodic stripping voltammetry at present seems to be the most suitable technique (Nürnberg, 1979; Sipos et al., 1979) because of its high sensitivity and because it allows metal speciation to be followed. However, this technique cannot be used for mercury. Low concentrations of mercury can be measured accurately with atomic fluorescence spectrometry (Ferrara et al., 1980).

In the literature, few papers report on the behaviour of metals in culture media (see Seritti et al., 1981; see Sunda and Lewis, 1978), and concentrations are commonly considered only at the beginning of a test. The results of one of our experiments using batch cultures of <u>Dunaliella salina</u>, spiked with 5 µg/l of ionic mercury, are schematized in Figure 2. It can be observed that, immediately after spiking, 50% of the metal is present in ionic form, approximately 40% is already organically-associated, about 7% is associated with cells (most likely adsorbed) and 3% is probably adsorbed on the container wall. This Figure (2) emphasizes the importance of the above considerations regarding change of chemical speciation and concentration of available chemical species.

CHOICE OF THE ORGANISM

The choice of biological species to be studied is a fundamental problem and obviously depends on the aims of the research. For example, cultures of bacteria or microscopic algae are very useful in evaluating effects at the population level and during several successive generations. Algal cultures may also be used for measuring the influence of pollutants on primary production, photosynthetic activity and population growth. In some species of metazoa (e.g., fish, insects, molluscs), the distribution of a metal and its biochemical fate in different tissues and organs can be studied. The ratio of the concentration of a given heavy metal in the organism to that of the same metal in the medium is called the concentration factor (or

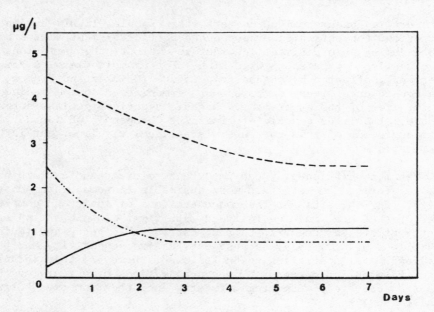

Figure 2. Distribution of mercury in a batch culture of <u>Dunaliella salina</u> spiked with 5 µg/l of ionic mercury. (Barghigiani, personal comm.). (-----) total mercury; (-··-··) ionic mercury; (———) mercury associated with the cells.

c.f.). This is a useful index in both pure and applied research, one example being studies on the transfer of a metal through succesive rings of a food-chain. If equilibrium between the concentration of metal inside and outside the organism is attained, the concentration factor is called "actual" to distinguish it from the "observed concentration factor" calculated when equilibrium is not attained. It is clear that any variation of the metal concentration in a medium modifies that in the body of an organism. The aliquot of metal adsorbed on the cell wall of microorganisms or on the body surface of metazoa influences the value of the concentration factor. Unfortunately, to distinguish between the amount of metal adsorbed and the amount of metal incorporated into cells and tissues is a very difficult problem.

To obtain a satisfactory evaluation of metal toxicity on aquatic biota, several taxonomical groups must be tested. In some groups, the number of species tested is relatively high (e.g., fish) but in others it is very low and, consequently, the rearing and testing of additional species should be investigated. Because the sensitivity to a particular metal generally varies with age, experiments must be carried out using the adult and all other developmental stages for each species.

Discrepancies between results obtained from laboratory experiments on the effects of a given concentration of a given metal on a single species may be very large for reasons such as differences in the acclimatization conditions and in the physiological and genetic characteristics of the organism tested as well as differences in handling of data.

If unicellular organisms are used in batch culture, cell concentrations must be measured since the metal-to-cell concentration ratio strongly decreases in long-term experiments. For the same reason, if metazoa are used, the ratio between biomass and volume of the medium cannot be higher than a certain value.

Laboratory experiments must generally be planned to reproduce, at least partially, the conditions in nature. For this reason, the choice of organism density and metal concentration should not exceed 10 to 100 times their natural background levels. As a consequence, using these concentrations for experiments on microorganisms, the following difficulties arise: (1) the cell count at low cell concentrations is never accurate and is always time-consuming; (2) the metal associated with cells is difficult to evaluate because of the low biomass present in the cultures. Furthermore, low metal concentrations decrease rapidly in the medium because of the fraction adsorbed onto the cells and container wall. In addition, analytical problems arise when determining the metal concentrations and their chemical forms in the medium. To overcome the above mentioned difficulties, at least partially, radioisotopes with a carrier may be used (a carrier being defined here as a stable metal of the same radionuclide) (Skaar et al., 1974; Davies and Sleep, 1979). In any case, the effect of metals is strongly dependent on the initial population density. This fact is probably responsible for the large range of inhibitory concentrations of metals reported in the literature (e.g., Berland et al., 1976; Thomas et al., 1977; Wilson and Freeberg, 1979) for planktonic algae. More homogeneous data are available on natural population experiments.

BIOLOGICAL EFFECTS

Commonly used biological parameters and the relevant experimental techniques are reported in Table 2.

Experimental responses may be divided into two groups: (1) <u>acute effects</u> (short-term effects occurring over hours or a few days); and (2) <u>chronic effects</u> (long-term effects occurring over more than a week). Some authors (Overnell, 1976; Venrick, 1977; Berland et al., 1980) prefer short-term experiments in order to minimize the accumulation of metabolic products which might complex the metals and thus affect their "availability" (Leland et al., 1978). This problem may be overcome by using a continuous flow system. Moreover, short-term experiments allow natural populations with a very rapid demographic turnover (such as phytoplankton) to be studied while avoiding changes in community structure (Hollibaugh et al., 1980). However, the metal concentrations used must be rather high in order for one to observe significant effects in a short time. The simpler experiments on metal toxicity employ mortality criteria; percentage mortality in a group of tested organisms, after a given exposure (such as 24, 48 or 96 hours) to a single metal or a mixture of heavy metals in solution, is

Table 2. Commonly used biological parameters and relevant experimental techniques.

Parameters	Experimental Techniques
- mortality	- counting
- growth and reproduction	- growth curves
- bioaccumulation phenomena	- metal determination in organisms
- photosynthetic activity	- oxygen evolution, $^{14}CO_2$ uptake
- heterotrophic activity	- ^{14}C glucose uptake
- respiration	- oxygen uptake
- pigmentogenesis	- pigment determination
- biochemical response	- enzymatic activity, ATP determination, etc.
- modification of cellular and subcellular structure	- optical and electron microscopy
- mutagenic effects	- mutagenic tests

commonly used. The time necessary for 50% of the organisms to be
killed by a given concentration of a pollutant is another estimation
of the toxicity (mean biological time: m.b.t.). Different indices
are adopted to estimate the influence of a given pollutant on the
mortality of a given species; for example there is the TL (tolerance
limit), the LD (lethal dose), the LC (lethal concentration) and the
ST (survival time). The lethal concentration (LC) concept is preferable when the pollutant concentration is measured in the medium, whereas the lethal dose (LD) concept should be used when the pollutant concentration concerning the body of the organism is known. Several of
these criteria are routinely used to establish maximum acceptable
values of pollutants.

The experiments referred to above must be considered complemen-
tary to long-term investigations on metal accumulation in organisms;
these latter investigations, as well as those on effects upon growth
rate, need a relatively long incubation time (days). Long-term experiments present, in general, more difficulties than do short-term
ones and, therefore, they need more extensive experimental planning.

To evaluate sub-lethal effects, other kinds of responses must
be obtained (for example, effects on growth, reproduction, physiological activities and biochemical characteristics). In addition,
some genetic, teratogenic and behavioural effects may be useful
criteria for evaluating heavy metal toxicity.

The accumulation of heavy metals in specific organs and tissues
(critical organs and tissues) is also used to predict the influence
of a given pollutant on the species tested. By using radionuclides,
the turnover-time of the metal assimilated may be accurately measured
for the whole organism as well as for its organs, tissues and biochemical compartments.

The few studies carried out on the effects produced by more than
one metal have shown the importance of the multi-stress approach.
Some effects are merely additive, whereas others may be synergistic
or antagonistic. In some cases, the influence of a given metal or
metals may be very complex.

CONCLUSIONS

It is always difficult to extrapolate results obtained in the
laboratory to the natural environment. However, the influence of
heavy metals on behavioural, physiological and biochemical parameters
(such as the uptake and the loss rate of the metal, and its fate after
assimilation) may be revealed by laboratory experiments. Also in the
laboratory, toxicity and detoxification mechanisms may be studied at
biochemical, cellular and individual levels. Experiments on heavy
metal transfer through short food chains may yield useful information

provided that the natural sequence of predator and prey is maintained in the chain.

It must be pointed out strongly that the biological response obtained in a test is related to the physical and chemical forms of the metal found in the medium; these depend not only on the form of the spiked metal but also on the complexing capacity of the medium and on the nature of the activities of the test organism. At present, most of the biological experiments reported in the literature on the toxic effects of metals do not take into account the speciation of the element. On the other hand, the physico-chemical studies of metals in aquatic environments often do not consider the influence of living matter on metal speciation.

In conclusion, the study of metal speciation in the culture medium situation allows a superior evaluation of a metal's toxic effects and also contributes to the understanding of metal speciation in aquatic environments.

REFERENCES

Batley, G.E., and Gardner, D., 1978, A study of copper, lead and cadmium speciation in some estuarine and coastal marine waters, Estuarine Coastal Mar. Sci., 7:59.

Berland, B., Bonin, D.J., Kapkov, V.I., Maestrini, S., and Arlhac, D., 1976, Action toxique de quatre métaux lourds sur la croissance d'algues unicellulaires marines, C.R. Hebd. Séanc. Acad. Sci., Paris, 282:633.

Berland, B., Chretiennot-Dinet, M., Ferrara, R., and Arlhac, D., 1980, Action à court terme du mercure sur les populations naturelles phytoplanctoniques et bactériennes d'eaux côtière de la Méditerranée nord-occidentale, Proc. V. Journées Etud. Pollutions, Cagliari, C.I.E.S.M., 721.

Davey, E.W., Gentile, J.H., Erickson, S.J., and Betzer, P., 1970, Removal of trace metals from marine culture media, Limnol. Oceanogr., 15:486.

Davies, A.G., and Sleep, J.A., 1979, Inhibition of carbon fixation as a function of zinc uptake in natural phytoplankton assemblages, J. Mar. Biol. Ass. U.K., 59:937.

Duinker, J.C., and Kramer, C.J.M., 1977, An experimental study on the speciation of dissolved zinc, cadmium, lead and copper in the River Rhine and North Sea water, by differential pulse anodic stripping voltammetry, Mar. Chem., 5:207.

Ferrara, R., Grassi, S., and Del Carratore, G., 1975, An automatic homocontinuous culture apparatus, Biotechnol. & Bioeng., 17:985.

Ferrara, R., Seritti, A., Barghigiani, C., and Petrosino, A., 1980, Improved instrument for mercury determination by atomic fluorescence spectrometry with a high frequency electrodeless discharge lamp, Anal. Chim. Acta, 117:391.

Fogg, G.E., 1977, Excretion of organic matter by phytoplankton, Limnol. Oceanogr., 22:576.

Gates, J.A., and Wilson, W.B., 1960, The toxicity of Gonyaulax monilata Howell to Mugil cephalus, Limnol. Oceanogr., 5:171.

Hart, B.T., 1981, Trace metal complexing capacity of natural waters: a review, Environ. Technol. Lett., 2:95.

Hollibaugh, J.T., Seibert, D.L.R., and Thomas, W.H., 1980, A comparison of the acute toxicities of ten heavy metals to phytoplankton from Saanich Inlet, B.C., Canada, Estuarine Coastal Mar. Sci., 10:93.

Huisman, J., and ten Hoopen, H.J.G., 1978, A mercury buffer for toxicity experiments with green algae, Water Air Soil Pollut., 10:325.

Kerrison, P.H., Sprocati, A.R., Ravera, O., and Amantini, L., 1980, Effects of cadmium on an aquatic community using artificial enclosures, Environ. Technol. Lett., 1:169.

Leland, H.V., Luoma, S.N., Elder, J.F., and Wilkes, D.J., 1978, Heavy metals and related trace elements, Journal WPCF, 50:1469.

Nürnberg, H.W., 1979, Polarography and voltammetry in studies of toxic metals in man and his environment, Sci. Total Environm., 12:35.

Overnell, J., 1976, Inhibition of marine algal photosynthesis by heavy metals, Mar. Biol., 38:335.

Premazzi, G., Ravera, O., and Lepers, A., 1978, A modified turbidostatic system for algal population studies, Mitt. Internat. Verein. Limnol., 21:42.

Premazzi, G., Bertone, R., Freddi, A., and Ravera, O., 1977, Combined effects of heavy metals and chelating substances on Selenastrum cultures, Proceedings of a Seminar on Ecological Tests Relevant to the Implementation of Proposed Regulations Concerning Environmental Chemicals: Evaluation and Research Needs, Berlin, Dec. 7-9, 169-187.

Provasoli, L., McLaughlin, J.J.A., and Droop, M.R., 1957, The development of artificial media for marine algae, Arch. Mikrobiol., 25:392.

Ravera, O., Cartisano, A., de Bernardi, R., and Guzzi, L., 1973, Effects of chelating agents (EDTA and SNTA) on the incorporation of radionuclides by freshwater filter feeding organisms (Copepod and Lamellibranch), in Atti 5° Coll. Int. Oceanogr. Med., Messina, 437-448.

Ravera, O., 1977, Effects of heavy metals (cadmium, copper, chromium and lead) on a freshwater snail: Biomphalaria glabrata Say (Gastropoda, Prosobranchia), Malacologia, 16:231.

Seritti, A., Ferrara, R., Barghigiani, C., Petrosino, A., Del Carratore, G., and Torti, M., 1981, A preliminary study on the distribution of ionic cadmium in batch cultures of Dunaliella salina by differential pulse anodic stripping voltammetry, Thalassia Jugosl., 17:55.

Sharp, J.H., 1977, Excretion of organic matter by marine phytoplankton: Do healthy cells do it?, Limnol. Oceanogr., 22:381.

Sipos, L., Valenta, P., Nürnberg, H.W., and Branica, M., 1979, Voltammetric determination of the stability constants of the predominant labile lead complexes in sea water, in: "Proc. Int. Experts

Discussion on Lead Occurrence, Fate and Pollution in the Marine Environment", Rovinj (October, 1977), M. Branica and Z. Konrad, eds., Pergamon Press, Oxford.

Skaar, H., Rystad, B., and Jensen, A., 1974, The uptake of ^{63}Ni by a diatom Phaeodactylum tricornutum, Physiologia Pl., 32:353.

Sunda, W.G., 1975, "Relationship Between Cupric Ion Activity and the Toxicity of Copper to Phytoplankton", (Ph.D. Thesis, Mass. Inst. Technol.), Cambridge, Mass., U.S.A., 167 pp.

Sunda, W.G., and Lewis, J.A.M., 1978, Effect of complexation by natural organic ligands on the toxicity of copper to a unicellular alga, Monochrysis lutheri, Limnol. Oceanogr., 23:870.

Takahashi, M., Thomas, W.H., Seibert, D.L.R., Beers, J., Koeller, P., and Parsons, T.R., 1975, The replication of biological events in enclosed water columns, Arch. Hydrobiol., 76:5.

Tan, T.L., 1980, Effect of long - term lead exposure on the seawater and sediment bacteria from heterogeneous continuous flow cultures, Microb. Ecol., 5:295.

Thomas, W.H., Seibert, D.L.R., and Takahashi, M., 1977, Controlled ecosystem pollution experiment: effect of mercury on enclosed water columns. III. Phytoplankton population dynamics and production, Mar. Sci. Commun., 3:331.

Venrick, E.L., 1977, Possible consequences of containing microplankton for physiological rate measurements, J. Exp. Mar. Biol. Ecol., 26:55.

Wilson, W.B., and Freeberg, L.R., 1979, (U.S.) E.P.A. Report No. R 801/511.

DISCUSSION: C. BARGHIGIANI, R. FERRARA, O. RAVERA AND A. SERITTI

H.W. NURNBERG   Your concluding statement, that physico-chemical investigations on speciation often neglect the influences of the biota in natural water systems, is not true. Sound investigations will try to account, at least, for all major substances contributing to speciation. Thus, inevitably, they will include the influences of living matter, including exudates, provided that the biological material of interest is present at a significant concentration level. There is no magic factor for living matter.

O. RAVERA   In some cases, metal speciation is considered only from a physico-chemical perspective. For example, some models of metal speciation are made purely in relation to physico-chemical factors such as pH and temperature. In other cases, the biota is considered from a point of view which is clearly in disagreement with basic ecological concepts.

G.E. BATLEY   I should like to correct your statement that A.S.V. cannot be used for mercury. As professor Nürnberg will verify, A.S.V. (using a gold or glassy carbon electrode) can provide a sensitive measure of mercury species.

R. FERRARA   I think our statement is correct. The gold electrode represents a promising technique, but, at the moment, it is not sensitive enough. This is true not only for speciation studies but probably also for the measurement of total dissolved mercury in sea water (2-10 ng/l).

F.H. FRIMMEL   The distribution of the originally-added inorganic Hg in a batch culture shows immediately a level of 40% "organically-associated" Hg as you call it. How do you explain that observation and how did you analyze it?

C. BARGHIGIANI It may be explained by the presence of cell metabolites in the culture medium. In fact, I wish to emphasize that the cultures are spiked with Hg^{2+} in the log phase of growth (about 5 to 7 days after the inoculum). The "organically-associated" Hg has been calculated as a difference between the total and the ionic Hg as measured by atomic absorption spectrometry. The total Hg figure is obtained by a previous photooxidation of the sample.

M. BRANICA   Your 5 µg/l is about 1000 times higher than the actual concentration of Hg in sea water. Have you done any experiments at lower concentration levels with regard to Hg?

C. BARGHIGIANI I wish to emphasize that, in our experiments, the cell concentrations are about 1000 times higher than those of phytoplankton in sea water. We have done experiments also at concentrations of 2.5 µg Hg/l and the transformations of Hg^{2+}, into the other chemical forms considered, followed approximately the same percentage distribution found earlier.
We have not yet done experiments with cells and Hg at concentrations close to the natural levels.

M. BRANICA   Experiments with radionuclides have one great drawback; you have to test the level of isotope dilution to be certain of it or else you will subject yourself to many possible artifacts.

O. RAVERA   There are several limitations using radioisotopes. On the other hand, I think we have no other method so accurate for following the fate of an element in the environment and for evaluating the turnover time of an element in the body of an organism.

FEASIBILITIES AND LIMITS OF FIELD EXPERIMENTS TO STUDY ECOLOGICAL

IMPLICATIONS OF HEAVY METAL POLLUTION

René Gächter and Jacques Urech

Federal Institute for Water Resources
and Water Pollution
Control (EAWAG)
CH-8600 Dübendorf, Switzerland

INTRODUCTION

Elevated metal concentrations in recent sediments indicate strongly that, during recent decades, heavy metal pollution of aquatic systems has increased practically all over the world. In spite of this, only a few cases of obviously adverse perturbations of aquatic populations have been reported. Nevertheless, it would be incorrect to conclude from this that widespread heavy metal pollution will be without effect on aquatic organisms. It must rather be assumed that changes in these ecosystems have evaded detection because of a lack of comparable unpolluted control systems or because such changes could not be related to increased heavy metal loadings for the reason that contamination with other substances changed simultaneously.

Most of the available information regarding the consequences of heavy metal pollution for aquatic organisms stems from: (1) experiments investigating short-term effects of increased heavy metal concentrations; (2) a few field observations made at extremely-heavily loaded sites. Although such investigations provide very valuable information about acute toxicity, it must be kept in mind that long-term exposure at much lower concentrations may eliminate sensitive species and thus cause major shifts in species composition. Due to interspecific relationships (for example, competition for nutrients, predator-prey interactions, etc.) it is conceivable that heavy metal tolerant species might also become negatively affected or even eliminated. On the other hand, a given population may react strongly in a short-term experiment but adapt metabolically during a long-term exposure so that no obvious effects can be detected afterwards. Thus, based on the available information, it is extremely difficult if not impossible to predict the ecological consequences of a continuously increased heavy metal discharge to a lake, especially if simultaneous interactions of different metals have to be considered. In the laboratory, it may be possible to study two or even more species in a defined medium or under defined conditions; however, these conditions usually are very remote from those of the natural environment with its multitude of diverse species living simultaneously within the same space. Hence, we believe that extrapolation of results obtained from such experiments to real ecosystems is very difficult, if not impossible.

One possible approach to carrying out experiments with natural populations is to impound or enclose the natural populations and then study the effects of added nutrients or pollutants on various species and on population interactions "in situ" (under conditions as natural as possible). This technique has been applied successfully to the study of various problems in aquatic biology (Strickland and Terhune, 1961; McAllister et al., 1961; Goldman, 1962; Antia et al., 1963; Gächter, 1968; Klussman and Inglish, 1968; Schelske and Stoermer, 1971; Lund, 1972; Jones, 1973; Bürgi, 1974; Jones, 1975; Davies et al., 1975; Lund, 1975; Jones, 1976; CEPEX, 1977; Lund, 1978; MELIMEX, 1979).

Enclosures which are open to the air and to the sediment, allowing interactions with the atmosphere and the sediments, are closer to natural conditions than completely closed systems. But even if they are very large, holding over 16,000 m^3 of water, there is a problem; development of phytoplankton populations in undisturbed tubes or corrals is markedly different from their development in the lake (Figure 1). Apart from quantitative changes towards lower densities, Lund (1972) reports that impoundment favoured the development of species which are commonly dominant in oligotrophic waters but uncommon or absent in eutrophic waters (Rodhe, 1948).

The reason for the differences in behaviour between an enclosed waterbody and the open lake is mainly the elimination of horizontal advection by mechanical barriers. This complete cut-off of lake external nutrient sources results finally in lower nutrient concentrations (Figure 2) and hence in productivity. Additionally, vertical enclosure walls increase the area for periphyton growth and, due to shading, diminish phytoplankton production. Thus, in artificial enclosures, as a result of the larger surface to volume ratio, the periphyton competes more effectively with phytoplankton than it does in natural lakes. To sum up, impoundment of a water body has an oligotrophying effect and therefore, in spite of the intention to simulate natural conditions as closely as possible, the results obtained from such studies cannot be transferred directly to lake conditions. Nevertheless, since large enclosures are able to sustain a multitude of diverse species of various trophic levels over months and since replicate enclosures have proven to behave similarly (Takahashi et

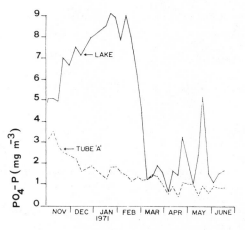

Figure 1. Fluctuations in phytoplankton, expressed in terms of chlorophyll a in Blelham Tarn and Tube "A" (Lund, 1972).

al., 1975; Gächter and Máreš, 1979; Urech, 1979; Baccini et al., 1979), limno-corrals can be very useful tools to study the effects of pollutants on complex ecosystems. This high degree of complexity, however, bears also disadvantages. As we shall demonstrate, such complicated systems do not always allow one to sort out clear cause-effect relationships. Necessarily, many attempts to explain observed results have too much of an hypothetical character. These hypotheses need to be tested, if at all possible, in simpler situations (less-natural laboratory experiments) which are easier to control. In other words, limno-corral studies are useful to observe long-term

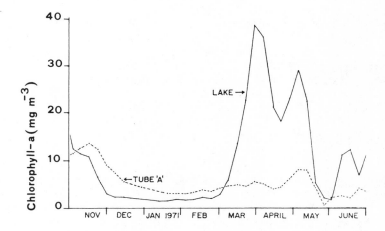

Figure 2. Phosphate concentration in Blelham Tarn and Tube "A" (Lund, 1972).

effects, but, additional laboratory experiments often will be necessary to fully understand the mechanisms causing these effects.

The goals of this paper are:

(1) to summarize the results obtained from two, long-term, heavy-metal, pollution studies, both of which having used large enclosures to isolate natural planktonic populations, with one having been conducted in the ocean and the other in a freshwater environment (CEPEX, 1977; MELIMEX, 1979);
(2) to compare and to discuss these results;
(3) to show the limitations of such large-scale long-term experiments;
(4) to present some additional results from laboratory studies necessary to the understanding of observations from enclosure studies;
(5) to draw some general conclusions.

RESULTS AND DISCUSSION

Field experiments in marine and freshwater environments

In CEPEX (Experiment 1), two containers served as controls and copper was added to a third enclosure 3 days after filling to provide a concentration of 10 µg/l. This concentration was maintained during the experiment by periodic copper analysis and addition. As Figure 3a shows, this increased total copper concentration resulted initially in a sharp decrease of in situ particulate photosynthesis. This is in agreement with observations made on natural freshwater phytoplankton (Gächter, 1976) which show, depending on the season, an extremely strong inhibition of CO_2 uptake when total copper concentration is increased up to the value above. On day 5, $^{14}CO_2$ uptake increased again (as the population recovered or adapted to copper stress) and finally it exceeded the values of controls at the end of the experiment. As a result, in the copper-treated enclosure, chlorophyll concentrations were lowest on day 5 and later reached or even exceeded those of the controls (Figure 3b). Figure 3c suggests that the recovery of phytoplankton occurred concomitantly with a drastic change in phytoplankton species composition.

Although with respect to microzooplankton, the two controls diverged markedly, there is an apparent trend to lower microzooplankton densities in the copper-loaded experiments (Figure 3d). A major phenomenon was a severe reduction in the abundance of the macrozooplankton in both the controls and the copper-polluted enclosures; however, the abundance of carnivorous ctenophores and medusae remained higher in the control than in the perturbed enclosures, thus indicating that these organisms also are adversely affected by a total copper concentration of 10 µg/l.

The MELIMEX study, which investigated the effects of an increased loading of a combination of various metals (Hg, Cu, Cd, Zn and Pb) (Gächter, 1979), yielded very similar results. The increased metal concentrations (Figure 4a) caused, in a first phase, a dramatic decrease of primary production in the metal-polluted corrals (Figure 4b), resulting in significantly lower chlorophyll concentrations from April, 1977, to April, 1978 (Figure 4c). In May and June of 1978, the chlorophyll concentration in the metal-loaded corrals (L1,L2) exceeded that of the control (C) by approximately tenfold. A depression in chlorophyll concentration was accompanied by a simultaneously-lowered zooplankton density (Figure 4d), mostly due to a strong inhibition of the filter feeding Daphnia longispina (Figure 4e). In order to get some qualitative information about the phytoplankton community structure, the pigment extract was measured at two different wave-lengths, $E^{\lambda 1}$ and $E^{\lambda 2}$. If the ratio, $E^{\lambda 1}/E^{\lambda 2}$, for two samples was different, then it might be assumed that the extracted phytoplankton communities of the two samples differed, either in species composition or in physiological state. According to Figure 4f in enclosures L1 and L2, the ratios were usually similar and either distinctly higher or lower than in the control (C). This strongly suggests that the species composition and/or the physiological state of the phytoplankton communities were similar within L1 and L2 but different in C.

When judging the results of phytoplankton enumerations, it must be kept in mind that, for technical reasons, the results of such enumerations are less precise and reproducible than those of some other measurements, such as those of chlorophyll concentrations or primary production rates. In addition, samples taken in bi-weekly intervals cannot exactly show the development and decline of short-lived population peaks. Small phase shifts in the development of such pulses might temporarily simulate differences in phytoplankton species composition and density, producing data which might be misleading. For this reason, with respect to algae observed only occasionally or at a very low density, it is not possible to determine if they were affected by the increased metal loading. Nevertheless, groups occurred in the control consistently with higher densities than in the metal-loaded limno-corrals: Aphanizomenon flos-aquae, Cyclotella spp., Fragillaria crotonensis, Fragillaria spp., Elakatothrix spp., Oocystis lacustris, Phacotus sp., Staurastrum sp., Cryptomonas ovata, Cryptomonas marsonii, Rhodomonas lacustris and microalgae. At the same time, the following groups were more abundant in the metal-loaded limno-corrals than in the control: Ankistrodesmus falcatus, Botryococcus braunii, Chlorella pyrenoidosa, Raphidium spp. and Schroederia setigera. Microalgae seemed to be growth-inhibited by increased metal concentrations during the first year of the experiment but, compared to the control, they developed higher densities in L1 and L2 in the spring of 1978. Other species or taxonomic groups were either not affected by the increased metal concentrations, or, effects could not be detected for reasons already mentioned. From

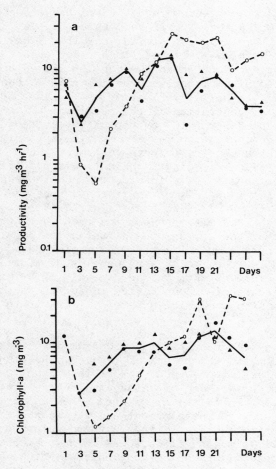

Figure 3. Response of a marine plankton community to an increase of total copper concentration to 10 µg Cu/l.
(a) Average (0 and 5m) particulate productivity in 2 controls (●,▲) and in a test enclosure with 10 µg Cu/l (o). Redrawn from Thomas et al., (1977).
(b) Concentration of chlorophyll a averaged between 0 to 10m depth in 2 controls (●,▲) and in a test enclosure with 10 µg Cu/l (o). Redrawn from Thomas et al., (1977).

April, 1977, to June, 1978, 51 different phytoplankton species or taxonomic groups were identified with a slight tendency towards higher species numbers in the control compared with metal-loaded limno-corrals. In about 70% of the observations, the species numbers of the control exceeded those of the metal-loaded enclosures.

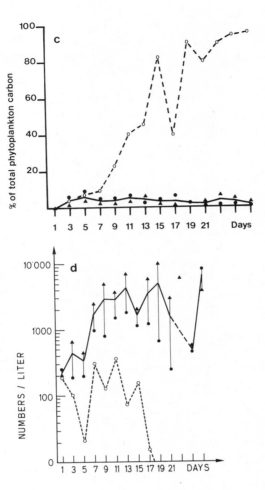

Figure 3. (continued)
(c) Microflagellates plus <u>Nitzschia delicatissima</u> as a percent of total phytoplankton. Values averaged between 0 and 10m depth in 2 controls (●,▲) and in a test enclosure with 10 μg Cu/l (o). Redrawn from Thomas and Seibert (1977).
(d) Average numerical abundance (surface to 13-14m) of tintinnid ciliates in 2 controls (●,▲) and in a test enclosure with 10 μg Cu/l (o). Redrawn from Beers et al., (1977).

To summarize briefly, increased metal concentrations caused shifts in phytoplankton species composition and, in general, the number of species observed in metal-loaded limno-corrals was lower than in the control.

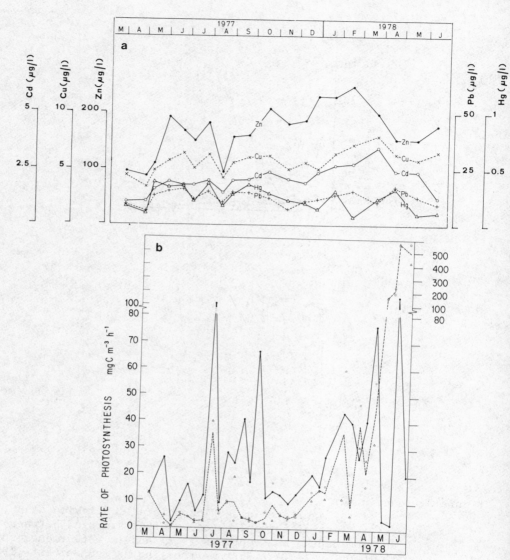

Figure 4. Response of a freshwater plankton community to increased heavy metal concentrations.
(a) Seasonal variation of zinc, copper, cadmium, mercury and lead concentrations in the epilimnion (surface to 5m depth) of a metal-loaded enclosure. Total metal concentrations in the control averaged 13 µg Zn/l, 1 µg Cu/l, 0.7 µg Cd/l, 0.0 µg Hg/l and 0.3 µg Pb/l.
(b) Seasonal variation of photosynthesis at optimum light intensity in metal-polluted corrals (o, △) and the control (●).

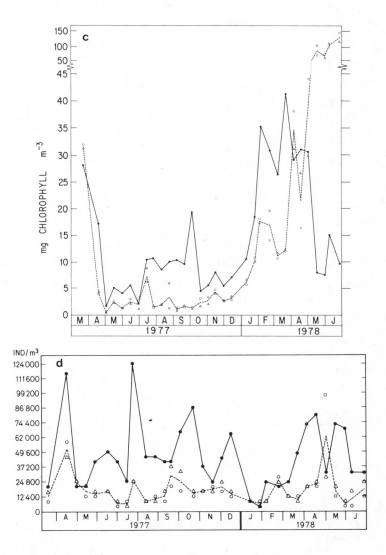

Figure 4. (continued)
(c) Seasonal variation of chlorophyll concentration in metal-polluted corrals (o,△) and the control (•).
(d) Abundance of crustacean plankton in metal-polluted corrals (o,△) and the control (•).

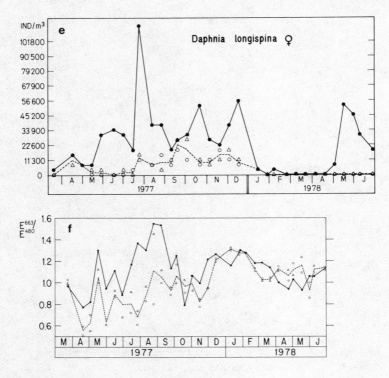

Figure 4. (continued)
(e) Abundance of Daphnia longispina in metal-polluted corrals (o,Δ) and the control (●).
(f) Ratio of light extinctions measured in the chlorophyll extract at 480 and 663 nm. Metal-polluted corrals (o,Δ). Control (●).

The susceptibility of L2 and C plankton towards copper is compared in Figure 5 which shows relative $^{14}CO_2$-assimilation rates ($^{14}CO_2$-assimilation rate in lake water without additional metals = 100%) of L2 and C plankton as a function of copper concentration in the medium. On eight occasions, the L2 plankton was found to be less susceptible, and on two occasions more susceptible towards copper, than was the C plankton. On two occasions no differences occurred. Zinc exhibited a similar result. The result might be interpreted in the following way. In the course of a year, phytoplankton communities are subject to continuous changes from a quantitative as well as from a qualitative point of view. The succession is most probably the result of many simultaneously-occurring variations of environmental conditions. Phytoplankton depend on abiotic factors, such as temperature, incident light, the concentration and speciation of

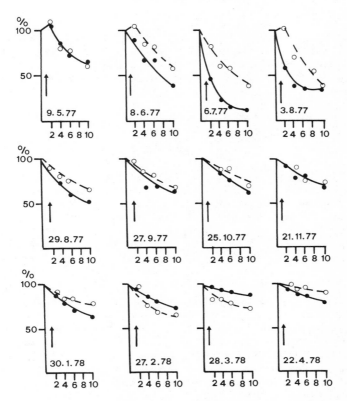

Figure 5. Depression of photosynthesis by copper for phytoplankton collected from a metal-polluted corral (o) and from the control (●). Phytoplankton collected with a 20 μm net were resuspended in filtered lake water (from Lake Baldegg). Subsamples were then spiked with increasing concentrations of copper. Arrows indicate actual metal concentrations in L2. The dimension of the X-axis is 10^{-7} mole/l.

nutrients and inhibitors, mixing and sedimentation. As well, they depend also on biotic factors such as grazing, infection by parasites, and the production and decomposition of growth promoting and inhibiting organic compounds. Obviously, the resultant phytoplankton population is adapted to actual environmental conditions. This means that species which momentarily profit the most from growth promoting factors and suffer the least from growth inhibiting factors will develop the best and reach the highest densities. If, in an ideal experiment, it would be possible to increase the metal concentration and keep all other regulating factors exactly at the level of the control, then, phytoplankton grown in the metal-polluted system would always be expected to be less susceptible to heavy metals than the control population. In the MELIMEX experiment, however, metal-loaded and control media differed not only in their metal content, but, due

to reduced primary production in L1 and L2, also in pH, alkalinity
and concentration of inorganic nutrients. Furthermore, as demonstrated with amino acids (Gächter and Máreš, 1979), they differed in
composition and concentration of organic compounds, as well as in
density and species composition of zooplankton and thus, also in
grazing pressure (Urech, 1979).

These concomitant changes of various growth-determining variables
might explain why the L phytoplankton were not always less susceptible to increased heavy metal concentrations than the control plankton.

It is well known that persistent, lipophilic, nonpolar substances
become accumulated in the liver and fatty tissue of higher organisms.
The following was done to answer the question, if, on a dry weight
basis, inorganic metal compounds really are biomagnified in an aquatic
food chain. Phytoplankton, periphyton and zooplankton samples, plus
chironomids and larvae of Sialis sp., as well as the fry of bream and
trout, were collected from unpolluted and metal-polluted limno-corrals
and then were analyzed for mercury, copper, cadmium, zinc and lead
(Gächter and Geiger, 1979). Table 1 shows that metal concentrations
were highest in phytoplankton and periphyton, lower in zooplankton,
chironomidae and Sialis and lowest in fish, thus indicating that, on
a dry weight basis, inorganic Hg, Cu, Cd, Zn and Pb are not accumulated through the food chain. This is in agreement with observations
of Topping and Windom (1977) who showed that copper to carbon ratios
were higher in fecal pellets than in phytoplankton. The ability of
zooplankton and crabs (Boothe and Knauer, 1972; Benayoun et al.,
1974) to concentrate metal in their faeces enables them to keep their
own metal content below the metal content of their food.

Table 2 suggests that increased metal concentrations favor phytoplankton species with a decreased metal sorption capacity. This
mechanism, which was also observed in the Flin Flon region in Canada
(Jackson, 1978), may play an important ecological role. The lower
the metal content in organisms at the lowest trophic level, the less
is the danger that organisms of higher trophic levels are adversely
affected by feeding upon them.

Need for additional experiments in laboratory

Figure 6 shows, strongly simplified, some important interdependencies regulating aquatic ecosystems. Obviously, heavy metals might
interact with a variety of these compartments independently of the
target-compartment. Also, as can be seen by the sketched interrelationships, the whole net-work might be affected at one time.

In CEPEX, as well as in MELIMEX, the applied metal concentrations
caused a decrease in phytoplankton as well as in zooplankton density.
Since depressed zooplankton density most likely is equivalent to a
lowered grazing pressure, it seems reasonable to assume that increased

Table 1. Average metal content of bream and trout fry, phytoplankton, zooplankton, Sialis sp. and chironomidae (10^{-8} M/g), as well as average dissolved metal concentration (10^{-8} M/l) in metal-polluted (L) and control (C) limno-corrals. (n.d. = not detectable)

Object	Cu L	Cu C	Cd L	Cd C	Pb L	Pb C	Zn L	Zn C	Hg L	Hg C
Phytoplankton	280	100	60	5	420	22	4000	1400	7	n.d.
Periphyton	70	12	60	1	500	8	5000	400	4	0.1
Chironomidae	104	12	17	0.2	64	2	2200	600	–	–
Zooplankton	70	15	25	2	20	3	650	250	1.5	n.d.
Sialis sp.	22	26	6	3	3	n.d.	500	300	1	n.d.
Bream (fry)	9	7	1	0.5	1	n.d.	400	400	0.7	0.2
Trout (fry)	4	5	1	0.7	4	n.d.	280	240	–	–
dissolved metal	8	1.3	1.8	0.07	2.8	0.1	230	17	0.04	n.d.

Table 2. Concentration factors (1/g) of phytoplankton collected in metal-polluted (L) and control (C) limno-corrals. n.d. = not detectable.

Cu		Cd		Pb		Zn		Hg	
L	C	L	C	L	C	L	C	L	C
35	77	33	71	150	220	17	82	175	n.d.

metal concentrations exerted a direct adverse effect on phytoplankton.

According to Figure 6, the observed decrease in zooplankton density in metal-loaded environments might have at least three reasons:
(1) a depressed food supply due to depression of primary production;
(2) a direct adverse effect of metals on zooplankton;
(3) a metal-induced increase in zooplankton predation.
Whereas, based on observations, the last possibility practically can be ruled out, it can hardly be distinguished whether the observed depression of primary consumers (such as Daphnia) was due to a direct adverse effect of metals on these organisms or due to an indirect metal-induced shortening in food supply.

To solve this problem, additional laboratory experiments with Daphnia longispina were conducted in which total metal concentration was the only variable and all other environmental factors, including food supply, were kept constant (Urech, 1982). In some of these experiments, the life history of animals from birth to death was recorded.

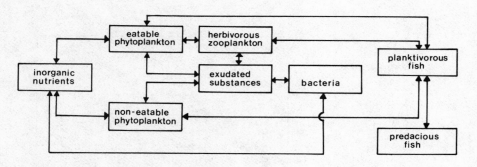

Figure 6. Interrelationships in an aquatic ecosystem.

Figure 7 shows that, up to a level of 50 μg Zn/l, the life span was not affected. At 100 μg Zn/l, animals became only half as old as in the control and, at a concentration of 200 μg Zn/l, juveniles died before becoming adult. The tendency towards an extension of the juvenile stage with increasing metal concentration was confirmed by additional experiments. Also, it can be deduced from Figure 7 that an increase of Zn concentration decreased not only the life span but also the rate of reproduction. On the average, control animals which became adult produced 59 eggs during their lifetime out of which 58 animals hatched. While almost not affecting the life span, 50 μg Zn/l lowered the egg production and increased the abortion rate. At 100 μg Zn/l, the average egg production was only 21% of the control and the abortion rate increased to 55%. At 200 μg Zn/l, animals did not reproduce because all died in the juvenile stage.

In the field study, the average epilimnic zinc concentration was on the order of 150 μg/l. Hence, with no doubt, the observed depression of <u>Daphnia</u> density was due mainly to a direct adverse effect of Zn on these animals, and was not (or at least not only) caused by a metal-induced shortening of the food supply.

SUMMARY AND CONCLUSIONS

From field studies like CEPEX and MELIMEX, it can be deduced that plankton communities are extremely susceptible to increased

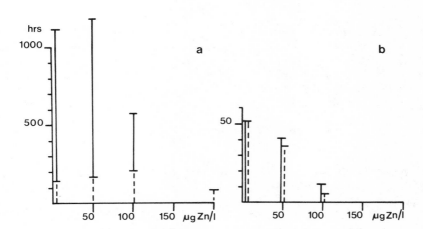

Figure 7. Effect of total zinc concentration on
(a) duration of juvenile stage (---) and total life span.
(b) egg production/animal (——) and production of juveniles per animal (---).
All results are averages from 25 records.

metal concentrations. If heavy metal concentrations exceed a certain limit, then freshwater and marine communities follow a similar pattern:

(1) a reduction of primary production followed by a decrease in chlorophyll concentration;
(2) a simultaneous decrease in zooplankton density;
(3) changes in phytoplankton community structure towards species which concentrate less metal and which are less susceptible to heavy metals;
(4) a recovery of the particulate primary production;
(5) a recovery of the phytoplankton density (chlorophyll concentration).

It was further demonstrated in both environments that the metal content of zooplankton did not exceed the metal content of phytoplankton and hence, on a dry weight basis, inorganic heavy metals were not accumulated in the aquatic food chain: phytoplankton-macrozooplankton-fish fry.

These results could hardly have been achieved by experimenting with cultures of single species under well-defined laboratory conditions. On the other hand, field experiments, working with complex communities, do not allow one to decide whether or not the organisms at higher trophic levels are affected directly or indirectly (via the organisms upon which they feed). Additional experiments with Daphnia longispina (Urech, 1982) indicate that increased heavy metal concentrations inhibit the egg production, increase the abortion rate and decrease the life span.

Based on present information, we conclude that 10 µg Cu/l or 50 µg Zn/l both exhibit deleterious effects on planktonic communities. If concentrations of different metals are increased simultaneously, they act additively or even synergistically (Gächter, 1976; Urech, 1982). Therefore, with no doubt, "safety limits" for Cu and Zn would have to be located much below 10 µg Cu/l and 50 µg Zn/l respectively.

REFERENCES

Antia, N.J., McAllister, C.D., Parsons, T.R., Stevens, K., and Strickland, J.D.H., 1963, Further measurements of primary production using a large-volume plastic sphere, Limnol. Oceanogr., 8:166.

Baccini, P., Ruchti, J., Wanner, O., and Grieder, E., 1979, Regulation of trace metal concentration in limno-corrals, Schweiz. Z. Hydrol., 41:202.

Baccini, P., and Suter, U., 1979, Chemical speciation and biological availability of copper in lake water, Schweiz. Z. Hydrol., 41:291.

Benayoun, G., Fowler, S.W., and Oregioni, B., 1974, Flux of cadmium through euphausiids, Mar. Biol., 27:205.

Beers, J.R., Steward, G.L., and Hoskins, K.D., 1977, Dynamics of microzooplankton populations treated with copper: Controlled ecosystem pollution experiment, Bull. Mar. Sci., 27:66.

Boothe, P.N., and Knauer, G.A., 1972, The possible importance of fecal material in the biological amplification of trace and heavy metals, Limnol. Oceanogr., 17:270.

Bürgi, H.R., 1974, Die Wirkung von NTA auf das Wachstum des Phytoplanktons unter besonderer Berücksichtigung des Eisens als Mikroelement, Schweiz. Z. Hydrol., 36:1.

CEPEX, 1977, Controlled ecosystem pollution experiment, (Various titles and authors), Bull. Mar. Sci., 27: pp. 1-145.

Davies, J.M., Gamble, J.C., and Steele, J.H., 1975, Preliminary studies with a large plastic enclosure, in: "Estuarine Research Vol. I.", E.L. Cronin, ed., Academic Press, New York.

Gächter, R., and Máreš, A., 1979, Effects of increased heavy metal loads on phytoplankton communities, Schweiz. Z. Hydrol., 41:228.

Gächter, R., 1968, Phosphorhaushalt und planktische Primärproduktion im Vierwaldstättersee (Horwer Bucht), Schweiz. Z. Hydrol., 30:1.

Gächter, R., 1976, Untersuchungen über die Beeinflussung der planktischen Photosynthese durch anorganische Metallsalze im eutrophen Alpnachersee und der mesotrophen Horwerbucht, Schweiz. Z. Hydrol., 38:97.

Gächter, R., and Davis, J.S., 1978, Regulation of copper availability to phytoplankton by macromolecules in lake water, Environ. Sci. Technol., 12:1416.

Gächter, R., 1979, MELIMEX, an experimental heavy metal pollution study: Goals, experimental design and major findings, Schweiz. Z. Hydrol., 41:169.

Gächter, R., and Geiger, W., 1979, Behaviour of heavy metals in an aquatic food chain, Schweiz. Z. Hydrol., 41:277.

Goldman, C.R., 1962, A method of studying nutrient limiting factors in situ in water columns isolated by polyethylene films, Limnol. Oceanogr., 7:99.

Imboden, D.M., Eid, B.S.F., Joller, T., Schurter, M., and Wetzel, J., 1979, Vertical mixing in a large limno-corral, Schweiz. Z. Hydrol., 41:177.

Jackson, T.A., 1978, The biogeochemistry of heavy metals in polluted lakes and streams at Flin Flon, Canada, and a proposed method for limiting heavy metal pollution of natural waters, Environ. Geol., 2:173.

Jones, J.G., 1973, Studies on freshwater bacteria; the effect of enclosure in large experimental tubes, J. Appl. Bact., 36:445.

Jones, J.G., 1975, Some observations on the occurrence of the iron bacterium Leptothrix ochracea in fresh water, including reference to large experimental enclosures, J. Appl. Bact., 39:63.

Jones, J.G., 1976, The microbiology and decomposition of seston in open water and experimental enclosures in a productive lake, J. Ecol., 64:241.

Klussman, W.G., and Inglish, J.M., 1968, Polyethylene tubes for studies of fertilization and productivity, Proc. SEast. Ass. Game Fish Commnrs., 22:415.

Lund, J.W.G., 1972, Preliminary observations on the use of large experimental tubes in lakes, Verh. Internat. Verein. Limnol., 18:71.

Lund, J.W.G., 1975, The uses of large experimental tubes in lakes, pp. 291-312, in: "The Effects of Storage on Water Quality", Medmenham.Water Research Centre.

Lund, J.W.G., 1978, Experiments with lake phytoplankton in large enclosures, Freshwater Biological Association Forty-Sixth Annual Report, p. 32.

McAllister, C.D., Parsons, T.R., Stevens, K., and Strickland, J.D.H., 1961, Measurements of primary production in coastal sea water using a large-volume plastic sphere, Limnol. Oceanogr., 6:237.

McLaren, I.A., 1969, Population and production ecology of zooplankton in Ogac Lake, a landlocked fiord on Baffin Island, J. Fish. Res. Board Can., 26:1485.

MELIMEX, 1979, An experimental heavy metal pollution study, (Various titles and authors), Schweiz. Z. Hydrol., 41:pp. 165-314. Separatum Nr. 770.

Rodhe, W., 1948, Environmental requirements of planktonic algae, Symb. Bot. Upsal., 10:149.

Schelske, C.L., and Stoermer, E.F., 1971, Eutrophication, silica depletion and predicted changes in algal quantity in Lake Michigan, Science, 173:423.

Strickland, J.D.H., and Terhune, L.D.B., 1961, The study of in situ marine photosynthesis using a large plastic bag, Limnol. Oceanogr., 6:93.

Takahashi,M., Thomas, W.H., Seibert, D.L.R., Beers, J., Koeller, P., and Parsons, T.R., 1975, The replication of biological events in enclosed water columns, Arch. Hydrobiol., 76:5.

Thomas, W.H., Holm-Hansen, O., Seibert, D.L.R., Azan, F., Hodron, R., and Takahashi, M., 1977, Effects of copper on phytoplankton standing crop and productivity: Controlled ecosystem pollution experiment, Bull. Mar. Sci., 27:34.

Thomas, W.H., and Seibert, D.L.R., 1977, Effects of copper on the dominance and the diversity of algae. Controlled ecosystem pollution experiment, Bull. Mar. Sci., 27:23.

Topping, G., and Windom, H.L., 1977, Biological transport of copper at Loch Ewe and Saanich Inlet: Controlled ecosystem pollution experiment, Bull. Mar. Sci., 27:135.

Urech, J., 1979, Effects of increased heavy metal load on crustacea plankton, Schweiz. Z. Hydrol., 41:247.

Urech, J., 1982, Experimentelle Untersuchungen im Freiland und im Labor über den Einfluss von Schwermetallen auf Crustaceen-Plankton, Thesis, ETH Zürich, (in preparation).

FEASIBILITIES AND LIMITS OF FIELD EXPERIMENTS

DISCUSSION: R. GACHTER AND J. URECH

M. BERNHARD — Of what material is the enclosure made? Is there a possibility that trace elements pass this material?

J. URECH — The outside wall was of rubber, the inside wall was of polyethylene. We believe that metal ions do pass the materials and it has been observed that they release some zinc. The periphyton growth on these materials absorbed heavy metal ions.

H. MUNTAU — I would like to comment on the question of Dr. Bernhard. It is likely that the polyethylene wall of the limno-corral is covered rapidly by periphyton. Hence, the periphyton will control the wall's behaviour towards heavy metals rather than having a control by the wall material *per se*.

J. URECH — There is no question about that!

Y.K. CHAU — Relating to Dr. Bernhard's question, we found that PVC materials adsorb trace metals. We also found that phytoplankton grow on the surface of PVC bags hung in lake water. Considering these phenomena, did you follow the concentration of your added metal during the course of your experiment?

J. URECH — Yes, we followed the metal concentration in all enclosures over the entire experimental time.

E.K. DUURSMA — Did you take into account a phytoplankton species composition change? As you know, the turbulence inside limno-corral containers is different from that of the surrounding lake water and some species cannot float without natural turbulence.

J. URECH — Yes, the containers were prepared long in advance.

R.F. VACCARO — I have three questions. Did you enrich the enclosures to maintain phytoplankton productivity? How did you handle sedimentable organic solids? How did you maintain water mixing within the enclosures?

J. URECH — The enclosures were not enriched with macronutrients but a system of pumps provided a continuous inflow of lake water into the corrals (resulting in an average residence time of about 100 days).

J. URECH
(continued)
Settled organic material was collected by sedimentation traps and it was analysed for organic carbon and the various heavy metals (Baccini et al., 1979). It was not intended in this experiment to influence natural stratification, and it was shown by Imboden et al. (1979) that the temperature structures of the water were very similar in the limno-corrals and in the surrounding lake.

R. FERRARA
How do you control the concentration of the metal under study in your open - flow system when you add fresh water?

J. URECH
The heavy metal concentrations in the inflow were kept constant, but they did fluctuate in the limno-corrals, due to seasonal fluctuations in sedimentation processes and mixing events. For the various metals, the following order of residence times has been estimated: $Zn > Cu > Cd > Pb \approx Hg$.

M. BRANICA
What was the composition of the added metal solution? Also, did you determine the concentration of metals actually present in the dissolved, particulate and complexed forms within the limno-corrals?

J. URECH
The inflow concentrations to the loaded corrals were: 5 µg Cd/l; 10 µg Cu/l; 200 µg Zn/l; 50 µg Pb/l; 1 µg Hg/l. All of the metals were added as nitrates. For Cu, Zn, Cd, and Pb, the dissolved metal (DM) and the total metal concentration (TM) were determined. The particulate fraction was calculated from (TM)-(DM). For Hg, three forms were measured: volatile mercury; dissolved, non-volatile mercury; and particulate mercury (Baccini et al., 1979). About 95% of the dissolved Cu was complexed by organic ligands having a molecular size within the range of 10^3 to 10^4 daltons (Baccini and Suter, 1979).

H.W. NURNBERG
Have the heavy metal levels been determined in various organs and parts of the fish of limno-corrals? There is to be expected a relatively large accumulation of Cd and Pb in kidney and liver while levels in muscle remain very small. This expectation is based on the investigation of thousands of fish specimens in my laboratories. Your conclusions based on the absence of accumulation effects in the food chain should be reconsidered in this context.

J. URECH — No, we did not study metal levels with regard to fish parts because we used fish fry as test organisms. For technical reasons we were unable to differentiate between various organs of these fish.

R. DE BERNARDI — Have you done experiments in enclosures without zooplankton? There is a means to discriminate between the metal effect and the effect of zooplankton grazing on phytoplankton populations, mainly in the recovery phase. Your conclusions are correct for what concerns the whole community but they are not the only possible conclusions when one considers single biotic compartments or species.

J. URECH — We conducted no experiments without zooplankton but since, during most of the time of the experiment, not only phytoplankton- but also zooplankton-density was lower in the loaded corrals, we concluded that phytoplankton depression in the loaded corrals was due to heavy metal stress and not to grazing. On the other hand, at the end of the experiment, the observed phytoplankton bloom in the loaded corrals might well be the result of lower zooplankton density.

O. RAVERA — I think that it is necessary in enclosure experiments to have replications because some non-conservative parameters (such as chlorophyll and nitrates) show large differences in different enclosures treated in the same way. Also, it seems very interesting that the metal concentration in zooplankton faeces is higher than that in zooplankton food. We have found the same for some radionuclides (for example Ru^{106}) in a fresh water prosobranch. This may be considered a protective mechanism of the organism against toxic substances.

J. URECH — It was observed during the pre-experimental phase of four and one half months that the three enclosure systems behaved very similarly. Again, during the experiment, the results from the two metal-loaded corrals were rather similar but quite different from those of the control.

M. SMIES — With regard to your Table 2, I am not certain that the difference in CF (concentration factor) is really a result of different phytoplankton species composition. It may well be that, for one species, different CF's are found on exposure to different metal concentrations. If this were the case, it would invalidate your conclusion about an ecological

M. SMIES
(continued)
adaptation of the phytoplankton community to metal stress.

J. URECH
We thought about this problem and made tests for various copper concentrations (Gächter and Geiger, 1979). For all of these concentrations the CF's were the same.

G.E. BATLEY
Limno-corrals obviously provide a valuable means for the study of metals on a range of organisms over a long term. Did you or do you intend to study: (1) synergistic effects of added metals; (2) measurements of the species distribution of added metals, both in dissolved and particulate forms; (3) the effect of the addition of specific forms of a given metal (for example, complexed, adsorbed on colloids, etc)? The results would provide much needed information but, to be properly done, they would require considerable research effort.

J. URECH
We observed synergistic effects in short term experiments with natural phytoplankton populations (Gächter, 1976) and couldn't see such effects with zooplankton (Urech, 1982). We also demonstrated that organic substances present in lake water are able to complex copper and to influence the bioavailability of copper (Gächter et al., 1978).

BIOLOGICAL RESPONSE TO TRACE METALS AND THEIR BIOCHEMICAL EFFECTS

V. Albergoni and E. Piccinni

Istituto di Biologia Animale
Università di Padova
Italy

TRACE AND ULTRATRACE METALS: PROBLEMS AND QUESTIONS.

It is well known that many elements, including metal ions, have a biological significance and are essential for plant, animal and human life. Some of the metals are required in very small amounts and thus are generally referred to as "trace metals". Until 20 years ago, few trace elements (such as Fe, I, Cu, Zn, Mn, Co, Cr, Se, F) were considered essential for animals. Recently, other elements, some regarded as environmental contaminants, were found to be beneficial in the diet of laboratory animals. Some of these are Ni, V, Si, and As. Very little is known concerning the specific biological functions of these newer trace and ultratrace elements. However, recent findings support the possibility that the interaction between one ultratrace element and another can be of nutritional significance. These results refer mainly to experiments (on mammals like rats or rabbits) which had been carried out to understand the dietary needs in humans. Data concerning the requirements for trace and ultratrace metals of unicellular organisms (protozoa and bacteria) grown on defined media are also available, but very little or even nothing is known about most animals.

In this paper, the emphasis will be on trace metals rather than on trace elements and the biological focus will be on animal life.

Certain elements, notably Cd, Pb and Hg are present in the environment in growing quantities. These have no known biological functions; they are regarded as undesirable and are considered as toxic trace elements at present. However, essential trace elements also may be harmful if present in great amounts in the environment. The toxicity of a trace element depends on the amount of the metal taken up from the environment in nearly all cases, and the toxic concentration greatly differs in different organisms. The toxicity is also influenced by abiotic and biotic factors; this will not be mentioned here since it has been reviewed recently (Bryan, 1976; Prosi, 1979).

The biological responses of organisms to trace elements are very complex. For instance, the same element exerts diverse toxicities for different animals; some metals can be concentrated or accumulated while other metals are maintained in low concentration even if they are present at a high level in the environment. Different chemical forms of trace metals may have different effects on animals. These effects may occur at the level of permeability (diffusion or active transport), or at the sites of accumulation and localization in animal tissues, but, there are no data on the effects of different chemical forms of metals on the detoxification and regulation systems. These and many other problems are connected with the biological response to trace metals. From the existing data, a lot of valuable but not always correlated information is available; so, it appears difficult to obtain a complete and satisfactory picture of response phenomena.

The problems of trace element effects must be treated and evaluated essentially in the framework of metabolic events for both essential and non-essential metals. This evaluation must be drawn comparatively using different animals. As for biological events, only the comparative approach will allow us to evaluate properly the physiological rules and the observed exceptions. We will discuss the problems related to the metabolism of trace metals according to the scheme of Table 1. This scheme does not take into account the interactions among elements.

RESPONSE TO METALS

We shall try to distinguish between the effects of essential and non-essential metals. As is true for several other physico-

Table 1. Complexities of the metabolism of trace metals.

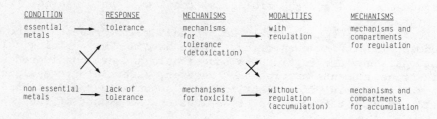

chemical parameters, animals can be tolerant or non-tolerant to metals; we refer to euryecious and stenecious organisms with respect to their response to environmental metals. The line separating tolerance from lack of tolerance to metals depends not only on the amount of metal present in the environment; it depends also on other parameters such as the rate of exchange, the salinity, the temperature, the kinds of ligands present in the environment, and so on. Tolerance also depends on parameters intrinsic to the animal itself. Moreover, even when the above parameters are the same, the extent of tolerance greatly differs across different groups of animals. Below, we refer to mechanisms for toxicity and to mechanisms for tolerance.

Mechanisms for toxicity

Firstly, it is not difficult to define whether a metal or an element in general is toxic or not. An element is toxic if it damages life functions (growth, reproduction, metabolism, etc.) of an organism. Normally, this definition takes in the concentration level which separates non-toxicity from toxicity. From the standpoint of environmental pollution, metals have been classified as (1) non-critical, (2) toxic but insoluble or rare, (3) very toxic and relatively accessible (Wood, 1974, as reported in Förstner and Wittmann, 1979).

We are inclined to consider the essential elements as less toxic than the non-essential ones. This is generally true, and an essential trace metal becomes toxic when the nutritional supply exceeds the optimal amount by a factor varying between 40 and 200 fold (Venugopal and Luckey, 1975). Some exceptions have been reported; for example, an essential metal, copper, is toxic at relatively low concentrations. Its toxicity seems to be higher for lower organisms than for higher organisms; however, there are exceptions as some protozoa and molluscs are highly tolerant. The high toxicity of copper, as for some other metals, has been tentatively correlated to the electronegativity (or to other physico-chemical properties) of its salts or of chelates (Bowen, 1966). The order of electronegativity (or of other properties) of different elements does not correspond to their order of toxicity for all organisms; the sequence of toxicity is different also for animals belonging to the same group.

It has been observed recently that most of the very toxic and relatively available metals are soft acceptors, according to the Pearson classification (hard and soft Lewis acids), as reported by Wittmann (1979). Soft acceptors (such as Cu^+, Ag^+, Hg^{2+}, CH_3Hg^+, Cd^{2+}, and others) prefer to bind to soft donors (such as SH^-, S^{2-}, alkyl or aryl-S compounds, CN^-, and others). At present, additional data are required for supporting the correlation between toxicity and soft-soft bonds. The characteristic coordination number and the coordination geometry sometimes have been considered in the evaluation of the metal effect upon some enzymes (Davis and Avram, 1980).

As for the mechanisms of toxicity, the most relevant is certainly the chemical inactivation of enzymes. The more electronegative metals and all the divalent transition metals are really more active in this respect. They react very promptly with the amino, imino and sulfhydryl groups of proteins; some of them (Cd, Hg) may compete with zinc and displace it in zinc-containing metalloenzymes. Some metals may also damage cells by acting as antimetabolites (for instance when arsenate substitutes for phosphate at specific sites), or by forming precipitates or chelates with essential metabolites (for instance when Al and Se react with phosphate or sulphate). Metals may also affect permeability by interacting with membranes and inhibiting specific transport sites (or by producing more extensive damages). A lesser degree of toxicity is due to the substitution of an element with another in some biological compartments (for example, Li for Na, or Br for Cl). Metals can be very dangerous for mutagenic effects. The literature on this subject is ever growing and the assessment of carcinogenic risk associated with metals has become a specialized branch of biology today. These topics are not referred to in the present paper (see, as an example of a recent review, Sunderman, 1979).

A large part of recent studies has been devoted particularly to ascertaining the effects of metal salts, in vitro and in vivo, on specific enzymatic systems and on specific metabolic pathways, many of which appear to be stimulated by some metals, while others are inhibited. According to the dose, different effects can be observed. Of importance are the effects of metals on cytochrome P_{450} which, as is well known, is particularly involved in the detoxification of xenobiotics, (Blumberg, 1978). Generally, the presence of metals (in vivo and in vitro) either lowers the content of or reduces the activity of cyt. P_{450}. Some data, however, indicate the contrary, and thus some more detailed studies are necessary to clarify this point. Other biochemical effects are discussed in this volume (see chapter by Vaccaro). Some metals (Pb and others) may manifest neurotoxic effects by interfering with the function of neurotransmitters or with the metabolism of drugs.

Other research is in progress to describe the ultrastructural modifications following metal exposure. One of the most significant results is the increase in the number and volume of lysosomes, especially in gills, liver and kidney. Mitochondrial swelling, altered vacuolar formations and nuclear changes are often reported.

Mechanisms for tolerance (detoxification)

We can define tolerance as the ability of being less responsive to the harmful effects of a metal by virtue of preventing its noxious interactions with reactive groups of enzymatic proteins, nucleic acids, and etc.

In the aquatic environment, the microbial community can react to the presence of some metals by detoxifying them. This protective response of a microbial population may produce compounds much more toxic towards other animals than the original impinging metal ion itself. Methylmercury and the Minamata disease is a well known example. The biomethylation of toxic metals was recently investigated from a biochemical point of view, and the conditions for methyl transfer in biological systems have been well-defined (Wood et al., 1978). Hg, As, Pb and Se are the most important of the metals which can be methylated. The high toxicity of these compounds is probably due to their liposolubility which facilitates a very rapid diffusion through cell membranes. The methyl derivatives interact mostly with phospholipids in the nervous system. Owing to the toxic effects, many studies have been done on the metabolism of these compounds. Recently, it has been demonstrated that mercury biomethylation may also occur in fish from the activity of intestinal bacteria (Rudd et al., 1980). Tolerant animals may protect themselves by lowering the toxicity of a metal ion and of its methyl derivatives. This detoxification may be performed in different ways. It has been suggested that methylmercury can be demethylated in the liver; such demethylation is secondary in the overall detoxification of this compound.

The first basic step of detoxification, performed just after entry of the offending metal into the cells, is the trapping of the metal ions by a large pool of linking substances. This pool is present both in insoluble and in particulate phases in some cellular systems and may be regarded as a "buffer". This "buffer" corresponds to the "insensitive sites" quoted by other authors such as Passow et al. (1961). The metals may be linked and detached very promptly from this pool. Therfore, the buffer system plays a role of temporary storage, with this mechanism being located mostly in liver or kidney. The toxicity, especially the acute toxicity, depends on the presence and the extent of this saturable compartment. Very few data are available on this buffer compartment; a comparative analysis of the situation in different groups of animals is particularly lacking.

For improved protection, tolerant animals have to utilize some other process and a more stable detoxification system must be employed. This latter system may consist of the production of a class of compounds equipped with a high affinity for metals. In fact, in most animals exposed to non-essential metals (such as Cd or Hg) or to an excess of essential ones (such as Zn or Cu), a class of proteins has been found to be particularly noteworthy, a class called metallothioneins. A primary role in detoxification has been ascribed in general to the metallothioneins. In spite of their presence, however, we remark that detoxification is performed by those metabolic events in which the metals are involved. In fact, the metabolic defence against metal damage consists of the consecutive transfer of the metal (or metals) through a series of subcellular compartments. The metallothioneins may represent a step in this overall sequence. Other steps of the detoxification process are either (1) the storage of metals in granules dispersed in the cytoplasm or (2) the accumulation of metals in vesicular bodies. The last process is the one of excretion.

The detoxification role ascribed to thioneins is certainly suggestive. The likely metallothionein roles are, according to Vallee (1979), "an enumeration of the most obvious possibilities", and should include transport, catalysis, storage, and detoxification. In reality their properties make such a type of protein "an ideal candidate for detoxification" but, in fact, "it was not proven possible to reach conclusions as to the function of this protein". These proteins have a high chelating capacity due to their high SH group content. The metal content of metallothionein amounts to 6 or 7 gram atoms per mole, and the metal ratio (namely Zn, Cu, Cd) varies with different animals, different treatments, and different tissues. Metallothioneins are present also in untreated animals and many metals may bind to them by zinc replacement. The typical molecular weight of these proteins ranges from 6,000 to 7,000. The metal ion is bound to the didentate site Cys.-X-Cys., and probably interacts also with a third cysteinyl residue. The ratio of metal ions to SH groups is 1/3 and the SH groups makeup 20-21 moles per metallothionein. Neither disulphide bridges nor free SH groups are present in metallothioneins. Other data cannot readily be summarized about these proteins. Many studies have been carried out and more are in progress. The various characteristics, properties and problems related to them were debated at the recent "Meeting on Metallothionein" (Zurich, 1979, see Kägi and Nordberg).

The mucous production that occurs in many animals (mostly in the gills) may be regarded as a protection from metals. However, this mucous production, associated with the detachment of the first cell layer in the gills, may interfere with respiratory gas exchange and cause rapid death. Other forms of rather unspecialized protection are all exclusion mechanisms. This topic will be considered in the section on regulation mechanisms.

MODALITIES FOR REGULATION AND ACCUMULATION

According to a kinetic evaluation of the presence of trace metals in living organisms, the input rate of a metal, when not balanced by an equal output rate, leads to accumulation. This may be a permanent or temporary event in tolerant animals. The capability of some animals to maintain a balance between input and output (such as between absorption and excretion) of some elements allows us to distinguish between regulator and conformer animals. This distinction may be applied to

BIOLOGICAL RESPONSE TO TRACE METALS

tolerant and non-tolerant animals in different ranges of external metal concentrations.

Regulation is a complex event, occurring in different ways in different animals. In this respect, accumulation too can be regarded as a form of regulation. In any case, the regulation and the storage of a metal require the previous formation of a complex with low toxicity. Now it seems convenient in a formal way to distinguish between (1) mechanisms and compartments for accumulation and (2) mechanisms and compartments for regulation, with the aim of clarifying the linkage between them.

Mechanisms and compartments for accumulation

The chemical affinity between trace elements and some sites on biological structures is the basis both for (1) the distribution of different metals in different localized compartments and for (2) metal storage. This is not quite true because some metals can apparently coprecipitate without any linking organic matrix. Nevertheless, even in this case, it is difficult to exclude a role for preexistent structures which catalyze the initial localization. The localization can therefore be regarded as the result of a site capability for interactions, both qualitative and quantitative.

The accumulation of metals can be regarded at the whole body level and at the subcellular level. Let us consider first the localization and accumulation at the body level. For Mollusca, many reports indicate localization of heavy metals; we note here the kidneys of Gasteropoda, as well as the gut, gills and kidneys of Bivalvia. The hepatopancreas of Octopus accumulates Cd, Cu, I and Mn, and the branchial hearts accumulate V, Fe and Cu. As for Crustacea, the hepatopancreas, gills and green glands are involved in Cd accumulation in Uca, the hepatopancreas for Cu accumulation in Carcinus, and the exoskeleton for Cd accumulation in three organisms (Callinectes, Carcinus, and the shrimp, Lysmata seticaudata) (Coombs and George, 1978; Bryan, 1976; Wright, 1978). Ostrea accumulates metals in amebocytes like the Ascidiacea accumulate V in blood cells. As for aquatic vertebrates, in the pisces the localization of metals appears to be the same (liver, kidneys, gills, gut). If we compare this localization to that of terrestrial vertebrates, the same picture can be seen except for the structures that are analogous to the gills (the lungs).

In the presence of very large amounts of metals and prolonged exposure, some other structures can be involved in metal accumulation in a limited way. Moreover, we notice that, according to the chemical properties of some metals, other localizations may exist. The lead in bone (substituted for calcium) and methylmercury in the nervous and muscular systems are some examples. The usual storage organs appear to protect other tissues, often those more vulnerable (such as nervous tissue). For our line of thought, we point out that such accumulating tissues are at the interface between the internal and external environments, and that they are involved in absorption or excretion (or both these functions). This subject will be referred to in the concluding remarks.

Now let us consider localization and accumulation at the subcellular level. The data on those subcellular compartments which localize metals comes from two different approaches, (1) the ultrastructural and (2) the biochemical. Analyses of subcellular fractions indicate that metals are distributed throughout both the particulate fractions (nuclear, large granule and microsomal) and the soluble phase. The presence of metals in the large granule fraction (which includes mitochondria, lysosomes and other membrane-bound vesicles) is characteristic. In marine invertebrates, this has been well-demonstrated by X-ray microanalysis for Hg, Pb, Fe and Cu. X-ray microprobe analysis of these compartments very often indicates the presence of sulphur. Recent data on copper granules in hepatopancreatic caeca of Amphipoda indicate that Cu is probably complexed with organic matter, and that granules can aggregate in multivesicular bodies (Icely and Nott, 1980). In the liver lysosomes of the toad, Bufo marinus, copper has been detected in a very large amount at the same level of concentration as in the liver of a man with Wilson's disease, or in the liver of newborn vertebrates (Goldfischer et al., 1970).

Often inorganic salts, like calcium phosphate, coprecipitate with metals. This mechanism seems to be very adequate to trap some metals. Granules derived in such a manner have been described for trapping Cd, Zn and Mn. Other data on the storage of metals in granular or vesicular structures are reported by Bryan (1976) and by Wright (1978). The deposition of metals in membrane-bound cytoplasmic bodies or lysosomes seems to be a general biological process. The soluble phase is the other intracellular compartment which in some tissues has a high capacity to link many metals. Nearly all the data about it show the presence of metallothioneins or similar compounds. A number of metallothionein-like proteins have been isolated from many organisms ranging from lower eukaryotes to higher vertebrates; at present they represent a broad class of compounds. Here, it is important to mention the very-low-molecular-weight protein found in Neurospora that has a striking sequence homology to the metallothioneins from vertebrates (Lerch, 1980). It is interesting to point out that metallothioneins can also interact very strongly with glucosaminoglycan so as to form a number of adducts with different molecular weights (Webb and Stoddart, 1978). One must remark that typical metallothioneins seem to be absent in some animals, and some metals (namely Cd and Pb) are linked to other compounds such as ATP, lysine, taurine and homarine in Ostrea (Coombs, 1974), glycopeptidic compounds in protozoa, (Albergoni et al., 1980; Piccinni and Coppellotti, in press), and high-molecular-weight proteins in barnacles (Rainbow et al., 1980). The metallothioneins seem to be inducible by an extra amount of many of the trace metals, but they have been detected in small amounts in unloaded organisms too. These proteins typically have a localization in the same organs that accumulate metals. They link metals according to (1) their presence in cells and (2) the affinity of the metals to SH groups.

Compartments and mechanisms for regulation

Regulation may be carried out at various levels. The first level is modification of the permeability of the membranes involved in the diffusion or in the transport of metals between the external and internal environments. A decrease in permeability as an adaptive modification has been suggested for organisms living in polluted waters (Bryan and Hummerstone, 1973; Foster, 1977). It has been postulated that metal ions may be trapped and complexed in the mucous layer which acts as a defensive barrier at the skin and gill levels (Pentreath, 1973; Coombs, 1979). At present, there are no satisfactory data about a control over absorption. Actually, a decreased permeability is a protective device and not a true regulation; this is so because it cannot promptly change according to the external levels of metals and, furthermore, it is not finely modifiable.

Similar devices for reducing the input of metal are seen in other exclusion mechanisms. Among them we can include the secretion products that bind metals extracellularly, thus reducing the concentration of free ions. A similar role may be ascribed to the fibrillar material described by Leppard (Leppard et al., 1977; Massalski and Leppard, 1979; see also the chapter of Leppard and Burnison in this volume). By such means there can be a change in the metal bioavailability because metals, in a new complexed form, may become unavailable for some animals, while also entering into new alimentary chains. An exclusion mechanism can be ascribed also to the corpuscles found in the intestinal lumen of marine fish. They bind metals directly from ingested sea water and limit the entry of these metals through the intestinal wall (Noel-Lambot, 1981).

A true regulation of the metal content can be performed only at the excretion level, provided that this can be varied according to absorption. There are three most important ways of excretion that can be used by animals. The first is a loss across the gills or the body surface. This way has been demonstrated for many metals in Crustacea and in fishes (Coombs, 1979); also, cadmium may be excreted via an increased mucous secretion (Bryan, 1976). The second way is excretion into the gut. There are many significant data about copper and iron granules formed in the hepatopancreas and extruded into the gut of Crustacea. It has been demonstrated that copper and zinc are usually excreted into the gut of many invertebrates (Bryan, 1976). Studies on the biliary excretion of copper have revealed that the excess copper is contained in numerous discrete granules arranged around bile canaliculi. These granules are rich in acid phosphatase

and are considered to be lysosomes (Lindquist, 1968). The third
important way for excreting metals is by way of kidneys or analogous
organs. Examples are represented by Homarus and Lobster and other
crustacean Decapoda which are generally able to excrete Cu, Zn, Co,
Mn, and inorganic Hg via urinary fluids(Bryan, 1976).

The excretion of granules (containing metals) by exocytosis is
well demonstrated in protozoa (lead and copper in Tetrahymena; Nilsson, 1979; Nilsson, in press). Granular excretions of zinc in the
kidney of Mytilus have also been reported (George and Pirie, 1980).
Many peculiar systems can also be present in animals. Cadmium and
other metals, having been accumulated in the carapace of Crustacea,
are removed by the moult. In some molluscs, such as Ostrea, copper-
accumulating amebocytes are directly excreted. From these data, it
appears that apart from the glomerular ultrafiltration of metals, the
mechanism of excretion of metals is supported by the intracellular
metabolism of them and realized by exocytosis or by a loss of accumulating cells.

Few data on the excretion of metallothioneins have been reported
and what exists is underutilized. The presence of metallothioneins
in urinary fluid was referred to by Nordberg in 1976; however, only
recently a urinary excretion of metallothioneins has been demonstrated
in man by a radioimmune assay, and, a linear relation has been shown
to exist between cadmium and metallothionein in urine (Chang et al.,
1980). The metallothionein excretion at the level of other excretory
compartments (like gills and liver, through biliary secretion) should
be tested.

A mechanism of regulation of metal content by metal excretion,
linked to detoxification and to metal metabolism, is suggested. We
will refer to this topic in the concluding remarks.

INTERACTIONS BETWEEN METALS

At present, the data concerned with this topic do not cover all
the possible interacting effects; the researchers have preferred
unifactorial experiments. This field certainly will be better explored in the future. Although most of the data are related to experiments or to observations on mammals, owing to their importance in
understanding human nutrition, a comparative approach to the problem
is lacking. The literature appears contradictory and unclear; this
is probably the result of unchecked experimental parameters. A conference on the interactions between various metals has been held recently by the New York Academy of Sciences (Levander and Cheng, 1980).

Interactions between an essential and other essential metals

Some of the most important interactions encompass iron, copper and zinc. In mammals, an over-administration of copper leads to toxic signs, which can be counteracted by an extra amount of iron or zinc. In this case, the copper accumulation in the liver is really decreased by zinc or iron. Our experiments on Euglena gracilis (Protozoa) show that a high amount of copper induces a decrease of the zinc content in the cells (Albergoni et al., 1980). This datum is supported by other authors who have demonstrated that the effect of zinc is to increase the excretion of copper. These interactions may occur at various levels; the most important are (1) the competition for the absorption sites, (2) the competition for metabolic sites and (3) the interaction on excretory systems.

Some researchers are now describing, in increasing detail, the interactions between trace and ultratrace elements. A synergism has been ascertained between nickel and iron. The nickel apparently converts the unavailable ferric ion into a form which can be absorbed. In every case, nickel seems to be involved in some mechanisms related to the active transport or to the passive diffusion of iron. This effect clearly appears when iron is low in the diet. An antagonistic interaction has been described between nickel and copper. Nickel supplementation tends to decrease the copper content in some tissues, especially when dietary copper is low. However, other data apparently do not agree with these.

Interactions between essential and non-essential metals

Interactions between many essential trace elements with non-essential ones are documented extensively in the literature. In general it seems that low levels of some essential metals exacerbate cadmium toxicity, whereas high levels are protective. Past observations and recent experiments confirm that the toxic effect of cadmium is due to copper deprivation when the dietary copper content is normal. An increase in copper or zinc intake counteracts these cadmium effects; higher doses of copper induce toxicity. As for the Cd-Zn interactions, on a chemical basis, cadmium should cause a decrease in tissue zinc content. On the contrary, cadmium causes zinc to accumulate in tissues. This is confirmed in Euglena gracilis (Albergoni et al., 1980).

SPECULATIONS ABOUT REGULATION, DETOXIFICATION, ACCUMULATION, AND INTERACTIONS

From the information above we intend to demonstrate a common mechanism that functions for the regulation, detoxification, and accumulation of metals in animals. The first point to consider is the relation (or the coincidence) between detoxification and regulation. The reported data tend to show that regulation is nearly always performed through excretory pathways. They also show that the trace metals are accumulated in organs that are clearly excretory or that may function also for excretion. All these organs contain metallothionein or analogous compounds, linked to zinc or to other essential trace metals. At the least, it has been shown that a very important point of the transfer of trace metals inside a cell is the storage of metals, sometimes temporary, in vesicular formations or in lysosomes. Additionally, many findings indicate that their contents can be discharged to the outside. The lysosomal step seems to be very important in the metabolism of metals.

A key point now is the relations between (1) cytoplasmic metallothioneins, (2) lysosomes and (3) excretion of metals. A polymeric form of Cu-Zn has been found in a lysosomal fraction of the liver of newborn mammals. Some X-ray microprobe analyses have revealed a peak of sulphur, one that may correspond to cysteine; this could mean the presence of metallothionein in a granular fraction. Recently, a cadmium thionein in lysosomes of the rat was demonstrated.

There is very little evidence on the excretion of metallothionein to permit a definite conclusion. The presence in the intestinal contents of such a protein has been related to epithelial desquamation of the intestinal mucosa. Also, the presence of Cd-thionein in urinary fluids has been related to damage owing to a cadmium effect. More careful investigations should be performed to control the regulative loss of metallothionein in the normal metabolism of essential trace metals. It is possible that, instead of metallothionein, fragments or aggregates of such protein are excreted. Here we must point out that an animal is advantaged by excreting the metals in a firmly bound form. Only in this case can reabsorption really be prevented. Metallothionein, owing to its structure and metal content, is barely digested by intestinal enzymes. This feature gives it the ideal role of an excretory product which could move metals through the digestive tract.

In our opinion, the metallothioneins may have a physiological role in the regulation of the essential metal content in the animal body, acting by binding and then by excreting the metals. During these steps, the metallothionein may also function as a donor of metals for normal metabolism. Many metals can be trapped by its SH groups and many may substitute for the essential ones.

The question now is: do the non-essential metals share the same secretion mechanism as do the essential ones? It could be so. Actually, the literature data indicate that non-essential metals are preferentially accumulated, although this event occurs at high levels of concentrations of the essential ones. This may mean that there is a saturation of the excretory system; it can be explained by the fact that the regulative system for trace metals normally regulates the metals present in trace amounts. The non-essential metals may contribute to saturate the cellular excretory pathways and may also interfere with the physiological regulation. Many recent data indicate that, in the soluble phase, cadmium or lead are accumulated, and are preferentially linked to high-molecular-weight compounds that cannot be promptly excreted. We have some evidence that these compounds are polymeric forms of metallothioneins or analogous compounds.

One must note that the lack of bioamplification of the metal content in the alimentary chain may be related, not only to the improvement of regulative mechanisms, but also to a lesser absorption of the metals chelated by metallothionein.

Finally, let us consider the last speculation. This regards the evaluation of some interactions between trace metals. In many cases, the regulation of an essential trace metal, when it is present in excess, induces a depauperization in another essential trace metal. This occurs between Mo and Zn, Cu and Zn, and so on. Considering the molecular basis of the regulation previously summarized, the excretion of metallothionein induces a concomitant excretion of the other linked metals. On the contrary, a lesser excretion, due to a non-essential metal, induces an increase in the content of the other co-linked metals. Many antagonistic effects can be so explained.

This mechanism of metal content regulation is supported by combinations of many experimental data and represents a proposal for well defined research to demonstrate the overall sequence of individual events. It never intends to explain all the biological events, but it represents an attractive tool, aiming to find the links between many physiological and pathological events on a molecular basis. One should mention that the fundamental pathways in biological regulation have a common basis in all animals, with secondary variations and improvements. Obviously, the proposed mechanism is not in opposition to other biological and chemical interactions which have been well-defined in some animals under some conditions. All forms of competition between the metals at the absorption level are very important in this respect.

Studies on chemical species of metals in the environment are really very important, but it is also necessary to know the fate of metals in organisms and the different metabolism-mediated transformations of chemical species. Only careful studies on these latter topics may explain the different effects of the same metal on different animals, and such studies can be very useful to find answers to many biological problems.

REFERENCES

Albergoni, V., Piccinni, E., and Coppellotti, O., 1980, Response to heavy metals in organisms. I. Excretion and accumulation of physiological and non-physiological metals in Euglena gracilis, Comp. Biochem. Physiol., 67C:121.

Blumberg, W.E., 1978, Enzymic modification of environmental intoxicants: the role of cytochrome P_{450}, Q. Rev. Biophys., 11:481.

Bowen, H.J.M., 1966, "Trace Elements in Biochemistry", Academic Press, New York.

Bryan, G.W., 1976, Heavy metal contamination in the sea, in: "Marine Pollution", R. Johnston ed., Academic Press, London.

Bryan, G.W., and Hummerstone, L.G., 1973, Adaptation of the polychaete Nereis diversicolor to estuarine sediments containing high concentrations of zinc and cadmium, J. Mar. Biol. Ass. U.K., 53:839.

Chang, C.C., Vander Mallie, R.J., and Garvey, J.S., 1980, A radioimmunoassay for human metallothionein, Toxic. Appl. Pharmac., 55:94.

Chang, C.C., Lauwerys, R., Bernard, A., Roels, H., Buchet, J.P., Garvey, J.S., 1980, Metallothionein in cadmium-exposed workers, Environ. Res., 23:422.

Coombs, T.L., 1974, The nature of zinc and copper complexes in the oyster Ostrea edulis, Mar. Biol., 28:1.

Coombs, T.L., 1979, Cadmium in aquatic organisms, in: "The Chemistry, Biochemistry and Biology of Cadmium", M. Webb, ed., Elsevier/ North Holland, Amsterdam.

Coombs, T.L., and George, S.G., 1978, Mechanisms of immobilization and detoxication of metals in marine organisms, in: "Physiology and Behaviour of Marine Organisms", D.S. McLusky and A.G. Berry, eds., Pergamon Press, Oxford.

Davis, J.R., and Avram, M.J., 1980, Correlation of the physico-chemical properties of metal ions with their activation and inhibition of human erythrocytic δ-aminolevulinic acid dehydratase (ALAD) in vitro, Toxic. Appl. Pharmac., 55:281.

Förstner, U., and Wittmann, G.T.W., 1979,"Metal Pollution in the Aquatic Environment",Springer-Verlag, New York.

Foster, P.L., 1977, Copper exclusion as a mechanism of heavy metal tolerance in a green alga, Nature, 269:322.

George, S.G., and Pirie, B.J.S., 1980, Metabolism of zinc in the mussel, Mytilus edulis (L.): a combined ultrastructural and biochemical study, J. Mar. Biol. Ass. U.K., 60:575.

Goldfischer, S., Schiller, B., and Sternlieb, I., 1970, Copper in hepatocyte lysosomes of the toad, Bufo marinus L., Nature, 228:172.

Icely, J.D., and Nott, J.A., 1980, Accumulation of copper within the "hepatopancreatic" caeca of Corophium volutator (Crustacea: Amphipoda), Mar. Biol., 57:193.

Kägi, J.H.R., and Nordberg, M., eds., 1979, "Metallothionein", Birkhäuser Verlag, Basel.

Leppard, G.G., and Burnison, B.K., in: this volume.

Leppard, G.G., Massalski, A., and Lean, D.R.S., 1977, Electron-opaque microscopic fibrils in lakes: their demonstration, their biological derivation and their potential significance in the redistribution of cations, Protoplasma, 92:289.

Lerch, K., 1980, Copper metallothionein, a copper binding protein from Neurospora crassa, Nature, 284:368.

Levander, O.A., and Cheng, L., eds., 1980, "Micronutrient Interactions: Vitamins, Minerals and Hazardous Elements", Ann. N.Y. Acad. Sci., Vol.355, New York.

Lindquist, R.R., 1968, Studies on the pathogenesis of hepatolenticular degeneration. 3. The effect of copper in rat liver lysosomes, Am. J. Path., 53:902.

Massalski, A., and Leppard, G.G., 1979, Morphological examination of fibrillar colloids associated with algae and bacteria in lakes, J. Fish. Res. Board Can., 36:922.

Nilsson, J.R., 1979, Intracellular distribution of lead in Tetrahymena during continuous exposure to the metal, J. Cell Sci., 39:383.

Nilsson, J.R., 1981, Effects of copper on phagocytosis in Tetrahymena, Protoplasma, in press.

Noel-Lambot, F., 1981, Presence in the intestinal lumen of marine fish of corpuscles with a high cadmium, zinc and copper binding capacity: a possible mechanism of heavy metal tolerance, Mar. Ecol. Prog. Ser., 4:175.

Nordberg, G.F., ed., 1976, "Effects and Dose-Response Relationships of Toxic Metals", Elsevier, Amsterdam.

Passow, H., Rothstein, A., Clarkson, T.W., 1961, The general pharmacology of heavy metals, Pharmac. Rev., 13:185.

Pentreath, R.J., 1973, The accumulation from water of ^{65}Zn, ^{54}Mn, ^{58}Co and ^{59}Fe by the mussel Mytilus edulis, J. Mar. Biol. Ass. U.K., 53:127.

Piccinni, E., and Coppellotti, O., 1982, Response to heavy metals in organisms. II. Effects of physiological and non-physiological metals in Ochromonas danica, Comp. Biochem. Physiol., in press.

Prosi, F., 1979, Heavy metals in aquatic organisms, in: "Metal Pollution in the Aquatic Environment", U. Förstner and G.T.W. Wittmann, eds., Springer-Verlag, New York.

Rainbow, P.S., Scott, A.G., Wiggins, E.A., and Jackson, R.W., 1980, Effect of chelating agents on the accumulation of cadmium by the barnacle Semibalanus balanoides, and complexation of soluble Cd, Zn and Cu, Mar. Ecol. Prog. Ser., 2:143.

Rudd, J.W.M., Furutani, A., and Turner, M.A., 1980, Mercury methylation by fish intestinal contents, Appl. Environ. Microbiol., 40:777.

Sato, M., Nagai, Y., 1980, Form of cadmium in rat liver subcellular particle, Toxicol. Lett., 7:119.

Sunderman, F.W., Jr., 1979, Mechanisms of metal carcinogenesis, Biol. Trace Element Res., 1:63.

Vaccaro, R.F., in: this volume.
Vallee, B.L., 1979, Historical review and perspectives, in: "Metallothionein", J.H.R. Kägi and M. Nordberg, eds., Birkhäuser Verlag, Basel.
Venugopal, B., and Luckey, T.D., 1975, Toxicology of non-radioactive heavy metals and their salts, in: "Heavy Metal Toxicity, Safety and Hormology", T.D. Luckey, B. Venugopal, D. Hutcheson, eds., Thieme, Stuttgart.
Webb, M., Stoddart, R.W., 1978, Isoelectric focusing of the cadmium ion binding protein of rat liver: interaction of the protein with a glycosaminoglycan, Biochem. Soc. Trans., 2:1246.
Wittmann, G., 1979, Toxic metals in: "Metal Pollution in the Aquatic Environment", U. Förstner and G.T.W. Wittmann, eds., Springer-Verlag, New York.
Wood, J.M., Fauchiang, Y.T., and Ridley, W.P., 1978, The biochemistry of toxic elements, Q. Rev. Biophys., 11:467.
Wright, D.A., 1978, Heavy metal accumulation by aquatic invertebrates, in: "Applied Biology, Vol. III", T. Coaker, ed., Academic Press, London.

DISCUSSION: V. ALBERGONI AND E. PICCINNI

G.G. LEPPARD   What happens to the lysosomal organic complexes of toxic metals when a small animal rich in these is eaten and passed through the gut of a larger predator animal?

V. ALBERGONI   This kind of event has not been investigated. One must consider that metallothioneins are not digested by trypsin and pronase. However, at very low pH, as is found in the stomache of some animals, the metals could be detached from the thioneins and therefore they could be absorbed by the animal.

F.H. FRIMMEL   How does mercury relate to thioneins?

V. ALBERGONI   Only inorganic mercury influences the formation of thioneins. It has been demonstrated that phenyl mercury can stimulate the formation of thioneins but this apparently occurs when phenyl mercury becomes broken down to inorganic mercury first. It has been reported that methyl mercury does not influence the formation of thioneins but, certainly after demethylation in the liver of fish, some mercury-thioneins can be found.

Y.K. CHAU   Biomethylation has been speculated to be a detoxification process. The methylated elements are often more toxic than the same elements prior to methylation. Can you comment on this?

V. ALBERGONI   Biomethylation permits a quick removal of some
                      toxic elements from niches occupied by bacteria. In
                      fact, methyl derivatives readily diffuse outwards
                      from the cell membrane but only a microbial popula-
                      tion is advantaged by this detoxification system.

M. BERNHARD   You showed that particles are eliminated by
                      protozoa. Could you briefly tell us something about
                      the reverse process?

V. ALBERGONI   Metal-bearing particles can enter small organ-
                      isms by phagocytosis.

G.G. LEPPARD   Many excellent studies have been done on the
                      immobilization of heavy metals by animal cells. How
                      much related work is done on algal cells and higher
                      plant cells?

V. ALBERGONI   If we consider unicellular algae, there have
                      been many studies done on the accumulation of metals
                      in vacuolar formations as well as on mechanisms of
                      excretion. In the roots of plants, clusters of
                      metals have been demonstrated to be linked to com-
                      pounds resembling metallothioneins. Some algae are
                      able to trap metals in their polysaccharidic wall.
                      An exclusion mechanism seems to be the one most used
                      to protect such organisms from metal entry into
                      cells.

J.P. GIESY   Cadmium is only slowly eliminated from animals.
                      If cadmium is bound to metallothioneins, it appears
                      the complex would not be excreted. Could the stra-
                      tegy of animals be to simply sequester cadmium to
                      metallothionein?

V. ALBERGONI   At this moment, I think it is possible that
                      metallothioneins linked to essential metals could
                      be excreted normally. In our opinion, cadmium
                      linked to high-molecular-weight metallothioneins
                      can be accumulated for a long period in lysosomes.
                      This may be regarded as an "interference" with the
                      homeostatic mechanisms that operate in the presence
                      of essential trace metals. As previously shown,
                      cadmium therefore may be lost preferentially by
                      cellular disintegration or by desquamation.

BIOLOGICAL ASPECTS OF TRACE ELEMENT SPECIATION IN THE AQUATIC ENVIRONMENT

Maarten Smies*

Delta Institute for Hydrobiological Research
Vierstraat 28, 4401 EA Yerseke, The Netherlands

INTRODUCTION

That different forms of a chemical element are likely to exert different biological effects in the natural environment would appear somewhat of a truism. Considering, however, that heavy metals are well established as environmental pollutants (Förstner and Wittmann, 1979), it is surprising that relatively little attention has been paid to the question of metal speciation in relation to biological effects. Recent reviews of metal pollution in the aquatic environment (Bryan, 1976; Prosi, 1979) have touched upon metal speciation as related to toxicity, but the aquatic toxicological literature by and large ignores it. An exception must be made for phytoplankton research, where worthwhile efforts have been made to properly define culture media (Morel et al., 1979) and to apply chemical speciation models to observations on phytoplankton growth (Jackson and Morgan, 1978). Otherwise, it would appear, few publications specifically address the above question in a more than rudimentary way. This is not meant to imply criticism of scientists working on heavy metals; it is merely an appraisal of the current situation.

The purpose of this paper is to assess the present state of the art and to indicate which research approaches appear most promising. It is clear from the previous papers in this volume that progress in the biological field is strongly dependent upon

---

*Present address: Shell Internationale Research Maatschappij BV,
                   Group Toxicology Division, PO Box 162,
                   2501 AN Den Haag, The Netherlands.

advances in chemical analyses and upon the possibility of providing biological interpretations of the effects of given chemical species, as identified by analytical techniques. However, the reverse must also be true; if the biologist is prepared to adapt his experimental design so that existing or newly available techniques for the identification of chemical species can be used, then he is most likely to obtain results that are meaningful, and in view of recent geochemical progress, worthwhile.

CATEGORIES OF CONSEQUENCES

Before discussing the consequences of trace element speciation for aquatic life, it is useful to take a brief look at the sort of consequences which will be considered.

First, the direct interaction between a chemical element and biota is of interest and this has received appreciable attention. Environmental factors such as pH, Eh, alkalinity, and chlorinity play an important part in the biological availability of trace element species and hence affect their apparent toxicity. Element uptake, binding, turnover and excretion by biota follow if the chemical species is available, so that effects on these processes also fall into this first category. Given sufficient data on bioavailability and the apparent toxicity of chemical forms to biota, a measure of the hazard to groups of organisms can be devised. Ideally, this would enable one to reach conclusions about the response of populations of different biological species, and consequently, to appraise the implications for the biological community. It will become clear that we have not come anywhere near this stage yet, while other experimental approaches, such as limnocorrals (butyl-rubber or polythene bags, enclosing a part of the water column in a lake or coastal lagoon - Gächter and Urech, this volume; Vaccaro, this volume), may provide better tools to study the effects in ecological systems.

Second, one may consider the effects of biological processes upon trace element speciation in the form of biotransformation and then study the fate and behaviour of the biotransformed products. Some trace elements, such as As (Waslenchuk, 1978; Andreae, 1978, 1979; Sanders, 1980) and Se (Measures and Burton, 1978, 1980; Measures et al., 1980) show biological control of inorganic chemical species composition, while a number of them, for example As, Hg, Se, and Te (Saxena and Howard, 1977; Summers and Silver, 1978) undergo biomethylation, or at least environmental methylation.

Third, besides the effects of living organisms upon trace element speciation, organic compounds of biological origin are clearly known to influence environmental chemical behaviour. Humic and fulvic acids and related organics are certainly the best known and probably the most important natural complexants (Mantoura et al., 1978; Mantoura, 1981), although other organic compounds may also play a role in complexing or adsorbing trace metals.

This third category effectively closes the circle, since dissolved and particulate organic matter (DOM and POM) are normal constituents of natural waters, so that we return to consequences in the first category.

STATE OF THE ART

In order to provide an overview of current research, literature data must be brought together in some specific way. Although it would seem preferable to order the data according to observed biological consequences or biological processes that may be affected, I have opted to present them by chemical element. In this way the interaction of biota and the element's geochemical cycle are the main subject of interest, while cross references will be made afterwards. The choice of elements is purely pragmatic and is based on the availability of sufficient information only.

Arsenic (As)

Arsenic is a very suitable element to start with because quite a lot of information on its biogeochemical behaviour has recently become available. In inorganic form it occurs in two oxidation states, As(V) and As(III), the latter being the product of biological reduction by phytoplankton (Andreae and Klumpp, 1979; Sanders and Windom, 1980; Wrench and Addison, 1981), probably as a protection against inhibition of phosphate uptake (Andreae, 1978; Sanders, 1979). Besides the production of arsenite, biota are also involved in the conversion of As into organic compounds, such as by methylation (Wood, 1974; Andreae, 1978, 1979; Andreae and Klumpp, 1979; Sanders, 1979; Sanders and Windom, 1980; Wrench and Addison, 1981). Other organic arsenic compounds, such as arseno-sugars (Edmonds and Francesconi, 1981) and arseno-lipids (Klumpp and Peterson, 1981; Wrench and Addison, 1981) are also produced. Both arsenite and the methylated arsenic compounds are excreted into the aquatic environment and are oxidized and demethylated there. The arseno-lipids can be transferred along the food chain (Wrench et al., 1979; Klumpp and Peterson, 1981). The biotic conversion may result in a 15-20% fraction of dissolved As in coastal water to be in As(III) or methylated form (Waslenchuk, 1978; Sanders, 1980), but in the deep sea As(V) becomes the sole species below the photic zone (Andreae, 1979). Conversion of As(V) by phytoplankton apparently acts as a detoxifying mechanism, since As(III) and methyl-arsenic compounds are less phytotoxic. As(III), however, is more toxic to most other life forms than is As(V).

Cadmium (Cd)

Cadmium is well researched as regards its toxicity to aquatic organisms and its occurrence in the environment. Contrary to many other trace metals, Cd is not strongly complexed by organic complexants (Mantoura, 1981), but its environmental toxicity has been reported to decrease with salinity (Engel and Fowler, 1979; Phillips, 1980), apparently because of reduced Cd^{2+}-ion activity by chloride complexation. George and Coombs (1977) reported faster uptake and higher accumulation of organically complexed Cd in the common mussel. In aquatic invertebrates, Cd is bound to proteins (Wright, 1977; Noel-Lambot et al., 1978; Jennings et al., 1979; Rice and Chien, 1979) which detoxify the element, while some fish have been reported to have Cd-binding corpuscles, in the intestinal lumen, which may serve the same purpose (Noel-Lambot, 1981). In the oceans, dissolved Cd concentrations are correlated with nutrient levels (Boyle et al., 1976; Bruland et al., 1978), indicating biological control of environmental concentrations.

Chromium (Cr)

Dissolved chromium occurs as Cr(III) and Cr(VI), with the latter normally predominating. Under anoxic conditions Cr(III) is formed, but in oxic conditions this is oxidized to Cr(VI) again (Emerson et al., 1979). Since oxidation may be slow, Cr(III) can exist metastably under oxic conditions (Cranston and Murray, 1980; Nakayama et al., 1981). Cr(III) is less toxic than Cr(VI) in general, which in fish appears to be the result of a totally different toxicity mechanism established by each of the two Cr species (van der Putte, 1981).

Copper (Cu)

Hart (1981) has reviewed the chemistry of copper in the aquatic environment. Cu toxicity to phytoplankton and other biota has been shown to depend on Cu(II)-ion activity or inorganically complexed Cu (Jackson and Morgan, 1978; Anderson and Morel, 1978; Young et al., 1979). Under natural circumstances, an appreciable fraction of Cu (65-90%) is bound by organic complexants, mostly of the fulvic acid type (see Hart, 1981, and references therein), although other complexant types, such as fractions of the dissolved organic nitrogen are also able to bind Cu (Tuschall and Brezonik, 1980). In fact, nominal ambient Cu concentrations may be sufficiently high to produce toxic effects in biota, if all Cu would be available biologically (Jackson and Morgan, 1978; Engel and Sunda, 1979). As regards detoxification, Seeliger and Edwards (1979) report that Cu is bound to organic compounds in marine benthic algae, and in the periwinkle (Littorina littorea) Cu is bound to a single low-molecular-weight complexant (Howard and Nickless, 1978). On the other hand, McKnight and Morel (1980) who demonstrated strong Cu-binding by siderophores excreted by blue-green algae, concluded that these were not excreted as a detoxification mechanism. Cu addition to marine and fresh water limnocorrals (CEPEX and MELIMEX) has been reported to cause a shift in phytoplankton species composition (Thomas and Seibert, 1977; Gächter and Máreš, 1979), but these results are non-replicated observations only.

Iron (Fe)

In contrast to copper, iron is an essential element for photosynthetic life forms and due to the insolubility of its dominant form, $Fe(OH)_3$, the Fe(III)-ion activities are extremely low. It has been postulated that organic chelators serve to make Fe available to phytoplankton, but the evidence is inconclusive (Jackson and Morgan, 1978). It is clear, however, that "dissolved" Fe concentrations comprise largely colloidal organic complexes (Eaton, 1979; Moore et al., 1979; Hoffmann and Eisenreich, 1981).

Mercury (Hg)

Mercury distinguishes itself from the heavy metals discussed before by its extensive environmental conversion, in which biota play an important role. Inorganic Hg, mainly as Hg^{2+} which is the normal natural and anthropogenic input into the environment, is to a large extent rendered biologically unavailable by complexation with humic compounds in fresh water (Mantoura et al., 1978), by chloride in estuarine and sea water (Dyrssen and Wedborg, 1980), and by sorption to particulates, especially bed sediments (Kudo et al., 1978). Methylmercury, although only a tiny fraction of the total Hg in the abiotic environment (Kudo et al., 1978; Bartlett and Craig, 1981), is much more important biologically not only because of its high toxicity but also because of its bioaccumulative properties (Mercury, 1976). It is produced as a by-product of microbial activity, predominantly in the bottom sediment (Summers and Silver, 1978) but also in the water column (Topping and Davies, 1981). Because of its high bioavailability and small clearance it represents a large, and in fish even predominant (Davies et al., 1979), fraction of the total Hg in biota. Besides methylation of Hg in natural systems, reduction to metallic Hg occurs in the environment, and this provides a release route from the aquatic compartment. Since a large fraction of Hg in aquatic systems is in the bottom sediment with low bioavailability (Luoma, 1977), it is interesting that biota may play a significant role in Hg clearance from the upper portion of the bottom sediment (Boddington et al., 1979; Bothner et al., 1980). Although this is not likely to affect significantly the Hg distribution in the abiotic environmental compartments, it may be of importance to the local aquatic organisms.

Manganese (Mn)

Manganese speciation in the aquatic environment (especially in the estuarine environment) has received appreciable attention in recent years. Mn cycles through estuaries, changing from dissolved to particulate species and back again (Duinker et al., 1979; Morris and Bale, 1979; Yeats et al., 1979). Very little is known apparently of the possible biological consequences. Sanders (1978) reports stimulated carbon uptake by estuarine phytoplankton under the influence of dissolved Mn and relates this to the concurrent high DOM concentrations. On the other hand, low oxygen concentrations in estuaries, caused by freshwater phytoplankton death (Morris et al., 1978) and sewage inputs, promote Mn dissolution (Wollast et al., 1979), so that there is a biological mediation of speciation. Boulègue and Renard (1980) point out that the oxidation of Mn in natural waters is catalyzed by bacteria and that this may explain high oceanic Mn deposition rates.

Lead (Pb)

Recently, the biogeochemistry of lead in the marine environment has been discussed extensively (Branica and Konrad, 1980), so that this element can perhaps be dealt with more summarily here. From various studies it can be concluded that Pb occurs mainly as $PbCO_3^0$ and $Pb(OH)$ in seawater (Sipos et al., 1980a,b), while in inshore and inland waters Pb is often associated with high-molecular-weight substances and colloidal organic matter (Giesy et al., 1978; Mantoura, 1981). A controversy has recently arisen on whether or not Pb undergoes biomethylation. This of course is of great interest because of the much higher toxicity of alkyllead compounds compared to inorganic Pb (Maddock and Taylor, 1980), and because of the volatility of alkyllead compounds. Reisinger et al. (1981a) have recently reported that they could not demonstrate biomethylation of inorganic Pb(II), but traces are produced in the presence of micro-organisms in the natural environment (Baker et al., 1981). Also, organic Pb(IV) salts may be converted into leadalkyls chemically. Tetraethyllead and tetramethyllead which are used as anti-knock agents in petrol are rapidly broken down in the environment to the corresponding trialkyllead compounds; these latter compounds are much more persistent (Grove, 1980) and more toxic to aquatic life forms (Röderer, 1981). It would, therefore, appear that alkyllead compounds found in the environment are essentially of anthropogenic origin.

Selenium (Se)

Selenium is biomethylated in the aquatic environment (Chau et al., 1976), but this has little biological consequence, because neither the inorganic nor the organic forms of Se have been reported to pose pollution problems in the environment (IRPTC, 1978). Se concentrations in marine mammals, birds, and fish are correlated with Hg burdens (Mercury, 1976) and selenite has been reported to detoxify Hg in the body. Since Se also increases the retention of Hg, it is not clear whether or not this detoxification is eventually effective (Beijer and Jernelov, 1978). The change of Se(IV) – Se(VI) concentration ratios in the sea indicates biological control of Se speciation (Measures and Burton, 1978, 1980; Measures et al., 1980).

Tin (Sn)

Little is known about tin speciation in the natural environment. This is regrettable because organotin compounds are used widely as biocides, especially for antifouling purposes (IRPTC, 1978), and this must mean that they are introduced into the aquatic environment. Organotin compounds are broken down in the environment into inorganic Sn which is considered non-toxic. As regards biomethylation of Sn, this has been reported to occur (Summer and Silver, 1978), but it is not at all certain (Craig, 1981). Recent papers (Blair et al., 1981; Chau et al., 1981) report its production in the presence of microbial activity, but this is almost certainly not an important conversion route.

Zinc (Zn)

The speciation and biological effects of zinc resemble those of cadmium. It is not strongly complexed by natural organic complexants (Mantoura, 1981) and occurs mainly in ionic or inorganically complexed form (Dyrssen and Wedborg, 1980). Biological availability of Zn is correlated with the free Zn^{2+}-ion activity (Anderson et al., 1978). Zn toxicity is usually less than that of Cd (IRPTC, 1978), and it becomes similarly bound to protein fractions in marine invertebrates (Howard and Nickless, 1978; Noel-Lambot et al., 1978) and to intestinal corpuscles in marine fish (Noel-Lambot, 1981). Under conditions of Zn limitation, silicic acid uptake is decreased in diatoms (Rueter and Morel, 1981). It is of interest that limiting and toxic Zn concentrations are only between two and three orders of magnitude different, indicating that both phenomena may well occur under natural or slightly polluted conditions (Anderson et al., 1978).

Synthesis

Trace element speciation may be subdivided into a number of classes of greater and lesser importance with regard to biota in the aquatic environment. Inorganic speciation in the dissolved phase does not seem to have profound consequences with the exception of chloride/carbonate complexing effects upon bioavailability/ apparent toxicity (Prosi, 1979). Biological control of inorganic speciation is relatively rare, although of great interest as regards the geochemical element cycles of As and Se. Organic complexation in the dissolved and colloidal phase has profound effects upon element availability (Mantoura, 1981). Some trace metals occur naturally at concentrations that in free ionic form would be toxic, but which by complexation are rendered unavailable. Removal from the dissolved phase by adsorption to suspended particles or organisms, followed by removal to the bottom sediment are also important in this respect, but this must be assessed against the probability of remobilization by redox changes in the sediment, by complexants or by biological or chemical conversion. Inorganic/organic conversions, whether or not biologically mediated, are of most direct importance biologically since organo-metals, at least the alkyl-metals, tend to be more toxic than the ionic forms, while they have a greater tendency towards bioaccumulation because of their lipid solubility. It would appear that environmental methylation needs to be studied in more detail now that it has become evident that, for most trace elements, it is not truly a biochemical process in the usual sense (Reisinger et al., 1981a, b). Detoxification and the biological fate of detoxification products also deserve attention. Although considerable information is available on heavy metal binding by proteins in a range of organisms (Roesijadi, 1981), the fate of the protein-metal in the foodweb (for example, the immediate fate upon death of the organism) has not been investigated. Biological consequences of trace element speciation above the level of individual organisms have not been seriously investigated yet. Limnocorral studies have been set up to specifically address this problem but, as far as I can see, the speciation aspect has not really been elucidated.

FUTURE

With regard to future research in this field, one must let oneself be guided both by the need to increase our understanding of the biogeochemical processes involved and by a number of practical considerations. It has, in my opinion, become irrelevant to continue producing bio-assay type data that show the effects of chlorinity, complexants, etc., upon the toxicity of added inorganic metal species, as is still done frequently for aquatic animals. Here, one should either choose a defined-medium approach (as outlined by Morel et al., 1979) or one should employ the relevant analytical techniques for chemical species and interpret the results on that basis. In fact, advanced analytical techniques and biological research on trace elements appear still to be far apart, and this gap urgently needs to be narrowed. This situation applies both to laboratory and field studies.

There is still a major gap in our knowledge of the availability of certain trace elements to biota, those which appear to have to be unavailable for reasons of solubility. The work by Jackson and Morgan (1978) indicates a possible approach to bridging this gap and, indeed, chemical analysis is unlikely to be of assistance here as one would expect to have to deal with extremely low levels of the bioavailable species.

Bioconversion and the fate and behaviour of bioconversion products form a subject that offers scope for further work. In view of the recent evidence on environmental alkylation of inorganic metal species, there is a need to establish how important it is as a conversion process. Since it is not a biochemical process, the reaction conditions and kinetics of formation and breakdown of methylated species must be investigated to determine its biogeochemical significance. Since the larger part of most aquatic foodwebs is detritus-based, the toxicity of metal-proteins to detritus feeders should be established, and the consequences of fungal and bacterial breakdown for the incorporated metals should be elucidated. The decomposition pathways of organic material with associated trace elements warrant attention more generally; appreciable amounts of trace elements are associated with DOM and POM and these materials are either decomposed in the water column or after sedimentation in the sediment. It is of utmost importance to find out in which chemical form these associated trace elements emerge from the decomposition process, and especially whether they will be bioavailable. In anoxic waters, such as the hypolimnion of deep lakes, and in the sediment the redox changes must also be considered. As sedimentation in limnocorrals is usually greater than in the parent water body, accelerated removal of trace element spikes is usually observed. It is essential that limnocorral/micro-ecosystem designs are such that biogeochemical cycling can closely correspond to that of the natural situation; the normal functional roles of the biota and of biological materials in general are essential for determining realistically the fate and behaviour of trace element species. At the same time in limnocorral work, more attention should be given to organic decomposition processes with regard to trace element studies.

As has already taken place in chemically oriented research on the fate of trace elements, there must be, by biologists, a move away from the use of artificial media, such as artificial seawater, to the use of natural water in combination with the use of suitable chemical analyses. This is of special relevance if the experimental design involves dilution or mixing-series, when the use of artificial media as diluters or the end members of mixing-series is totally inappropriate.

Finally, I would like to advocate experiments for which natural transient conditions form the basis. Estuarine field studies or mixing-series of estuarine origin, such as epi-/hypolimnion transitions, etc., appear to me to be more worthwhile than static steady-state systems. The natural change of physicochemical conditions must have consequences for trace element speciation and associated biological processes.

SUMMARY

Research on biological consequences of trace element speciation has focussed on bioavailability in relation to complexation by inorganic and natural organic compounds. Bioconversion and especially biologically mediated alkylation are receiving attention, and their role in biogeochemical cycling deserves continued research input. The environmental fate of those metal-protein compounds which appear to act as detoxifiers in aquatic animals is insufficiently known to assess the consequences for the detritus-based community.

Advanced chemical analytical methods should be employed rigorously in biological trace element research, rather than placing reliance upon nominal trace element concentrations. Natural conditions should be employed whenever possible, while natural transient environments may well be most worthwhile for the study of the biological consequences of trace element speciation in aquatic habitats.

ACKNOWLEDGEMENTS

I am greatly indebted to Miss M.J. de Dreu for her sustained technical assistance in the production of typescripts.

REFERENCES

Anderson, D.M., and Morel, F.M.M., 1978, Copper sensitivity of Gonyaulax tamarensis, Limnol. Oceanogr., 23:283.
Anderson, M.A., Morel, F.M.M., and Guillard, R.R.L., 1978, Growth limitation of a coastal diatom by low zinc ion activity, Nature, 276:70.
Andreae, M.O., 1978, Distribution and speciation of arsenic in natural waters and some marine algae, Deep-Sea Res., 25:391.

Andreae, M.O., 1979, Arsenic speciation in seawater and interstitial waters: The influence of biological-chemical interactions on the chemistry of a trace element, Limnol. Oceanogr., 24:440.

Andreae, M.O.,and Klumpp, D., 1979, Biosynthesis and release of organoarsenic compounds by marine algae, Environ. Sci. Technol., 13:738.

Baker, M.D., Wong, P.T.S., Chau, Y.K., Mayfield, C.I.,and Innis, W.E., 1981, Methylation of lead, mercury, arsenic and selenium in the acidic aquatic environment, in: "Heavy Metals in the Environment", CEP Consultants, Edinburgh.

Bartlett, P.D.,and Craig, P.J., 1981, Total mercury and methyl mercury levels in British estuarine sediments - II, Water Res., 15:37.

Beijer, K.,and Jernelov, A., 1978, Ecological aspects of mercury-selenium interactions in the marine environment, Environ. Health Perspect., 25:43.

Blair, W.R., Jackson, J.A., Olson, G.J., Brinckman, F.E.,and Iverson, W.P., 1981, Biotransformation of tin, in: "Heavy Metals in the Environment", CEP Consultants, Edinburgh.

Boddington, M.J., DeFreitas, A.S.W.,and Miller, D.R., 1979, The effect of benthic invertebrates on the clearance of mercury from sediments, Ecotoxicol. Environ. Safety, 3:236.

Bothner, M.H., Jahnke, R.A., Peterson, M.L.,and Carpenter, R., 1980, Rate of mercury loss from contaminated estuarine sediments, Geochim. Cosmochim. Acta, 44:273.

Boulègue, J.,and Renard, D., 1980, Catalyse bactérienne de l'oxydation du manganèse manganeux dans les eaux. Conséquences géochimiques, C.R. Acad. Sc. Paris, D, 290:1165.

Boyle, E.A., Sclater, F.,and Edmond, J.M., 1976, On the marine geochemistry of cadmium, Nature, 263:42.

Branica, M.,and Konrad, Z., eds., 1980, "Lead in the Marine Environment", Pergamon Press, Oxford.

Bruland, K.W., Knauer, G.A.,and Martin, J.H., 1978, Cadmium in northeast Pacific waters, Limnol. Oceanogr., 23:618.

Bryan, G.W., 1976, Heavy metal pollution in the sea, in: "Marine Pollution", R. Johnston, ed., Academic Press, London.

Chau, Y.K., Wong, P.T.S., Kramar, O.,and Bengert, G.A., 1981, Methylation of tin in the aquatic environment, in: "Heavy Metals in the Environment", CEP Consultants, Edinburgh.

Chau, Y.K., Wong, P.T.S., Silverberg, B.A., Luxon, P.L.,and Bengert, G.A., 1976, Methylation of selenium in the aquatic environment, Science, 192:1130.

Craig, P., 1981, Biomethylation: Pollution amplified, New Sci., 90:694.

Cranston, R.E.,and Murray, J.W., 1980, Chromium species in the Columbia River and estuary, Limnol. Oceanogr., 25:1104.

Davies, I.M., Graham, W.C.,and Pirie, J.M., 1979, A tentative determination of methylmercury in seawater, Mar. Chem., 7:111.

Duinker, J.C., Wollast, R.,and Billen, G., 1979, Behaviour of manganese in the Rhine and Scheldt estuaries. II. Geochemical cycling, Estuarine Coastal Mar. Sci., 9:727.

Dyrssen, D., and Wedborg, M., 1980, Major and minor elements, chemical speciation in estuarine waters, in: "Chemistry and Biogeochemistry of Estuaries", E. Olausson and I. Cato, eds., Wiley, Chichester.

Eaton, A., 1979, Removal of 'soluble' iron in the Potomac river estuary, Estuarine Coastal Mar. Sci., 9:41.

Edmonds, J.S., and Francesconi, K.A., 1981, Arseno-sugars from brown kelp (Ecklonia radiata) as intermediates in cycling of arsenic in a marine ecosystem, Nature, 289: 602.

Emerson, S., Cranston, R.E., and Liss, P.S., 1979, Redox species in a reducing fjord: Equilibrium and kinetic considerations, Deep-Sea Res., 26A:859.

Engel, D.W., and Fowler, B.A., 1979, Factors influencing cadmium accumulation and its toxicity to marine organisms, Environ. Health Perspect., 28:81.

Engel, D.W., and Sunda, W.G., 1979, Toxicity of cupric ion to eggs of the spot Leiostomus xanthurus and the Atlantic silverside Menidia menidia, Mar. Biol., 50:121.

Förstner, U., and Wittmann, G.T.W., 1979, "Metal Pollution in the Aquatic Environment", Springer-Verlag, Berlin.

Gächter, R., and Máreš, A., 1979, MELIMEX, an experimental heavy metal pollution study: Effects of increased heavy metal loads on phytoplankton communities, Schweiz. Z. Hydrol., 41:228.

Gächter, R., and Urech, J., in: this volume.

George, S.G., and Coombs, T.L., 1977, The effects of chelating agents on the uptake and accumulation of cadmium by Mytilus edulis, Mar. Biol., 39:261.

Giesy, J.P., Briese, L.A., and Leversee, G.J., 1978, Metal binding capacity of selected Maine surface waters, Environ. Geol., 2:257.

Grove, J.R., 1980, Investigations into the formation and behaviour of aqueous solutions of lead alkyls, in: "Lead in the Marine Environment", M. Branica and Z. Konrad, eds., Pergamon Press, Oxford.

Hart, B.T., 1981, Trace metal complexing capacity of natural waters: A review, Environ. Technol. Lett., 2:95.

Hoffmann, M.R., and Eisenreich, S.J., 1981, Development of a computer-generated equilibrium model for the variation of iron and manganese in the hypolimnion of Lake Mendota, Environ. Sci. Technol., 15:339.

Howard, A.G., and Nickless, G., 1978, Heavy metal complexation in polluted molluscs. 3. Periwinkles (Littorina littorea), cockles (Cardium edule) and scallops (Chlamys opercularis), Chem.-Biol. Interact., 23:227.

IRPTC, 1978, "Data Profiles for Chemicals for the Evaluation of their Hazards to the Environment of the Mediterranean Sea, Vol. I", (IRPTC Data Profile Series 1), UNEP, Geneva.

Jackson, G.A., and Morgan, J.J., 1978, Trace metal-chelator interactions and phytoplankton growth in seawater media: Theoretical analysis and comparison with reported observations, *Limnol. Oceanogr.*, 23:268.

Jennings, J.R., Rainbow, P.S., and Scott, A.G., 1979, Studies on the uptake of cadmium by the crab *Carcinus maenas* in the laboratory. II. Preliminary investigation of cadmium-binding proteins, *Mar. Biol.*, 50:141.

Klumpp, D.W., and Peterson, P.J., 1981, Chemical characteristics of arsenic in a marine food chain, *Mar. Biol.*, 62:297.

Kudo, A., Miller, D.R., Akagi, H., Mortimer, D.C., DeFreitas, A.S., Nagase, H., Townsend, D.R., and Warrock, R.G., 1978, The role of sediments on mercury transport (total- and methyl-) in a river system, *Prog. Water Technol.*, 10:329.

Luoma, S.N., 1977, The dynamics of biologically available mercury in a small estuary, *Estuarine Coastal Mar. Sci.*, 5:643.

McKnight, D.M., and Morel, F.M.M., 1980, Copper complexation by siderophores from filamentous blue-green algae, *Limnol. Oceanogr.*, 25:62.

Maddock, B.G., and Taylor, D., 1980, The acute toxicity and bioaccumulation of some lead alkyl compounds in marine animals, in: "Lead in the Marine Environment", M. Branica and Z. Konrad, eds., Pergamon Press, Oxford.

Mantoura, R.F.C., 1981, Organo-metallic interactions in natural waters, in: "Marine Organic Chemistry", E.K. Duursma and R. Dawson, eds., Elsevier, Amsterdam.

Mantoura, R.F.C., Dickson, A., and Riley, J.P., 1978, The complexation of metals with humic materials in natural waters, *Estuarine Coastal Mar. Sci.*, 6:387.

Measures, C.I., and Burton, J.D., 1978, Behaviour and speciation of dissolved selenium in estuarine waters, *Nature*, 273:293.

Measures, C.I., and Burton, J.D., 1980, The vertical distribution and oxidation states of dissolved selenium in the northeast Atlantic Ocean and their relationship to biological processes, *Earth Planet. Sci. Lett.*, 46:385.

Measures, C.I., McDuff, R.E., and Edmond, J.M., 1980, Selenium redox chemistry at GEOSECS 1 reoccupation, *Earth Planet. Sci. Lett.*, 49:102.

Mercury, 1976, "Environmental Health Criteria 1", WHO, Genève.

Moore, R.H., Burton, J.D., Williams, P.J. Le B., and Young, M.L., 1979, The behaviour of dissolved organic material, iron and manganese during estuarine mixing, *Geochim. Cosmochim. Acta*, 43:919.

Morel, F.M.M., Rueter, J.G., Anderson, D.M., and Guillard, R.R.L., 1979, Aquil: A chemically defined phytoplankton culture medium for trace metal studies, *J. Phycol.*, 15:135.

Morris, A.W., and Bale, A.J., 1979, Effect of rapid precipitation of dissolved Mn in river water on estuarine Mn distributions, *Nature*, 279:318.

Morris, A.W., Mantoura, R.F.C., Bale, A.J., and Howland, R.J.M., 1978, Very low salinity regions of estuaries: Important sites for chemical and biological reactions, *Nature*, 274:678.

Nakayama, E., Tokoro, H., Kuwamoto, T., and Fujinaga, T., 1981, Dissolved state of chromium in seawater, Nature, 290:768.

Noel-Lambot, F., 1981, Presence in the intestinal lumen of marine fish of corpuscles with a high cadmium-, zinc-and copper-binding activity: A possible mechanism of heavy metal tolerance, Mar. Ecol. Prog. Ser., 4:175.

Noel-Lambot, F., Bouquegneau, J.M., Frankenne, F., and Disteche, A., 1978, Le role des métallothioneines dans le stockage des métaux lourds chez les animaux marins, Rev. Int. Océanogr. Méd., 49:13.

Phillips, D.J.H., 1980, "Quantitative Aquatic Biological Indicators", Applied Science Publishers, London.

Prosi, F., 1979, Heavy metals in organisms. in: "Metal Pollution in the Aquatic Environment", U. Förstner and G.T.W. Wittmann, Springer-Verlag, Berlin.

Reisinger, K., Stoeppler, M., and Nürnberg, H.W., 1981a, Evidence for the absence of biological methylation of lead in the environment, Nature, 291:228.

Reisinger, K., Stoeppler, M., and Nürnberg, H.W., 1981b, On the biological methylation of lead, mercury, methylmercury and arsenic in the environment, in: "Heavy Metals in the Environment", CEP Consultants, Edinburgh.

Rice, M.A., and Chien, P.K., 1979, Uptake, binding and clearance of divalent cadmium in Glycera dibranchiata (Annelida: Polychaeta), Mar. Biol., 53:33.

Röderer, G., 1981, Fate and toxicity of tetraalkyl lead and its derivatives in aquatic environments, in: "Heavy Metals in the Environment", CEP Consultants, Edinburgh.

Roesijadi, G., 1981, The significance of low molecular weight, metallothionein-like proteins in marine invertebrates: Current status, Mar. Environ. Res., 4:167.

Rueter, J.G., and Morel, F.M.M., 1981, The interactions between zinc deficiency and copper toxicity as it affects the silicic acid uptake mechanisms in Thalassiosira pseudonana, Limnol. Oceanogr., 26:67.

Sanders, J.G., 1978, Enrichment of estuarine phytoplankton by the addition of dissolved manganese, Mar. Environ. Res., 1:59.

Sanders, J.G., 1979, Effects of arsenic speciation and phosphate concentration on arsenic inhibition of Skeletonema costatum (Bacillariophyceae), J. Phycol., 15:422.

Sanders, J.G., 1980, Arsenic cycling in marine ecosystems, Mar. Environ. Res., 3:257.

Sanders, J.G., and Windom, H.L., 1980, The uptake and reduction of arsenic species by marine algae, Estuarine Coastal Mar. Sci., 10:555.

Saxena, J., and Howard, P.H., 1977, Environmental transformation of alkylated and inorganic forms of certain metals, Adv. Appl. Microbiol., 21:185.

Seeliger, U., and Edwards, P., 1979, Fate of biologically accumulated copper in growing and decomposing thalli of two benthic red marine algae, J. Mar. Biol. Assoc. U.K., 59:227.

Sipos, L., Raspor, B., Nürnberg, H.W., and Pytkowicz, R.M., 1980, Interaction of metal complexes with coulombic ion-pairs in aqueous media of high salinity, Mar. Chem., 9:37.

Sipos, L., Valenta, P., Nürnberg, H.W., and Branica, M., 1980, Voltammetric determination of the stability constants of the predominant labile lead complexes in seawater, in: "Lead in the Marine Environment", M. Branica and Z. Konrad, eds., Pergamon Press, Oxford.

Summers, A.O., and Silver, S., 1978, Microbial transformations of metals, Ann. Rev. Microbiol., 32:637.

Thomas, W.H., and Seibert, D.L.R., 1977, Effects of copper on the dominance and diversity of algae: Controlled ecosystem pollution experiment, Bull. Mar. Sci., 27:23.

Topping, G., and Davies, I.M., 1981, Methylmercury production in the marine water column, Nature, 290:243.

Tuschall, J.R., and Brezonik, P.L., 1980, Characterization of organic nitrogen in natural waters: Its molecular size, protein content, and interactions with heavy metals, Limnol. Oceanogr., 25:495.

Vaccaro, R.F., in: this volume.

van der Putte, I., 1981, "An Assessment of the Environmental Toxicity of Hexavalent Chromium in Fish", (Dissertation Agric. Univ.), Pudoc, Wageningen.

Waslenchuk, D.G., 1978, The budget and geochemistry of arsenic in a continental shelf environment, Mar. Chem., 7:39.

Wollast, R., Billen, G., and Duinker, J.C., 1979, Behaviour of manganese in the Rhine and Scheldt estuaries. I. Physico-chemical aspects, Estuarine Coastal Mar. Sci., 9:161.

Wood, J.M., 1974, Biological cycles for toxic elements in the environment, Science, 183:1049.

Wrench, J.J., and Addison, R.F., 1981, Reduction, methylation, and incorporation of arsenic into lipids by the marine phytoplankton Dunaliella tertiolecta, Can. J. Fish. Aquat. Sci., 38:518.

Wrench, J., Fowler, S.W., and Ünlü, M.Y., 1979, Arsenic metabolism in a marine food chain, Mar. Pollut. Bull., 10:18.

Wright, D.A., 1977, The uptake of cadmium into the haemolymph of the shore crab, Carcinus maenas (L.). The relationship with copper and other divalent ions, J. Exp. Biol., 67:147.

Yeats, P.A., Sundby, B., and Brewer, J.M., 1979, Manganese recycling in coastal waters, Mar. Chem., 8:43.

Young, J.S., Gurtisen, J.M., Apts, C.W., and Crecelius, E.A., 1979, The relationship between the copper complexing capacity of sea water and the copper toxicity in shrimp zoeae, Mar. Environ. Res., 2:265.

DISCUSSION: M. SMIES

M. BRANICA — After your excellent overview of different trace metals in aquatic environments, can you summarize which of the trace elements is relatively more dangerous for aquatic life (consider concentrations, species present, added contributions from pollution)?

M. SMIES — It is, of course, impossible to make valid general statements in view of the great variety of environmental conditions and discharge conditions. It would appear that those trace elements of special interest are those which are either not extensively bound to stable complexes, or which undergo extensive environmental alkylation to more toxic organo-metal compounds.

O. RAVERA — To calculate the biological risk, one must understand, in addition to the availability and " apparent toxicity " of a given chemical form, also the effects on reproduction in relation to the intrinsic rate of natural increase. In this context: what do you mean by " apparent toxicity "?; do you agree with the importance of relating to the intrinsic rate of natural increase?

M. SMIES — The term " apparent toxicity " is used in this paper in response to a chronic biological problem of interpretation. The problem is that, in many test cases, the actual chemical species to which the test organism is exposed is not known. Consequently, it is not altogether correct to speak of toxicity proper. To your second question, I would agree that effects of environmental chemicals on reproductive potential are of great importance for environmental risk assessment. There is a tendency to try to obtain information on the effects of environmental chemicals over the entire life cycle of test organisms. Because of practical difficulties in exposing organisms during an entire life time to defined concentrations of chemicals, this is usually approached by short-term experiments at different life stages, including the reproduction stage.

R.F. VACCARO — Does methylation of metals occur in the water column as well as in sediments?

M. SMIES   Environmental alkylation has been reported to occur in the water column, but the contribution of the sediment is probably much greater.

M. BERNHARD   There should occur methylation in the water column, and not only in the sediments. How otherwise can phytoplankton and other pelagic organisms contain methyl mercury?

R.F. VACCARO   What is the fate of methylated metals in natural waters; how are they recycled?

M. SMIES   Because there is no accumulation of alkylated metals in the abiotic environment, it must be concluded that they are broken down again to produce inorganic species.

H.W. NURNBERG   Biomethylation of lead in organisms can be excluded according to the cited proof of Reisinger et al., 1981. Biomethylation of mercury happens only to a secondary extent. There occurs methylation of lead from organic species such as lead acetate, etc., and of mercury, but, this occurs by a chemical mechanism mediated by sulfide or free SH- groups of some amino acids.

Y.K. CHAU   There is no sharp and clear definition to differentiate between biomethylation and biologically-mediated or biologically-induced methylation. They are all chemical reactions basically.

BIOLOGICAL CONTROL OF TRACE METAL EQUILIBRIA IN SURFACE WATERS

John P. Giesy

Pesticide Research Center, and
Department of Fisheries and Wildlife
Michigan State University
East Lansing, Michigan 48824 USA

It has long been known that aquatic organisms can release organic compounds to their environment. While there are organics present from the decomposition of animals and, through excretion, as waste products from animals (Ferrante, 1976), most of the organics released to the aquatic environment are from plants (Fogg, 1951; Fogg, 1963; Gessner, 1965; Hellebust, 1965; Fogg, 1966; Forsberg and Taube, 1967; Khailov and Burlakova, 1969; Sieburth, 1969; Anderson and Zeutschel, 1970; Khailov and Finenko, 1970; Zajic, 1970; Berman, 1976). The compounds released by phytoplankton which form complexes with metals include amino acids, polypeptides, proteins, porphyrins, pterins, and purines (Khailov, 1964). The littoral marine alga, Fucus vesiculosus, releases as much as 40% of the carbon which it fixes. Khailov and Burlakova (1969) found as much as 39% of brown algal (and 38% of red algal) production was released as soluble extracellular products. Phytoplankton can release as much as 50% of their carbon which has been photosynthetically fixed (Fogg, 1951; Berman, 1976). Fogg and Westlake (1955) speculated that the polypeptides extensively released by blue-green algae may form complexes with metals and have important effects on ecology. Most of the organic complexing capacity of sea water is due to autochthonous production of organic compounds (Davey et al., 1973) by phytoplankton (Duursma, 1963; Anderson and Zeutschel, 1970; Thomas, 1971; Daumas, 1976). In a study of copper release from the thalli of benthic red algae, Seeliger and Edwards (1979) found that 22% of the copper released from living thalli was bound to dissolved organic matter, and 80-90% of the copper released from decomposing thalli was associated with dissolved organic compounds.

Hellebust (1965) determined the composition and quality of excreted photoassimilated carbon for 22 species of marine phytoplankton. In this study, compounds were characterized by solubility in extractants with particular attention paid to the quantity of glycolic acid excreted. This study also determined the proportion of charged and uncharged compounds in the excreted organics. The compounds in the neutral fraction were further characterized by chromatography. The isolated compounds included mannitol, aspartic acid, arabinose, glutamic acid, lysine, glycerol, glucose, proline, and a number of unidentified compounds which yielded a number of amino acids upon hydrolysis. The extracellular products of Cricosphaera elongata have been characterized by ultrafiltration (Gnassia-Barelli et al., 1978).

Early studies of the possible interactions of metals with extracellular products were circumstantial because researchers were unable to isolate metal-organic complexes from water or to show a change in the form of trace metals attributable to association with extracellular products. As early as 1939, Harvey had found that diatom growth could be stimulated by soil extracts. Johnson (1955) isolated organic constituents of sea water but was unable to characterize qualitatively the compounds. He was, however, able to show that the isolates were bioactive. The isolated compounds caused both increases and decreases in the growth rate of phytoplankton. In 1964, Johnson conducted studies with the synthetic chelating agent EDTA and concluded that, by analogy, natural marine phytoplankton populations required a concentration of natural chelating agents equivalent to an EDTA concentration of 10^{-8}M. From these studies, Johnson (1964) concluded that powerful natural chelators were present in sea water in small concentrations.

Extracellular products from aquatic organisms have been much studied and have been implicated in many important ecological processes, especially those with nutritive and antibiotic properties (Lucas, 1946). It is often difficult to separate these effects from those of changes in trace metal equilibria in aquatic systems. Analysis is complicated further by the circumstantial nature of some of the evidence for biological control of trace metal equilibria.

Biological control of the speciation of trace metals has been implicated in decreasing the inhibitory effect of some poisonous metals such as copper (Steemann-Nielsen and Wium-Andersen, 1970; Gächter et al., 1973; Gächter et al., 1978; Gächter and Máreš, 1979; Davies and Sleep, 1980) and in enhancing the availability of nutrients such as iron, a nutrient which is very insoluble under most conditions in the photic zone of aquatic environments and, thus, by being in short supply is potentially limiting to primary productivity (Giesy, 1976). The inhibitory effects of copper on phytoplankton, such as lengthened lag period and reduced growth rate (Guy and Kean, 1980), have been attributed to the free (Cu^{++}) form of copper (Sunda and Guillard, 1976; Sunda et al., 1978; Sunda and Gillespie,

1979; Jackson and Morgan, 1980).

Reduced productivity in waters of marine upwellings has been observed even though the nitrogen and phosphorus concentrations in these waters were greater than the surrounding surface waters (Barber and Ryther, 1969). These latter authors observed that the lag phase was elongated. This fact, coupled with the observation that copper concentrations in these waters were high (and that after phytoplankton had grown in the water, the effect was reduced), led to the conclusion that the phytoplankters were releasing a substance which was reducing the inhibitory effects of copper. To test the hypothesis that the observed effects were due to some organic compounds in the water, investigators added EDTA, glycine, and an uncharacterized zooplankton extract to culture media (Barber and Ryther, 1969; Barber, 1973; Huntsman and Barber, 1975). All three of these organics enhanced the growth rate of phytoplankton. It was concluded that the marine upwelling water was rich in free Cu^{++} inhibitory to phytoplankton, and that the water needed to be conditioned by the extracellular products of algae.

Studies of the extracellular products of Chlorococcum ellipsoideum showed that these substances stimulated growth of Chlamydomonas globosa (Kroes, 1972). The extracellular products were presumed to be polysaccharides and proteins which elicited responses similar to those of EDTA. The author suggested that the observed response may be due to keeping iron in solution or chelating other metals. When Huntsman and Barber (1975) separated, by a filter membrane, the extracellular products of a dense marine phytoplankton population from a less dense population of mixed phytoplankton in upwelling water, the lag phase of the less dense population was decreased and carbon fixation was increased relative to populations which were not grown in proximity to the phytoplankton filtrate. The effects observed for the phytoplankton filtrate were similar to those observed for additions of EDTA.

In a laboratory culture study, Hardstedt-Romeo and Gnassia-Barelli (1980) found that natural phytoplankton exudates of Cricosphaera elongata decreased copper and cadmium accumulation by this marine phytoplankter. The polymeric phenols released as extracellular products from the marine brown algae, Ascophyllum nodosum (L.) and Fucus vesiculosus (L.), decreased the toxicity of Zn to both Skeletonema costatum (Greve) and Phaeodactylum tricornutum (Ragan et al., 1980). In these studies, in the presence of 0.5 mg Zn/l, all concentrations of polymeric phenols between 100 and 2000 µg/l increased the rate of cell division. This effect was not observed in the absence of Zn additions. At concentrations of polyphenols greater than 2000 µg/l, the rates of cell division were decreased relative to controls. This effect may have been due to the available zinc concentration being decreased below the concentration which is limiting for growth. Low-molecular-weight extracellular products

(<500 MW) of Cricosphaera elongata did not reduce copper toxicity; however, extracellular products with molecular weights > 500 MW decreased copper toxicity to this species (Gnassia-Barelli et al., 1978). The greatest decrease in toxicity was observed in the presence of extracellular products ranging from 500 to 1000. Similar reductions in copper toxicity were observed when C. elongata was grown in sea water containing extracellular products from Prorocentram micans, Dunaliella primolecta, and Chaetocaeros lauderi.

Khailov (1964) found metal-binding extracellular products were excreted from the marine phytoplankters, Dunaliella salina and Pontosphaera huxley, and three marine macrophytic algae (Fucus serratus, Ascofillum nodosum, and Rhodymenia palmota). In a study of eight species of marine phytoplankton, Swallow et al. (1978) found that only one species, Gloeocystis gigas, produced enough extracellular product to reduce the cupric ion activity when the total copper concentration was 10^{-6}M. These researchers concluded that this copper binding was simply the result of the mucilaginous extracellular products of this species. The other eight species studied all created copious quantities of mucilaginous material and none of them affected the speciation of copper. A large proportion of the soluble organic nitrogen in aqueous systems is proteinaceous and this material can be important in the complexation of heavy metals (Tuschall and Brezonik, 1980). These authors reported stability constants for copper-organic complexes ranging from 1.6×10^6 to 7.9×10^6 at a pH of 7.5 and from 3.2×10^6 to 1.3×10^7 at a pH of 4.5.

These field and culture studies were only circumstantial evidence that complex formation with copper was responsible for the observed reduction in inhibition. It still remained to isolate a copper-organic complex in natural waters or demonstrate a reduction in copper activity due to actual extracellular products. Many studies on the effects of synthetic chelating agents on phytoplankton have been conducted. Avakyan and Rabotnova (1971) found that natural products, such as oxalic, pyruvic, malic, tartaric, and citric acids, which are known to chelate copper, also reduce the toxicity of copper to the yeast Candida utilis. Slowey et al. (1967) found indirect evidence of copper-organic complexes in sea water by extracting the phospholipids, amino lipids, and porphyryn lipids into chloroform and measuring the copper content of this fraction of the Hirsch-Ahrens separation method.

Further indirect evidence of copper-organic complexes has been found by measuring free copper before and after oxidation. After oxidation of the organic compounds, the total measurable copper concentration increased (Corcoran and Williams, 1964; Williams, 1969; Foster and Morris, 1971). Rona et al. (1962) found that zinc and manganese in sea water were in a filterable, yet undializable, form and suggested that this was probably due to formation of an organic complex. This may not necessarily be true since colloidal inorganic forms of zinc and manganese could do the same thing. Burton (1966) observed that vanadium in sea water was not precipitated with Fe(OH)$_3$ and suggested that the vanadium may be complexed by organic matter. Albert (1950) inferred, from potentiometric studies, that copper and zinc are bound by amino acids. Fukai and Huynh (1975) found a high-molecular-weight fraction of dissolved organic matter in sea water, one which contained zinc and could not be measured potentiometrically or colorimetrically until the organic matter was oxidized with persulfate. Khailov (1964) and Koshy et al. (1969) studied the changes of color of ferric iron in solution and inferred that complexes were being formed. Similar results were found by Laevestu and Thompson (1958) when they oxidized sea water and observed an increase in measurable iron which they attributed to the iron being bound to organic complexes.

Studies of boron in sea water have also indicated that boron is released from organic complexes after oxidation (Gast and Thompson, 1958; Noakes and Hood, 1961). These authors suggest that the boron is bound to polyhydroxy organic compounds. However, Williams and Strack (1966) concluded that there was insufficient organic matter available in sea water to complex boron if stability constants for a boron-mannitol complex are assumed. Sillén (1961) and Goldberg (1965) have noted that, on a theoretical basis, concentrations of organic compounds at 10^{-4} to 10^{-5} M, in sea water, can form strong complexes with trace metals at low concentrations (10^{-6} to 10^{-7} M).

Besides the reductions in toxicity of trace metals which have been attributed to extracellular products, a number of authors have attributed nutrient availability to algal extracellular products. Early workers were not sure why algae released extracellular products; however, it was thought that the release of polypeptides may be due to cell wall formation processes. Fogg (1951) noted that the concentration of extracellular polypeptides increased when cells were iron deficient. Fogg (1951) suggested that iron was required for respiration and that energy was required to keep polypeptides inside living cells. Thus, iron limitation would result in increased release of polypeptides. Lange (1974) and Hunter (1972) suggested that the extracellular products of blue-green algae were as effective as EDTA in keeping iron and trace metal nutrients from precipitating at pH's above 8, and suggested that this could be ecologically important in maintaining the availability of trace metal nutrients. Lange (1974) did not characterize the extracellular products but speculated that they were probably polysaccharides. Spencer et al. (1972) studied the properties of extracellular products and their interaction with iron by infrared spectra, and suggested that the extracellular products were chelators specific for ferric-iron. Estep et al. (1975) investigated the growth-promoting activity of extracts from marine algal mats and sea grass on the siderochrome auxotroph <u>Arthrobacter</u> JG-9. Growth promotion of this particular species indicates the presence of secondary hydroxamic acids which are specific iron chelators. These investigations found that the stimulatory compound was resistant to autoclaving and was not desferal; it was suggested that the compound was a secondary hydroxamate.

Neilands (1967) described hydroxamic acids and their role in iron metabolism by microorganisms. The hydroxamic acids are oxidized peptides with the general structure R-CON(OH). These compounds are formed by yeasts, bacteria, fungi and higher plants, and they act as growth factors, antibiotic antagonists, tumor inhibitors, or cell division factors. The complexes with ferric iron are of the form:

$$\left. \begin{array}{c} -C=O \\ \\ -N-O \end{array} \right\rangle_n Fe^{+3}$$

where n can range from 1 to 3. The 1:1 structure forms at low pH and is transformed to the 3:1 complex as the pH approaches neutrality. These compounds are considered to be iron-transfer agents, not electron transfer agents such as the hemes.

Blue-green algae have been found to excrete hydroxamate chelators which can enhance the growth of Anabaena species or inhibit growth of other species or both (Murphy et al., 1976). Murphy tested seven species of blue-green algae and 10 species of green algae for hydroxamate chelating activity and found only three species of blue-green algae (Microcystis aeruginosa, Phormidium autumnale, and Anabaena flos-aquae) which showed such activity. Murphy was also able to isolate three species of Pseudomonas and one species of Aerobacter which excreted hydroxamates. Some species of Scenedesmus are able to excrete polypeptides which solubilize iron, but these compounds do not form very strong complexes relative to the siderochromes of blue-green algae.

Murphy et al. (1976) suggest that the blue-green algae can assimilate iron-hydroxamate siderophore complexes, while eucaryotic algae cannot. This is ecologically important because it gives the blue-green algae a competitive advantage under low iron conditions. McKnight and Morel (1980) found that Anabaena cylindrica released strong copper complexing agents in response to iron-limited conditions. Under iron-rich conditions, only weak copper-organic complexes were formed. Competition studies between copper and iron indicated that the extracellular products were siderophores, especially trihydroxamates. They also found that only siderophores free of iron will bind copper, a fact which is in keeping with the great difference in formation constants for copper and iron complexes. The complex formation of copper with A. flos-aquae exudate was indistinguishable from that of the copper-desferal (CIBA-Geigy) complex, which is a hydroxamate sideramine.

The coordination numbers for copper (+4) and iron (+6) will result in different geometries for the copper- and iron-siderophore chelates. The iron will form an octahedral geometry while the copper will form a square planar complex (Emery, 1971; as cited by McKnight and Morel, 1980). McKnight and Morel (1980) state that it is unlikely that the copper-siderophore complex would be assimilated by blue-green algal cells. Studies of copper toxicity to A. flos-aquae indicated that hydroxamate was not released from cells when they were copper stressed. It was found that the hydroxamates will chelate copper but form a less stable complex than with iron, and it was concluded that, in the system studied by the latter authors, the release of hydroxamate siderophores was in response to low iron concentrations, not copper toxicity.

Few of the culture or field studies have actually identified the extracellular products responsible for chelating metals and even fewer have quantified the amount released or the stoichiometric relations and affinity strengths. In the following section, I present information on the characterization of extracellular products which have been isolated.

Khailov (1964) found that the extracellular products of Fucus serratus, which formed complexes with copper, could be separated into four fractions on a column of Dowex 50-2B. Khailov (1964) noted that hydrogen ions were released upon titration with copper and suggested the formation of copper-organic complexes by hydrogen bonding. Dialysis studies indicated that most of the copper binding was due to compounds other than proteins or polypeptides. Sunda and Gillespie (1979), using a ^{14}C-glucose assimilation technique, determined the reactive chelating concentration in sea water to be approximately 0.05 µM EDTA equivalents and the formation constant of the copper-organic complex to be $\leq 10^{10}$. Based on these values and the concentrations and stability constants for the inorganic constituents of sea water, Sunda and Gillespie (1979) calculated the cupric ion activity to be $\leq 10^{-11}M$ when the total copper concentration was $1.4 \times 10^{-2}M$ (0.9 µg Cu/l). These studies were conducted in the Newport River Estuary (USA) and the organics were probably best characterized as humic-type substances rather than as extracellular products.

Johnson (1964) concluded from experiments where he compared the effect of EDTA on the growth of Skeletonema costatus that, by analogy, natural populations in sea water would require $10^{-8}M$ EDTA equivalents to chelate existing trace metals. Thus, he concluded that the natural chelators exist in small concentrations and form strong complexes with trace metals. The speculation of Johnson (1964) is in good agreement with the results of Sunda and Gillespie (1979).

The copper complexing capacities and formation constants for the extracellular products of Anabaena cylindrica, Navicula pelliculosa, and Scenedesmus quadricauda have been determined in culture (Van den

Berg et al., 1979) (Table 1). Using these stability constants, the free Cu^{++} concentration was predicted and was found not to be significantly ($\alpha = 0.05$) different from the measured cupric ion concentration. Stolzberg and Rosin (1977) observed a copper binding capacity of 4×10^{-7}M in continuous cultures of the marine diatom Skeletonema costatum which were not stressed by copper. The conditional formation constant for the copper-organic complexes of Anabaena flos-aquae and Anaceptis nidulans have been determined to be 10^8 and 10^{10}, respectively (McKnight and Morel, 1979).

Anderegg et al. (1963) reported stability constants for hydroxamate complexes with lead, nickel, cobalt, and aluminum to be greater than 10^8. The formation constants for the iron and copper siderophore complexes, determined by McKnight and Morel (1980) were 2×10^{10} and 3.7×10^7, respectively. Hydroxamate concentrations of 5×10^{-6} to 5×10^{-5}M were found in cultures with chlorophyll concentrations similar to those observed in field conditions. McKnight and Morel (1980) speculated that the iron concentration which limits growth of nitrogen-fixing, blue-green algae would be approximately 10^{-7}M, and that, at iron concentrations in this range, these algae would excrete hydroxamates at concentrations between 10^{-7} and 10^{-5}M. These authors concluded that the copper-siderophore complex would be the major copper species in blooms of nitrogen-fixing, blue-green algae where iron is the micronutrient which is limiting primary production.

Gnassia-Barelli et al. (1978) measured the complexing capacity for copper of several ultrafilter fractions of extracellular products of Cricosphaera elongata. Unfortunately, they report their results

Table 1. Concentrations of extracellular products and conditional stability constants for copper-organic complexes of 10-day-old algal cultures and two natural waters (Van den Berg et al., 1979).

	Ligands produced (μM/L)	Ligands per mg dry weight (μM/mg)	Conditional stability constants (pH) 7.6	8.0
A. cylindrica	6.73	0.110	7.7 ± 0.1	8.1 ± 0.2
N. pelliculosa	2.86	0.036	8.1 ± 0.2	8.5 ± 0.4
S. quadricauda	0.66	0.010	8.6 ± 0.2	9.0 ± 0.3
Bay of Quinte	—	—	7.6 ± 0.2	8.0 ± 0.2
Lake Ontario	—	—	8.7 ± 0.5	9.1 ± 0.5

as mg Cu/l or percent of total carbon bound in each fraction. Therefore, we do not know the binding capacity on a weight basis for the extracellular products. The authors do, however, report that 60% of the copper is bound by that fraction of the extracellular products having a molecular weight between 500 and 10,000. By comparing the copper threshold toxicity of sea water in the presence and absence of extracellular products, Gnassia-Barelli et al. (1978) concluded that approximately 50% of the copper was complexed to organic extracellular products when they were present. Unfortunately, this does not allow much generality in calculating the speciation of copper under different situations.

Gillespie and Vaccaro (1978) measured the relative copper binding capacities of organic carbon in sea water by measuring the assimilation of ^{14}C-glucose by marine bacteria when copper was added. The binding capacities of water from Saanich Inlet and Vinyard Sound (British Columbia and the Sargasso Sea) were found to be 0.0030, 0.0035, and 0.0060 µg Cu/µg DOC, respectively. This technique assumes that only free Cu is able to inhibit assimilation of the glucose substrate.

In conclusion, it can be said that those specific extracellular products of aquatic organisms which can form complexes with trace metals have not been well characterized. When they have been identified they have not generally been quantified in a manner such that the molar concentrations, required for thermodynamic speciation models, are available. The extracellular products released by phytoplankton have been the most studied in relation to possible metal chelating properties. The metal chelating properties of the extracellular products of macrophytic algae also have been investigated, but to a lesser extent. The quantity and quality of extracellular products are dependent upon species (Briand et al., 1978), growth phase (McKnight and Morel, 1979), light conditions and nutrient concentrations.

If all of the possible organic extracellular products are quantified, the problem exists that stoichiometric and thermodynamic constants are known for some of the compounds but not all of them. In fact, the formation constants, when known, may be conditional upon pH and ligand-metal ratios, so that competitive distribution equations cannot be written to account for pH and competing cations. In some cases there will be a number of possible organic ligands such as amino acids, polysaccharides, polypeptides, or simple organic acids, but no single compound which is very important in overall speciation. Also, there may be compounds such as humic-like compounds, which are a poorly-characterized continuous spectrum of polymeric compounds, for which no single simple formation constant is available.

While, because of theoretical considerations, it would be nice to describe all of the metal-ligand interactions individually and solve all of the competitive interactions simultaneously, for practical reasons this may be impossible. This may be so because: there are many compounds with very similar formation constants; there is an error associated with the conditional stability constants; there is an error associated with measuring the total concentration of each individual ligand; or there is a lack of information about the stoichiometric and thermodynamic interactions of primary and competing trace metals. Even if all of the required information were available, it might be desirable to make simplifications to decrease the errors introduced by solving so many simultaneous equations.

Sposito (1981) has suggested that we use a quasiparticle model to represent a simplification of the complex situation of metal-organic interactions in surface waters. A quasiparticle model is a mathematical description of an aqueous solution, in which the actual assembly of organic compounds is replaced by an assembly of hypothetical identical macromolecules, whose mole mass is the number-average mole mass of the actual mixture and whose metal-complexation reactions closely mimic those of the real system (Sposito, 1981). An example of this conceptualization is the determining of overall conditional stoichiometric and formation constants by Scatchard analyses. Certain classes of metal-organic complexes may have similar stability constants and all of the sites which can be measured on polymeric molecules may not be available for binding metals. For these reasons, overall conditional stability constants and maximum binding capacities can be measured for organic ligands by the Scatchard analysis. Numerical solution of the partial derivatives which describe the hyperbolic Scatchard function results in overall conditional stability constants which can be used in quasiparticulate models (Giesy, 1980) to predict the forms of trace metals in solution. Giesy and Alberts (1981) have presented a discussion of humic compounds as quasiparticles which includes a discussion of functional group analysis and the problems of determining the molar concentrations of heterogeneous mixtures of organic ligands. They show that all of the measurable functional groups, carboxylic and phenolic, are not available for binding to metals such as copper, cadmium and lead.

It is often difficult to isolate the organic ligands from the inorganic ligands in natural waters. Sorption onto macroreticular resins, such as XAD-4, results in changes in the nominal molecular size distribution due to selective adsorption of the larger organics to the resin. Ultrafiltration may leave as much as 50% of the organic carbon in solution; this is due to the fact that often this much of the organic carbon in surface waters is less than the nominal 500 molecular weight cutoff of the smallest pore-size ultrafilters.

Because of the problems associated with isolation and quantification of organic ligands from surface waters, it is useful to measure the overall conditional stability constants in water before and after photo-oxidation. In this way, the relative effects of organic and inorganic ligands can be separated. This technique suffers from the problem of release of cations and particulates upon oxidation of the organics with which they were associated.

In summary, the quasiparticulate model is not a perfect, or even theoretically-appealing, alternative to the complete description of the metal-ligand interactions, but may be the only pragmatic method of describing complex natural systems in a manner which allows the prediction of free metal concentrations. From this synoptic review it can be seen that much more information will be required if we are going to be able to quantify the effects of aquatic organisms on trace metal equilibria.

REFERENCES

Albert, A., 1950, Quantitative studies of the avidity of naturally-occurring substances for metals. I. Amino-acids having only two ionizing groups, Biochem. J., 47:531.
Anderegg, G., L'Eplattenier, F., and Schwarzenbach, G., 1963, Hydroxamate complexes. 3. Iron III exchange between sideramines and complexones. A discussion of the formation constants of the hydroxamate complexes, Helv. Chim. Acta, 46:1409.
Anderson, G.C., and Zeutschel, R.P., 1970, Release of dissolved organic matter by marine phytoplankton in coastal and offshore areas of the northeast Pacific Ocean, Limnol. Oceanogr., 15:402.
Avakyan, Z.A., and Rabotnova, I.L., 1971, Comparative toxicity of free ions and complexes of copper with organic acids for Candida utilis, Mikrobiologiya, 49:305.
Barber, R.T., 1973, Organic ligands and phytoplankton growth in nutrient-rich seawater, in: "Trace Metals and Metal-Organic Interactions in Natural Waters", P.C. Singer, ed., Ann Arbor Science Publ., Ann Arbor, Michigan.
Barber, R.T., and Ryther, J.H., 1969, Organic chelators: factors affecting primary production in the Cromwell Current upwelling, J. Exp. Mar. Biol. & Ecol., 3:191.
Berman, T., 1976, Release of dissolved organic matter by photosynthesizing algae in Lake Kinneret, Israel, Freshwater Biol., 6:13.
Briand, F., Trucco, R., and Ramamoorthy, S., 1978, Correlations between specific algae and heavy metal binding in lakes, J. Fish. Res. Board Can., 35:1482.
Burton, J.D., 1966, Some problems concerning the marine geochemistry of vanadium, Nature, Lond., 212:976.
Corcoran, E.G., and Alexander, J.E., 1964, The distribution of certain trace elements in tropical sea water and their biological significance, Bull. Mar. Sci. Gulf Caribb., 14:594.
Davey, E.W., Morgan, M.J., and Erickson, S.J., 1973, A biological measurement of the copper complexation capacity of seawater, Limnol. Oceanogr., 18:993.

Davies, A.G., and Sleep, J.A., 1980, Copper inhibition of carbon fixation in coastal phytoplankton assemblages, J. Mar. Biol. Ass. U.K., 60:841.

Daumas, R.A., 1976, Variations of particulate proteins and dissolved amino acids in coastal sea water, Mar. Chem., 4:225.

Duursma, E.K., 1963, The production of dissolved organic matter in the sea as related to the primary gross production of organic matter, Neth. J. Sea Res., 2:85.

Emery, T., 1971, Role of ferrichrome as a ferric ionophore in Ustilago sphaerogena, Biochemistry, 10:1483.

Estep, M., Armstrong, J.E., and Van Baalen, C., 1975, Evidence for the occurrence of specific iron (III)-binding compounds in near-shore marine ecosystems, Appl. Microbiol., 30:186.

Ferrante, J.G., 1976, The characterization of phosphorus excretion products of a natural population of limnetic zooplankton, Hydrobiologia, 50:11.

Fogg, G.E., 1951, The production of extracellular nitrogenous substances by a blue-green alga, Proc. R. Soc. Lond. B., 139:372.

Fogg, G.E., 1963, The role of algae in organic production in aquatic environments, Br. Phycol. Bull., 2:195.

Fogg, G.E., 1966, The extracellular products of algae, Oceanogr. & Mar. Biol. Ann. Rev., 4:195.

Fogg, G.E., and Westlake, D.F., 1955, The importance of extracellular products of algae in freshwater, Verh. Int. Verein. Theor. Angew. Limnol., 12:219.

Forsberg, C., and Taube, O., 1967, Extracellular organic carbon from some green algae, Physiologia Pl., 20:200.

Foster, P., and Morris, A.W., 1971, The seasonal variation of dissolved ionic and organically associated copper in the Menat Straits, Deep-Sea Res., 18:231.

Fukai, R., and Huynh, N.L., 1975, Chemical forms of zinc in sea water. Problems and experimental methods, J. Oceanogr. Soc. Japan, 31:179.

Gächter, R., Davis, J.S., and Máreš, A., 1978, Regulation of copper availability to phytoplankton by macromolecules in lake water, Environ. Sci. Technol., 12:1416.

Gächter, R., Lum-Shue-Chan, K., and Chau, Y.K., 1973, Complexing capacity of the nutrient medium and its relation to inhibition of algal photosynthesis by copper, Schweiz. Z. Hydrol., 35:252.

Gächter, R., and Máreš, A., 1979, Effects of increased heavy metal loads on phytoplankton communities, Swiss J. Hydrobiol., 41:228.

Gast, J.A., and Thompson, T.G., 1958, Determination of the boron concentration of sea water, Anal. Chem., 30:1549.

Gessner, F., 1955, The importance of extracellular products of algae in freshwater, Discussion on page 232 of G.E. Fogg and D.F. Westlake (above), Verh. Int. Verein. Theor. Angew. Limnol., 12:219.

Giesy, J.P., 1976, Stimulation of growth in Scenedesmus obliquus (Chlorophyceae) by humic acids under iron limited conditions, J. Phycol., 12:172.

Giesy, J.P., 1980, Cadmium interactions with naturally occurring organic ligands, in: "Cadmium in the Environment. Part One: Ecological Cycling", J.O. Nriagu, ed., John Wiley and Sons, New York.

Giesy, J.P., and Alberts, J.J., 1981, Trace metal speciation: the interaction of metals with organic constituents of surface waters, in preparation.

Gillespie, P.A., and Vaccaro, R.F., 1978, A bacterial bioassay for measuring the copper-chelation capacity of water, Limnol. Oceanogr., 23:543.

Gnassia-Barelli, M., Romeo, M., Laumond, F., and Pesando, D., 1978, Experimental studies on the relationship between natural copper complexes and their toxicity to phytoplankton, Mar. Biol., 47:15.

Goldberg, E.D., 1965, Minor elements in sea water. Chapter 5, in: "Chemical Oceanography Volume 1", J.P. Riley and G. Skirrow, eds., Academic Press, London.

Guy, R.D., and Kean, A.R., 1980, Algae as a chemical speciation monitor. I. A comparison of algal growth and computer calculated speciation, Water Res., 14:891.

Hardstedt-Romeo, M., and Gnassia-Barelli, M., 1980, Effect of complexation by natural phytoplankton exudates on the accumulation of cadmium and copper by the haptophyceae Cricosphaera elongata, Mar. Biol., 59:79.

Harvey, H.W., 1939, Substances controlling the growth of a diatom, J. Mar. Biol. Ass. U.K., 23:499.

Hellebust, J.A., 1965, Excretion of some organic compounds by marine phytoplankton, Limnol. Oceanogr., 10:192.

Huntsman, S.A., and Barber, R.T., 1975, Modification of phytoplankton growth by excreted compounds in low-density populations, J. Phycol., 11:10.

Hunter, S.H., 1972, Inorganic nutrition, Ann. Rev. Microbiol., 26:313.

Jackson, G.A., and Morgan, J.J., 1978, Trace metal chelator interactions and phytoplankton growth in sea water media: theoretical analysis and comparison with reported observations, Limnol. Oceanogr., 23:268.

Johnson, R., 1955, Biologically active compounds in the sea, J. Mar. Biol. Ass. U.K., 34:185.

Johnson, R., 1964, Sea water, the natural medium of phytoplankton. Trace metals and chelation, J. Mar. Biol. Ass. U.K., 44:87.

Khailov, K.M., 1964, The formation of organometallic complexes with the participation of extracellular metabolites of marine algae, Dokl. Akad. Nauk SSSR., Biol. Sci., 155:237.

Khailov, K.M., and Burlakova, Z.P., 1969, Release of dissolved organic matter by marine seaweeds and distribution of their total organic production to inshore communities, Limnol. Oceanogr., 14:521.

Khailov, K.M., and Finenko, Z.Z., 1970, Organic macromolecular compounds dissolved in sea-water and their inclusion into food chains, in: "Marine Food Chains", J. H. Steele, ed., University, of California Press, Berkeley.

Koshy, E., Desai, M.V., and Ganguly, A.K., 1969, Studies on organometallic interactions in the marine environment. I. Interaction
of some metallic ions with dissolved organic substances in sea
water, Curr. Sci., 38:555.

Kroes, H.W., 1972, Growth interactions between Chlamydomonas globosa
Snow and Chlorococcum ellipsoideum Deason and Bold: The role of
extracellular products, Limnol. Oceanogr., 17:423.

Laevestu, T., and Thompson, T.G., 1958, Soluble iron in coastal
waters, J. Mar. Res., 16:192.

Lange, W., 1974, Chelating agents and blue-green algae, Can. J. Microbiol., 20:1311.

Lucas, C.E., 1958, External metabolites and productivity, Rapp. P.-
v. Réun. Cons. Perm. Int. Explor. Mer, 144:155.

McKnight, D.M., and Morel, F.M.M., 1979, Release of weak and strong
copper-complexing agents by algae, Limnol. Oceanogr., 25:62.

McKnight, D.M., and Morel, F.M.M., 1980, Copper complexation by siderophores from filamentous blue-green algae, Limnol. Oceanogr.,
25:62.

Murphy, T.P., Lean, D.R.S., and Nalewajko, C., 1976, Blue-green algae:
their excretion of iron-selective chelators enables them to
dominate other algae, Science, 192:900.

Neilands, J.B., 1967, Hydroxamic acids in nature, Science, 156:1443.

Noakes, J.E., and Hood, D.W., 1961, Boron-boric acid complexes in sea
water, Deep-Sea Res., 8:121.

Ragan, M.A., Ragan, C.M., and Jensen, A., 1980, Natural chelations in
sea water: detoxification of Zn^{+2} by brown algal polyphenols,
J. Expl. Mar. Biol. Ecol., 49:261.

Rona, E., Hood, D.W., Muse, L., and Buglio, B., 1962, Activation analysis of manganese and zinc in sea water, Limnol. Oceanogr.,
7:201.

Seeliger, V., and Edwards, P., 1979, Fate of biologically accumulated
copper in growing and decomposing thalli of two benthic red
marine algae, J. Mar. Biol. Ass. U.K., 59:227.

Sieburth, J.M., 1969, Studies on algal substances in the sea. III.
The production of extracellular organic matter by littoral
marine algae, J. Expl. Mar. Biol. Ecol., 49:261.

Sillén, L.G., 1961, The physical chemistry of sea water, in: "Oceanography", M. Sears, ed., Publ. No. 67 of the American Association
for the Advancement of Science, Washington, D.C.

Slowey, J.F., Jeffrey, L.M., and Hood, D.W., 1967, Evidence for
organic complexed copper in sea water, Nature, 214:377.

Spencer, L.J., Barber, R.T., and Palmer, R.A., 1972, The detection
of ferric specific organic chelators in marine phytoplankton
cultures, Mar. Technol. Soc. Proc., p. 203.

Sposito, G., 1981, Trace metals in contaminated waters, Environ. Sci.
Technol., 15:396.

Steemann-Nielsen, E., and Wium-Andersen, S., 1970, Copper ions as
poison in the sea and in freshwater, Mar. Biol., 6:93.

Stolzberg, R.J., and Rosin, D., 1977, Chromatographic measurement of
submicromolar strong complexing capacity in phytoplankton media,

Anal. Chem., 49:226.

Sunda, W.G., Engel, D.W., and Thuotte, R.M., 1978, Effects of chemical speciation on toxicity of cadmium to grass shrimp Palaemonetes pugio: importance of free cadmium ion, Environ. Sci. Technol., 12:409.

Sunda, W.G., and Gillespie, P.A., 1979, The response of a marine bacterium to cupric ion and its use to estimate cupric ion activity in sea water, J. Mar. Res., 37:761.

Sunda, W.G., and Guillard, R.R.L., 1976, The relationship between cupric ion and the toxicity of copper to phytoplankton, J. Mar. Res., 34:511.

Swallow, K.C., Westall, J.C., McKnight, D.M., Morel, N.M., and Morel, F.M.M., 1978, Potentiometric determination of copper complexation by phytoplankton exudates, Limnol. Oceanogr., 23:538.

Thomas, J.P., 1971, Release of dissolved organic matter from natural populations of marine phytoplankton, Mar. Biol., 11:311.

Tuschall, J.R., and Brezonik, P.L., 1980, Characterization of organic nitrogen in natural waters: its molecular size, protein content, and interactions with heavy metals, Limnol. Oceanogr., 25:495.

Van den Berg, C.M.G., Wong, P.T.S., and Chau, Y.K., 1979, Measurement of complexing materials excreted from algae and their ability to ameliorate copper toxicity, J. Fish. Res. Board Can., 36:901.

Williams, P.M., 1969, The association of copper with dissolved organic matter in sea water, Limnol. Oceanogr., 14:156.

Williams, P.M., and Strack, P.M., 1966, Complexes of boric acid with organic cis-diols in sea water, Limnol. Oceanogr., 11:401.

Zajic, J.E., ed., 1970, "Properties and products of algae", Proceedings of the Symposium on the culture of algae, New York, 1969, sponsored by the Division of Microbial Chemistry and Technology of the American Chemical Society, Plenum Press, New York.

DISCUSSION: J.P. GIESY

P. VALENTA   In our voltammetric measurement of the complexation of Cd with humic acids in sea water (not yet published), we could state that there is only one kind of site available for cadmium. Can you comment on this and cadmium complexation in general?

J.P. GIESY   We have observed cadmium to have a greater affinity for some humic materials than even copper but the total number of sites available is generally less. Also, there are seasonal changes in the number of types of sites. During some times of the year we can measure only one type of organic-cadmium binding site. Thus our data are not in disagreement with your findings.

Y.K. CHAU   Do you take into account the hydrolysis constants, solubility product constants, etc., in your model for the calculation of species?

J.P. GIESY   Yes the model includes solubility products and redox reactions; they don't happen to be very important in this system because of pH and redox considerations but are considered. Also, let me say that all solubility products and stability constants, as well as the activities of the constituents, are corrected for ionic strength.

VOLTAMMETRIC STUDIES ON TRACE METAL SPECIATION IN NATURAL WATERS
PART II: APPLICATION AND CONCLUSIONS FOR CHEMICAL OCEANOGRAPHY
AND CHEMICAL LIMNOLOGY

Hans Wolfgang Nürnberg

Chemistry Department, Institute of Applied Physical
Chemistry, Nuclear Research Center (KFA)
Juelich, Federal Republic of Germany

INTRODUCTION

This paper will focus on the heavy metals which occur in natural waters. They usually occur at the trace level; this situation holds even for the more elevated concentrations in polluted areas. Generally, the total heavy metal concentration is distributed between the dissolved state and the surface zones of suspended inorganic and organic matter. In this context, films of organic materials on the surface of inorganic particles will play a significant, although not exclusive, role in the binding of metals to suspended matter. In addition, metals will be taken up by aquatic organisms. Among them, the phytoplankton are of particular significance from the biomass aspect and the accumulation potential. In rather shallow waters, typically down to 100 m, the surface layers of sediments are also involved more or less significantly as depot sites in the dynamics of metal transfer and exchange between the components and various phases of a natural water system.

Although the concentrations of heavy metal traces in the dissolved state are, as a rule, orders of magnitude lower than in any other phase of a natural water system, the dissolved state is nevertheless of central significance. It is the state in which occurs, to an overwhelming extent, the transfer of metals to and from the other phases of natural water systems. Uptake and release of metals by suspended matter or sediments occurs usually from the dissolved state and to a major extent this is also the case for organisms. This holds for phytoplankton, the most important component in terms of biomass, exclusively. Organisms at higher trophic levels have, in addition, an alternative pathway along the food chain but, for some of them (fish), a direct uptake from the dissolved state via the gills remains very important. Furthermore, certain classes of organisms, such as filter feeders, are able to take up metals immediately from incorporated suspended matter.

Thus, it is doubtless an important contribution to the understanding of the heavy metal situation in a natural water mass to determine reliable data on the metal concentrations in the various aquatic phases, such as in the dissolved state (Mart, 1979; Mart et al., 1980a, 1982a, 1982b), in suspended matter (Mart, 1979; Mart et al., 1982a, 1982b), and in sediments and organisms (Stoeppler and Brandt, 1979; Stoeppler and Nürnberg,1979; Ahmed et al., 1981). For higher organisms, the distribution of heavy metals over essential organs and body fluids is of particular interest (Stoeppler and Brandt, 1979; Martincic et al., 1982). Modern trace analytical methods, if applied with appropriate expertise, care and scrutiny, provide excellent potentialities in this respect. As a consequence of its general features and its inherently low accuracy risks, voltammetry, mainly in the mode of differential pulse stripping voltammetry (Nürnberg, 1981b), constitutes a particularly attractive approach among the choices of those instrumental methods which are applicable to the determination stage of the analytical procedure for the respective phase or component of a natural water system (Nürnberg, 1979; Nürnberg, 1982a). However, for the most interesting tasks connected with speciation problems in organisms, immediate applications of voltammetry are not possible, although indirect applications in combination with other methods (such as the usage of voltammetric detectors in HPLC) offer interesting future potentialities.

The situation becomes quite different if attention is focussed on heavy metals in the particularly important dissolved state; here, suitable modes of voltammetry provide one of the hitherto most powerful and illuminating methodological approaches for speciation studies (Nürnberg, 1979, 1981a, 1982a, 1982b; Nürnberg and Valenta, 1982). Thus, further treatment here of the subject of speciation will be restricted to the dissolved state. The capabilities of voltammetry in this respect stem from the fact that it is basically a species-sensitive method, in contrast to the purely element-sensitive methods, such as atomic absorption spectroscopy (AAS), emission spectroscopy (AES), X-ray fluorescence (RFA) or neutron activation analysis (NAA).

The distribution of the overall (dissolved) concentration of a given heavy metal trace depends on its physical properties and the other dissolved chemical constituents of a natural water type according to the physicochemical laws relevant for aqueous solutions. This is by no means a trivial statement; it emphasizes that, basically, the speciation of dissolved metals in natural waters has to be described and can be understood only in terms of physical chemistry. In this respect, for instance, sea water is an aqueous electrolyte, although a rather complicated one. As will be featured by the examples to be discussed, this physicochemical treatment of metal speciation provides not only a sound basis for the interpretation and explanation of experimental findings but it permits also to draw conclusions of general significance and validity and also to formulate prognostic predictions on the speciation of dissolved metals in the various types of natural waters. Voltammetry certainly has proven to be a very versatile methodological tool for obtaining to a considerable extent the necessary experimental data in a very reliable manner. Nevertheless, it would be unrealistic to expect from voltammetry the answer to all speciation problems regarding dissolved metals.

The strongly growing interest in the speciation of dissolved heavy metal traces is, to a considerable extent, connected with the basically correct realization that the respective processes of metal uptake or release by suspended matter, sediments and organisms depend strongly on the dissolved chemical species of the respective metals interacting with the respective interface types. This remains relevant despite the fact that, after interfacial contact in the course of the interfacial reaction (which will involve for organisms the transfer through biological membranes or other modes of incorporation or the corresponding inverse processes in exudation), the speciation of metals will usually be altered. Consequently, the metal speciation within organisms, including toxicologically-active metal species, will probably be different from that found dissolved in the natural water milieu. Nevertheless, it is obvious that one significant and fundamental objective of both chemical oceanography and chemical limnology, with respect to heavy metal traces, is to elucidate the speciation of dissolved metals in the various types of natural waters. In this context, a somewhat short-sighted view, nurtured in certain quarters, needs to be corrected; it is a viewpoint derived from the exaggerated conclusion that only the speciation, but not the overall concentration, of the dissolved heavy metal traces is of ecochemical and ecotoxicological relevance. This is, of course, not true for a simple reason; the quantities of the various species of a metal, in a given natural water type, will depend on the overall concentration of the respective metal in the dissolved state. Accurate and precise data on the overall concentrations of dissolved metals are of great significance and provide one of the first prerequisites to quantify the species distribution in any natural water type under consideration. Voltammetry provides, in addition to its suitability for speciation studies on most metals of toxicological relevance, the optimal approach for carrying out reliable analytical determinations of overall (dissolved) concentrations (Nürnberg, 1979, 1982a). As this task can be performed at an analytically-optimal pH, it is in principle easier to do than are speciation studies, and, the sensitivity in such analytical applications of voltammetry is substantially higher, reaching the ng/l level or even less (Mart, 1979; Mart et al., 1980a, 1982a, 1982b; Nürnberg, 1979, 1982a).

The methodological aspects of various kinds of speciation studies by voltammetry have been treated in detail in part I (Valenta, this volume). Therefore, the following discussion can be restricted to established findings and conclusions, mentioning only the respective method applied.

RESULTS AND DISCUSSION

A comprehensive literature review on all speciation studies carried out by the application of voltammetry is beyond the scope of this paper. Therefore, the subject will be restricted to some important examples plus the emerging essential conclusions on some metals under study in the ongoing research program of the author's institute.

Labile complexes

For most parts of the sea, the salinity and the ionic composition are homogeneous within a rather narrow range. The same applies to the components of the carbonate buffer system and, therefore, to the pH of about 8. In most areas of the open sea, areas which constitute the major portion of the seas and oceans, the level of dissolved organic matter (DOM) remains well below 1 mg/l (Duursma, 1965); the reason is that there is a low biological productivity in vast parts of the oceans except for upwelling regions, divergencies, coastal zones and estuaries (Bunt, 1975). Moreover, the amount of suspended particulate matter is usually very small in the open oceans (Mart et al., 1982a). Consequently, an overwhelming portion of the usually very low concentrations of heavy metals in the open sea is present in the dissolved state. There, the heavy metals are speciated to at least a considerable extent (depending on the species distribution of the respective heavy metal) as complexes with inorganic anionic ligands. These ligands, X, are mainly Cl^-, OH^-, CO_3^{2-}, HCO_3^-, SO_4^{2-}. They form labile mononuclear complexes of the type MeX_j. Labile means, in this context, that the rate constants of complex formation, k_f, and dissociation, k_d, will be rather high and, therefore, the complex

$$Me + X \underset{k_d}{\overset{k_f}{\rightleftharpoons}} MeX \quad (1)$$

(The charge signs have been omitted for simplicity.)

$$MeX_{j-1} + X \underset{k_d}{\overset{k_f}{\rightleftharpoons}} MeX_j$$

equilibria will be rather mobile. As a consequence of this behaviour, the heavy metal central ion in such labile complexes will always undergo a reversible electrode reaction. Then, the stability constants, β_j, and the ligand numbers, j, of the complexes belonging to a consecutive series, MeX_j, with a given ligand, X, can be evaluated. They are evaluated from the relationship between the reversible half wave potential, $E_{1/2}$, or peak potential, E_p, and the logarithm of the ligand concentration, (X), because these potentials are closely related to the redox potential of Me and this is thermodynamically well-defined (for details, see Valenta, this volume).

For a number of reasons, particularly to avoid the side effects of relatively small concentrations of other trace constituents present in a natural water, it is desirable to make measurements of overall concentrations of the considered trace metal (in its dissolved state) which resemble as closely as possible the levels existing in reality. Then, the pseudo-polarogram approach, utilizing anodic stripping voltammetry (ASV), has to be applied. In this manner, studies on the inorganic speciation of Pb and Cd in sea water have been performed in the 10^{-9}M range for the overall concentration of these metals. In the oceans, the actual overall concentrations of both metals can often be one or two orders of magnitude lower than this level (Mart et al., 1982a, 1982b), such as is the case for Cd in the surface region (above 30 m depth) and for Pb in deeper waters. Such small concentrations are at present beyond the capabilities of voltammetric speciation measurements, but not of a voltammetric determination of the overall dissolved concentration. Nevertheless, this is not of fundamental relevance for the validity of the results of speciation measurements; one can say this because the β_j- and j-values are, in principle, independent of the overall concentration of the considered heavy metal if this is in the trace range. As stated above, doing the speciation measurements at rather low overall concentrations of Me, at or close to the realistic trace level, is primarily a practical safeguard so as not to overlook eventual specific side effects significant at the trace level.

The measurements for Pb and Cd have been performed with the ligands Cl^-, CO_3^{2-} and HCO_3^- and the results (Sipos et al., 1980) are represented in Table 1. The β_j-values for complexes with the ligands OH^- and SO_4^{2-} have been taken from Sillén and Martell (1971) and from Dyrssen and Wedborg (1974). It should be noted that the β_j- values are (in the terminology of coordination chemistry) conditional values. That means that they contain the activity coefficients of the reactants and refer, with respect to salt concentration and to composition, to the ionic strength of the medium which is sea water. To avoid interferences by complexing components of DOM, the measurements have been done at first in artificial sea water. Later, the data for Cd have been reproduced in genuine Ligurian Sea water which had a negligible DOM level.

With the aid of the β_j-values, the percentage species distribution has been evaluated for Pb and Cd, taking into account the major inorganic ligands operative in the sea. A significant improvement has been achieved when the ion pairing effects, operative between the major cationic and anionic salinity components, were not neglected in the evaluation of β_j, as was commonly the case in the past (Table 1).

Table 1. Stability constants and species distribution for Cd and Pb in sea water without and with a correction for ion pairing between major salinity components. Reference A is Dyrssen and Wedborg (1974) and reference B is Sillén and Martell (1971).

Species	β_j	Without ion pairing correction Distribution (%)	β_j	With ion pairing correction Distribution (%)	
Cd^{2+}	–	2.7	–	1.9	
$CdCl^+$	27.25	34.6	32.69	29.1	
$CdCl_2$	61.66	36.8	88.74	37.2	
$CdCl_3^-$	91.20	25.6	157	31.0	
other CdX_j	–	0.3	–	0.8	
among them					
$CdHCO_3^-$	1.52	7.4×10^{-3}	1.8	6.1×10^{-3}	
$Cd(HCO_3)_2$	24.9	2.2×10^{-4}	35.0	2.2×10^{-4}	
$CdSO_4$	10.0	0.12	–	0.66	A
Pb^{2+}	–	3.0	–	1.8	
$PbCl^+$	8.57	12.0	10.28	8.6	
$PbCl_2$	22.24	14.7	32.00	12.6	
$PbOH^+$	7.9×10^6	50.0	7.9×10^6	30.0	
$Pb(OH)_2$	6.3×10^{10}	0.9	6.3×10^{10}	0.5	B
$PbCO_3$	4.2×10^5	18.9	1.6×10^6	43.0	
$Pb(CO_3)_2^{2-}$	6.25×10^8	3.7	9.1×10^9	3.7	

The essential results of these investigations are as follows.

Cd is predominantly present as chlorocomplexes (97%) in the sea. The three chlorocomplexes occur in comparable quantities, although there is a slight preference for the neutral $CdCl_2$. All other inorganic species, including free hydrated Cd^{2+}, contribute to the total only about 3%. The contribution of carbonato and bicarbonato complexes amounts only to traces; however, it is interesting to note that, in the trace range, bicarbonato complexes of Cd do exist in sea water.

For Pb, a completely different inorganic speciation pattern emerges. The most abundant species are carbonato complexes with a distinct predominance of $PbCO_3$, which alone amounts to 43%. The second most important species type is $PbOH^+$ followed by the chloro-complexes. All other species, including free hydrated Pb^{2+}, contribute much less. Even in traces, no bicarbonato complexes are formed.

A further important general result comes from the determined Pb-speciation which showed that the relevance of the sea water model of Pytkowicz and Hawley (1974) could be experimentally confirmed; this is not the case for controversial species distributions emerging from the models of other authors (see Sipos et al., 1980). Such a success emphasizes the necessity for the experimental verification of models, which, as such, are to be regarded merely as strategical aids in the screening of possibilities for the speciation distributions existing in a given natural water type.

The inorganic speciation patterns elucidated for Pb and Cd are of general validity for the major part of a sea. They ignore contributions made by the formation of organic complexes or chelates with the various constituents of DOM. As pointed out before (and to be emphasized in later sections), this is permissible for major parts of the open sea and even for significant parts of coastal waters, such as large zones of the Ligurian and Tyrrhenian coast in the Mediterranean Sea (Mart, 1979, 1980a). This can be attributed to low levels of DOM, of which only a fraction has complexing (or more generally-speciating) capabilities. However, in areas where DOM-constituents contribute to the speciation, as a consequence of elevated DOM-levels, a more complicated situation of special speciation arises. In addition, the percentage distribution of the inorganic species of both metals will be altered. Such specific situations will be, however, relatively localized (such as to upwelling zones, to certain coastal areas and frequently to estuaries). Moreover, in particular microenvironments, such as the interstitial pore waters of certain sediment types or the immediate milieu of planktonic particles affected by the release of organic exudates, there are speciation contributions by organics to be expected. For the major part of the seas, however, the elucidated inorganic speciation patterns will prevail for Pb and Cd.

Although the labile complex species discussed are formed mainly with major inorganic anions, organic ligands too may be labile. Examples are the recently studied complexes of Cd with the amino acid anion glycinate (Simoes Goncalves et al., 1982). Experiments have been performed in artificial sea water and on genuine samples from the Pacific which had been spiked with an overall concentration of 6 x 10^{-7}M Cd. The Pacific water samples had been subjected previously to

UV-irradiation (Mart et al., 1980b) to exclude any interference from natural DOM components already present. The results on both media agreed and increasing glycine concentrations were investigated. Up to the rather high glycine concentration of 3.6×10^{-3} M, the Cd is complexed by this ligand to less than 5%. At substantially higher glycine concentrations, two Cd-glycinates with one and two glycinate ligands exist. The glycine levels required for any noticeable speciation of Cd in sea water are far above the concentration levels occurring under any imaginable conditions in the sea, both for this and other amino acids of a similar type. Thus, it can be safely concluded that glycine must be excluded as a significant natural ligand for Cd in the sea, a result which was to be expected a priori from the literature data on the stability constants (Sillén and Martell, 1971). Nevertheless, it was regarded as desirable (with respect to the potential importance of amino acids of the glycine type as DOM components) to provide an experimental confirmation for some cases.

For the same reason also, the speciation of Zn with glycine in sea water has been studied (Simoes Goncalves and Valenta, 1982), yielding similar results as for Cd. The ligand level required for the onset of Zn-speciation is well above the value occurring for glycine in the sea; the stability constant is about an order lower than for Cd. Studies on sea water with Zn and the ligand, L-aspartic acid, regarded as another potential candidate for amino acid speciation, yielded ligand concentration requirements much too high for the reaction to be of any significance (Sugawara and Valenta, 1982). In this context, it has to be mentioned that the investigations with Zn have not been based on shifts of the reversible half wave or peak potentials with the amino acid concentration but, instead, on the voltammetric titration of the trace metal with the ligand, an approach to be applied to more stable complexes (see also next section of this chapter). As Zn forms more stable complexes with the studied amino acids than does Cd, the Zn-complexes with these amino acids form a border-line case where the voltammetric titration approach is still applicable provided that a voltammetric mode with a short time constant, as has differential pulse polarography (DPP), is applied; the Cd-glycinates are already so labile that the half wave or peak potential shift approach has to be applied for them.

The results of the investigations of amino acid speciation of Cd and Zn in sea water are summarized in Table 2.

Although the above mentioned investigations with Cd and Zn confirm that, in the sea for both metals, the speciation by amino acids remains usually negligible, this might be different in special cases (such as in the microscopic exudate zone in the immediate vicinity of phytoplankton cells). Moreover, special situations might arise in areas with substantial DOM-levels (due to high planktonic productivity), particularly for the Zn species having higher stability constants. Similar conclusions should be forthcoming in another special

Table 2. Complexation of Cd and Zn by amino acids in sea water.

Sea Water Type	Metal	lg β_1	lg β_2	Ligand	Required ligand concentration for a given complexation degree
North Sea	Cd	4.30	-	L-aspartic	1.0×10^{-2} M for 10%
North Sea	Zn	5.44	10.2	L-aspartic	8.8×10^{-4} M for 10%
Pacific	Cd	4.03	7.29	glycine	3.6×10^{-3} M for 5%
Pacific	Zn	4.29	7.82	glycine	2.0×10^{-4} M for 2%

case, still to be investigated, which is represented by interstitial and pore waters of sediments containing substantial amounts of relevant organic materials. Furthermore, particularly for Zn, a decrease of salinity in estuaries and (particularly) in fresh waters will lower the ligand concentration requirements, as a result of the concomitant decrease of competing salinity components (such as Cl^-); it may lower them so as to make effective the amino acid levels which are possible under certain conditions in natural waters. Then at least, some contribution to Zn speciation can no longer be excluded. Thus, particularly with respect to low salinity media, the amino acid speciation seems to merit further research. This situation refers also to other heavy metals of ecotoxicological significance (such as Cu, Pb and Hg), in low salinity media and in sea water, because these metals form more stable complexes with amino acids than with some other common ligands. A field of special importance will arise from future investigations of the complexation of these trace metals (as well as of Cd and Zn) with sulphur-containing amino acids in sea water, estuarine waters and various inland water types; a particularly important basis for this statement is the fact that such amino acid complexes of higher stability really are to be expected.

High stability complexes with organic ligands

Besides the amino acids already discussed, there are certainly other DOM-components (among them both smaller molecules and polymeric entities, including proteins, humic materials, nucleic acids and their degradation products) which can act as rather inert complexes or chelates. Obviously such DOM-components act not only in fresh water but also in sea water, as is evidenced by the necessity to subject sea water samples, for the determination of overall concentrations of dissolved heavy metals prior to voltammetry, to UV-irradiation in acid milieu (pH 2) if the samples had been taken in areas of high phytoplankton productivity (Mart, 1979; Mart et al., 1980a, 1982b). Also, finite complexation capacities with Cu or Pb as reference metals point to the same conclusion (see later section). In organically-polluted coastal or estuarine areas, additional ligands of anthropogenic origin might occur.

A question of general significance has been stated as follows; what amount of organic ligand is required in different water types to contribute significantly to the speciation of a given trace metal? An important consequent question is as follows; what are, in the various water types, the general common specific features of the formation of such rather inert complex species? To tackle these general problems of inert complex formation with dissolved organics, voltammetric investigations have been carried out for Cd, Pb and Zn with the well-defined ligand, NTA. They have primarily an experimental model character. However, it should be noticed that NTA is appearing already in the aquatic environment as an anthropogenic ligand. This trend would become substantially larger for rivers, lakes, estuaries and coastal waters if NTA would be used in the future on a large scale in many countries as a phosphate substitute in detergents.

The approach applied in these studies was the voltammetric titration of a heavy metal with NTA as described in part I (Valenta, this volume). One determined, after the addition of a certain amount of NTA, the percentage of the metal remaining unchelated, via its corresponding reversible voltammetric response. As the natural solution contains also its salinity components, the unchelated portion of the dissolved heavy metal trace will be speciated as labile inorganic complexes, MeX_j. Quoted heavy metals have been studied in sea water and in Lake Ontario water, the latter being used as an example of an inland water type with a certain hardness (Raspor et al., 1980a, 1981; Nürnberg and Raspor, 1981). To disentangle the various common specific effects of sea water constituents on the chelation equilibria with NTA, some additional investigations have been made on model solutions of important relevant sea water components (such as 0.55 M NaCl, 1 x 10^{-2} M $CaCl_2$, 5.36 x 10^{-2} M $MgCl_2$), always adjusted to an ionic strength of 0.7 (if necessary by the addition of some $NaClO_4$) and a pH around 8 (using borate buffer or carbonate buffer). Unequivocal tests have ensured that no noticeable perturbation of the complex equilibria with NTA were caused by the voltammetric measurements (see part I - Valenta, this volume). Table 3 contains, for sea water and for Lake Ontario water, the conditional stability constants, K, of the respective NTA-chelates and also the NTA concentrations required for a chelation degree of 20%, which would produce a noticeable contribution to the speciation of those heavy metals in the dissolved state. The required NTA concentrations are considerable despite the rather high K-values and are orders of magnitude higher than in pure inert $NaClO_4$-solution adjusted to the same ionic strength and pH. The reason is that the alkaline earth metals, in particular Ca, exert significant specific side effects; they compete strongly, due to their substantial excess over the heavy metal traces, with these traces for the ligand NTA. Much smaller, additional, specific side effects stem

Table 3. Chelation of heavy metal traces and alkaline earth metals with NTA in sea water (and Lake Ontario water).

Metal	Concentration (M)	lg K	Required NTA conc. for 20% chelation degree (M)	Water type
Cd	3.0×10^{-9}	9.0	1.0×10^{-4}	sea
Pb	2.0×10^{-8}	10.75	4.3×10^{-6}	sea
Zn	3.2×10^{-8}	10.5	4.0×10^{-7}	sea
Zn	4.8×10^{-7}	10.94	1.84×10^{-7}	Lake Ontario (ionic strength 0.007)
Ca	1.0×10^{-2}	5.76	3.4×10^{-7}	sea
Mg	5.36×10^{-2}	4.76	3.4×10^{-7}	sea

in natural waters from the inorganic anionic ligands, X, competing with NTA for the heavy metal trace; they tend to speciate it, as described before, in labile complex species, MeX_j. The fact that the ligand requirements in artificial and genuine sea water are practically the same emphasizes the idea that no other constituents of genuine sea water exert significant specific side effects on the heavy metal chelation by NTA.

Looking at the rather substantial NTA concentrations required in both natural water types to achieve a not-negligible chelation degree, the following prognostic conclusions can be drawn. DOM-components contributing significantly to the speciation of Cd, Pb and Zn in sea water or lake water, of the quoted hardness, must be available at similar levels as NTA provided that the formed complexes are of similar stability. If the stability constants of heavy metal complexes with DOM-components are smaller, or even larger (with respect to NTA), then the required concentration of the respective DOM-component becomes correspondingly higher or lower than the NTA-values in Table 3.

Under the assumption that the complexing DOM-components would have a molecular weight of 100, one can estimate, based on the data in Table 3, the approximate amount of DOM-components in mg/l required as a function of lg K to achieve a complexation degree of 20% of the considered heavy metal traces (see Table 4). It should be noted that the average molecular weights of relevant DOM-components could be higher. This would lead to even higher requirements in ligand concentration if expressed in mg/l. Polymeric DOM-components, such as humates and proteins, will have several chelator sites per molecule. In this case, the estimates in Table 4 refer to monomeric complexing

Table 4. Prognostic correlation between the stability constant and the required concentrations of organic ligands (MW=100) for a 20% chelation degree of heavy metal traces ($10^{-8} - 10^{-9}$M) in sea water.

Metal	log K =	7	8	9	10	11	12
		\multicolumn{6}{c}{Ligand Concentrations in mg/l}					
Cd		1000	100	10	1	0.1	0.001
Pb		2600	260	26	2.6	0.26	0.026
Zn		130	13	1.3	0.13	0.013	0.0013

units of the polymeric ligand. The data in Table 3 and Table 4 show that, for Pb and Zn in sea water, the ligand requirements differ substantially, although the conditional stability constants for their NTA-chelates are almost equal. The reason is that the side effects exerted by the inorganic anionic constituents of sea water are quite different in strength for both heavy metals. As has been shown for Pb, the formation of the rather stable carbonato species plays a substantial role in its inorganic speciation pattern (Sipos et al., 1980), while Zn forms much weaker complexes with all relevant inorganic anionic constituents of sea water (Raspor et al., 1981; Nürnberg and Raspor, 1981).

The prognosis potential in Table 4, originating for the organic speciation of Cd, Pb and Zn from the experimental studies with NTA, can be utilized to predict, from stability constant data available in the literature (Sillén and Martell, 1971), if certain types of organic ligands are to be considered as a probable candidate for such speciation. Provided that the typical DOM-level in a natural water type is known, one sees immediately if there exists a chance at all for a certain ligand to be effective. If the required ligand concentration exceeds the total DOM-concentration, or a substantial amount of it, then the probability becomes zero or very small.

A good example is the widely discussed and controversial question of the contribution of dissolved (in sea water) humic material to the speciation of heavy metals. According to the literature, the total DOM level of the open sea remains usually below 1 mg/l with a humic content below 0.11 mg/l and certainly below 0.2 mg/l (Duursma, 1965; Ogura, 1972). Considering the K-value range reported for complexes of Cd, Zn and Pb with various types of humates (Mantoura et al., 1978; Whitfield and Turner, 1980), one concludes immediately from Table 4 that no noticeable contribution to the speciation of those heavy metals by dissolved humic material is to be expected in the open sea. There is simply not enough dissolved humic material present. This prediction has been made already several years ago

(Raspor et al., 1979, 1980a). Results of electrophoretic studies with Pb and humic material of marine origin support this conclusion (Musani et al., 1980).

Recently this prediction has been confirmed experimentally by the voltammetric titration approach with humic material of marine origin from various parts of the sea (Raspor et al., unpublished). At 0.3 to 0.4 mg/l of humate, the complexation degree for Cd and Pb reaches only 10% and, below 0.1 mg/l of humate, the complexation degree remains 0 to 5% in sea water. For Zn, a complexation degree of 10% requires about 1.0 mg/l of humates while, below 0.4 mg/l of humate, no complexation of Zn is detectable. With Zn, the same humate types have been tested also in Lake Ontario water. The ionic strength differs by a factor 10^3 and the concentration of particularly competitive Ca differs by a factor 750 less than in sea water (Raspor et al., 1981). The humate concentrations required for a Zn complexation degree of 10% range from 0.6 to 1.15 mg/l. These recent findings confirm that, at least in the open sea, the required dissolved humate concentrations are, at less than 0.1 mg/l, too small to produce any significant contribution to the speciation of Cd, Pb and Zn. A similar conclusion is drawn for the inland water type represented (with respect to ionic strength and composition) by Lake Ontario water.

For Cu and Hg, however, a speciation contribution of dissolved humic substances is to be expected in sea water, and particularly in inland waters, due to the significantly higher stability constants of humate complexes (Mantoura et al., 1978; Whitfield and Turner, 1980).

Kinetics and mechanisms

As has been described in part I (Valenta, this volume), the approach of the voltammetric titration (of dissolved heavy metal traces with an organic ligand forming inert complexes) can also be used to determine the kinetics of the formation of those inert complexes. Then one can add equal trace concentrations of the trace metal and ligand to the natural water to establish second order kinetics and follow the time function of the establishment of the complex equilibrium. To elucidate general features, the chelation kinetics of Cd, Pb and Zn with EDTA have been studied in sea water and in model solutions (adjusted to the sea water ionic strength of 0.7 and to a pH of 8) of those major ionic components of sea water relevant for significant specific side effects on the formation of inert heavy metal complexes with organic ligands (Raspor et al., 1977, 1980b, 1981). The resulting conditional formation rate constants, k_f, are listed in Table 5.

The data reveal the following important findings. The k_f-values parallel approximately the sequence of the stability constants (Sillén and Martell, 1971). Both are equal for the EDTA-chelates of

Table 5. Conditional rate constants, k_f, of chelate formation with EDTA in model media, sea water and Lake Ontario water.

Medium	k_f(moles/l/s)			Remark
	Cd	Pb	Zn	
0.55 M NaCl	unmeasurably fast			recombination mechanism
0.0536 M $MgCl_2$	2.8×10^3	5.0×10^4	3.3×10^3	
0.01 M $CaCl_2$	4.2×10^2	2.5×10^3	1.9×10^2	
Artificial sea water	3.3×10^2	4.0×10^3	3.2×10^2	uncertainty ± 20%
North Sea water	3.9×10^2	3.9×10^3	-	
Ligurian Sea water	-	2.0×10^3	-	
Lake Ontario water (ionic strength 0.007; 1.35×10^{-3}M Ca)	-	-	6.0×10^3	

Cd and Zn, while for Pb the stability is higher and the formation rate constant is larger. In all cases, the k_f-values in the 1×10^2M Ca-model solution coincide practically with those measured in artificial and genuine sea water. This finding features again clearly the dominant specific competitive influence of the Ca excess in sea water on organic ligands. The influence of Mg, although present in a more than fivefold larger concentration, remains comparatively marginal. The approximately 20-fold higher k_f-value for the chelation of Zn in Lake Ontario water, having a Ca concentration 7.5 times smaller than sea water, again emphasizes the significance of Ca as a competitor. Thus, in general the kinetic measurements reflect and confirm in their way the fundamental conclusions from the equilibrium studies on the complexation of the studied heavy metals by inert complex-forming organics. Such mutually-supporting general basic results, on the complexation as seen from equilibrium and kinetic studies, are to be expected in terms of physicochemical principles.

A most important finding on the mechanism of the formation of inert complexes with organics is, however, furthermore reflected by the k_f-values in Table 5. The heavy metal chelates are formed in sea water and hard inland waters by a ligand exchange mechanism.

$$MgL \rightleftharpoons CaL + MeX_j \xrightarrow{k_f} MeL + Ca^{2+} + jX^- \qquad (2)$$

with $X \equiv Cl^-, OH^-, CO_3^{2-}$, etc.

The reason is that the alkaline earth ions present in substantial excess consume, a priori, all the organic ligands, L. In this manner, a fraction of the overall alkaline earth concentration in sea water and hard inland waters becomes complexed by the respective components of DOM. If, however, a labile heavy metal species, MeX_j, encounters an alkaline earth complex such as MgL or CaL, then ligand exchange has to occur because the stability of the respective heavy metal complex, MeL, is usually higher. Due to the higher stability of CaL compared to MgL, the concentration of CaL will be almost immediately reestablished if CaL has been consumed in this ligand exchange mechanism. The alternative mechanism for the formation of inert heavy metal complexes, MeL, with organic ligands by simple recombination of a free hydrated heavy metal ion, Me^{2+}, and a free ligand particle, L, is thus practically not existent in both sea water and hard inland waters; this is a result of the specific role of Mg, and particularly of Ca. In such a recombination mechanism, not the elementary step of complex formation but the diffusional approach of both reactants, up to a critical reaction distance of several Å-units, is rate determining (Eigen, 1963); here, much higher k_f-values in the order of 10^9 1 mol^{-1} s^{-1} would be operative. They would be far above the kinetic resolution of the applied voltammetric method (Raspor et al., 1980b). In accordance with that, the formation kinetics for inert heavy metal complexes, MeL, with organic ligands become unmeasurably fast in the 0.55 M NaCl-solution (see Table 5); this is because, due to the absence of alkaline earth ions here, a recombination mechanism has to operate. This will be also the case in soft fresh waters containing only negligible amounts of Ca and Mg. It should be emphasized that the prevailing of the ligand exchange mechanism in sea water and hard inland waters is just a consequence of the particular concentration ratios of the substances involved, leading to an excess (by orders of magnitude) of the alkaline earth ions over the rather small concentration of DOM and the heavy metals present at an even lower trace level. Fundamental studies in aqueous solutions reported in the literature reveal that, when the concentrations of Ca and organic ligands become of comparable order with that of heavy metals, the recombination mechanism for the formation of heavy metal complexes takes over (Kuempel and Schaap, 1968).

Complexation capacity

The applications of voltammetry discussed in the preceding sections have emphasized the particular suitability of this method to obtain results and findings of general and fundamental significance on the speciation of dissolved heavy metals in natural waters in terms of the ultimately-ruling physicochemical principles relevant for aqueous solutions and electrolytes. Typically, those experiments have to be performed separately for one heavy metal with one defined inorganic or organic ligand.

Voltammetry is also a very efficient method to obtain diagnostic information on global speciation parameters of the dissolved state. An important example is the growing significance of the measurement of complexation capacity (Duinker and Kramer, 1977) of natural water types for certain reference metals such as Cu, Pb, Cd and Zn. The methodology is described in part I (Valenta, this volume). Essentially, one performs, for the selected natural water type, a voltammetric titration of the overall concentration of all dissolved ligands, L, capable of forming inert complexes, MeL, with the applied reference heavy metal. Suspended particulate matter is eliminated, at a prior step, from the natural water sample by filtration, usually through a membrane filter of 0.45 µm pore size. It is noted that any other manipulation of the natural water sample, such as pH changes, has to be avoided to maintain the original speciation in the dissolved state. In context with the findings on the specific role of Ca, and to an analogous but less pronounced extent on the role of Mg, in sea water and in hard inland waters, it is to be concluded that, in these water types, the majority of the amounts of the various ligands, L, capable of forming inert heavy metal complexes exist speciated by alkaline earth ions. A marginal remainder of those ligands, L, will exist as heavy metal complex species, due to the original heavy metal trace levels in the given natural water sample. The reference heavy metal (such as Cu or Pb), progressively added as titrant, will ultimately consume all the ligands, L, according to the outlined ligand exchange mechanism, provided it is capable of forming more stable complexes, MeL, than those existing with Ca.

Our recent systematic determinations of the complexation capacity at a number of stations in the Atlantic Ocean, Pacific Ocean, Weddell Sea and the western Mediterranean Sea, as well as in an ongoing investigation of the Lake of Constance, indicate that this entity is, in several aspects, an interesting diagnostic parameter of a natural water system. In general accordance with the findings reported earlier (in the section on heavy metal speciation as inert complexes by organic ligands), higher complexation capacities are found (in the oceans) for Cu than for Pb, while for Cd it remains zero. The data sets obtained at the stations above at different depths along the water column down to the sea bottom, in oceanic regions of low and moderate biological productivity and also in upwelling regions, suggest that correlations exist with phytoplankton productivity and the corresponding DOM-levels. Significantly higher complexation capacities are obtained in low salinity fresh water systems which are also well-supplied with nutrients such as is the case for the Lake of Constance. A more detailed discussion of these freshly-emerging findings and trends, which in terms of the present state of knowledge on speciation of dissolved heavy metals are not unexpected, seems, however, premature. It shall be emphasized only that it should become common practice to regard, in future field studies on natural water systems, the complexation capacity (preferentially for Cu and other heavy metals) as a further common mandatory parameter; this should include the investigation of vertical profiles along the water column.

Concluding remarks

A justified general basic question has been asked; we wish to know if, and in what respect, analytically-defined chemical speciation is a needed parameter (Baudo, this volume). The answer is, for the dissolved state, obviously a positive one; it is a prerequisite for a comprehensive understanding of the fate of heavy metals in natural waters, based on physicochemical principles, which fundamentally governs what happens. With respect to bioavailability, it will provide only a (necessary) portion of the needed answers within the frame of its potentialities (i.e., up to the interaction with biological interfaces). For speciation in the dissolved state, voltammetry is one of the most suitable experimental approaches, as has been concluded also by other experts (Florence and Batley, 1980). At biological interfaces, the speciation prevailing in the dissolved state usually will necessarily undergo alterations which require the application of other methods. The same applies also to detailed investigations on the speciation of heavy metals bound to or sorbed at the surface of suspended particulate matter and sediments.

REFERENCES

Ahmed, R., Valenta, P., and Nürnberg, H.W., 1981, Voltammetric determination of mercury levels in tuna fish, Mikrochim. Acta (Wien), p. 171.
Baudo, R., in: this volume.
Bunt, J.S., 1975, Primary productivity in marine ecosystems, in: "Primary Productivity of the Biosphere", H. Lieth and R.S. Whittaker, eds., Springer, Berlin.
Duinker, J.C., and Kramer, C.J.M., 1977, An experimental study on the speciation of dissolved zinc, cadmium, lead and copper in the River Rhine and North Sea water by differential pulse anodic stripping voltammetry, Mar. Chem., 5:207.
Duursma, E.K., 1965, Biological and chemical aspects of dissolved organic matter in sea water, in: "Chemical Oceanography,Vol.1",J.P. Riley and G. Skirrow, eds., Academic Press, London.
Dyrssen, D., and Wedborg, M., 1974, Equilibrium calculations of the speciation of elements in sea water, in: "The Sea, Volume 5", E.D. Goldberg, ed., Wiley, New York.
Eigen, M., 1963, Ionen- und Ladungsübertragungsreaktionen in Lösung, Ber. Bunsenges. Phys. Chem., 67:753.
Florence, T.M., and Batley, G.E., 1980, Chemical speciation in natural waters, CRC Crit. Rev. Anal. Chem., 9:219.
Kuempel, J.R., and Schaap, W.B., 1968, Cyclic voltammetric study on the rate of ligand exchange between cadmium ion and Ca-EDTA, Inorg. Chem., 7:2435.

Mantoura, R.F.C., Dickson, A., and Riley, J.P., 1978, The complexation of metals with humic materials in natural waters, Estuarine Coastal Mar. Sci., 6:387.

Mart, L., 1979, Ermittlung und Vergleich des Pegels toxischer Spurenmetalle in nordatlantischen und mediterranen Küstengewässern, Doctoral thesis, RWTH Aachen.

Mart, L., Nürnberg, H.W., and Dyrssen, P., 1982b, Low level determination of trace metals in Arctic sea water and snow by differential pulse anodic stripping voltammetry, in: "Trace Metals in Sea Water", C.S. Wong, ed., Plenum Press, New York, in press.

Mart, L., Nürnberg, H.W., and Valenta, P., 1980a, Comparative base line studies on Pb-levels in European coastal waters, in: "Lead in the Environment", M. Branica and Z. Konrad, eds., Pergamon Press, Oxford.

Mart, L., Nürnberg, H.W., and Valenta, P., 1980b, Voltammetric ultra trace analysis with a multicell system designed for clean bench working, Fresenius' Z. Analyt. Chem., 300:350.

Mart, L., Rützel, H., Klahre, P., Sipos, L., Platzek, U., Valenta, P., and Nürnberg, H.W., 1982a, Comparative studies on the distribution of trace metals in the oceans and coastal waters, in: "Trace Metals in Sea Water", C.S. Wong, ed., Plenum Press, New York, in press.

Martincic, D., Stoeppler, M., Nürnberg, H.W., and Branica, M., 1982, Toxic metals levels in bivalves and their ambient water from the Lim Fjord, Thalassia Jugosl., in press.

Musani, Lj., Valenta, P., Nürnberg, H.W., Konrad, Z., and Branica, M., 1980, On the chelation of toxic trace metals by humic acid of marine origin, Estuarine Coastal Mar. Sci., 11:639.

Nürnberg, H.W., 1979, Polarography and voltammetry in studies of toxic metals in man and his environment, Sci. Total Environm., 12:35.

Nürnberg, H.W., 1981a, Voltammetric studies on toxic metal speciation in natural waters, Proc. Int. Conf. Heavy Metals in the Environment, Amsterdam, p. 635, CEP Consultants, Edinburgh.

Nürnberg, H.W., 1981b, Differentielle Pulspolarographie, Pulsvoltammetrie und Pulsinversvoltammetrie in:Analytiker-Taschenbuch,R. Bock, W. Fresenius, H. Günzler, W. Huber, G. Tölg, eds.,Springer,Bd.2.

Nürnberg, H.W., 1982a, Voltammetric trace analysis in ecological chemistry of toxic metals, Pure Appl. Chem., in press.

Nürnberg, H.W., 1982b, Features of voltammetric investigations on trace metal speciation in sea water and inland waters, Thalassia Jugosl., in press.

Nürnberg, H.W., and Raspor, B., 1981, Applications of voltammetry in studies of the speciation of heavy metals by organic chelators in sea water, Environ. Technol. Lett., 2:457.

Nürnberg, H.W., and Valenta, P., 1982, Potentialities and applications of voltammetry in chemical speciation of trace metals in the sea, in: "Trace Metals in Sea Water", C.S. Wong, ed., Plenum Press, New York, in press.

Ogura, N., 1972, Rate and extent of decomposition of dissolved organic matter in surface sea water, Mar. Biol., 13:89.

Pytkowicz, R.M., and Hawley, J.E., 1974, Bicarbonate and carbonate ion-pairs and a model of sea water at 25°C, Limnol. Oceanogr., 19:223.
Raspor, B., Nürnberg, H.W., Valenta, P., and Branica, M., 1980a, The chelation of Pb by organic ligands in sea water, in: "Lead in the Environment", M. Branica and Z. Konrad, eds., Pergamon Press, Oxford.
Raspor, B., Nürnberg, H.W., Valenta, P., and Branica, M., 1980b, Kinetics and mechanism of trace metal chelation in sea water, J. Electroanal. Chem., 115:293.
Raspor, B., Nürnberg, H.W., Valenta, P., and Branica, M., 1981, Voltammetric studies on the stability of the Zn(II)-chelates with NTA and EDTA and the kinetics of their formation in Lake Ontario water, Limnol. Oceanogr., 26:54.
Raspor, B., Nürnberg, H.W., Valenta, P., and Branica, M., unpublished work.
Raspor, B., Valenta, P., Nürnberg, H.W., and Branica, M., 1977, Polarographic studies on the kinetics and mechanism of Cd(II)-chelate formation with EDTA in sea water, Thalassia Jugosl., 13:79.
Raspor, B., Valenta, P., Nürnberg, H.W., and Branica, M., 1979, Voltammetric studies on the potentialities of Cd(II)-chelate formation in sea water, Rapp. Comm. Int. Mer Médit., 25/26:31.
Sillén, G., and Martell, A.E., 1971,"Stability Constants of Metal Ion Complexes, Second Edition", Chem. Soc. Spec. Publ., No. 17.
Simoes Goncalves, M.L.S., and Valenta, P., 1982, Voltammetric and potentiometric investigations on the complexation of Zn(II) by glycine in sea water, J. Electroanal. Chem., in press.
Simoes Goncalves, M.L.S., Valenta, P., and Nürnberg, H.W., 1982, Voltammetric and potentiometric investigations on the complexation of Cd(II) by glycine in sea water, J. Electroanal. Chem., in press.
Sipos, L., Raspor, B., Nürnberg, H.W., and Pytkowicz, R.M., 1980, Interaction of metal complexes with coulombic ion-pairs in aqueous media of high salinity, Mar. Chem., 9:37.
Stoeppler, M., and Brandt, K., 1979, Trace metals in krill, krill products and fish from the Antarctic Scotia Sea, Z. Lebensm.Unters. Forsch., 169:95.
Stoeppler, M., and Nürnberg, H.W., 1979, Typical levels and accumulation of toxic trace metals in muscle tissue and organs of marine organisms from different European seas, Ecotoxicol. Environ. Safety, 3:335.
Sugawara, M., and Valenta, P., 1982, Voltammetric studies on the speciation of trace metals by amino acids in sea water, Rapp. Comm. Int. Mer Médit., in press.
Valenta, P., in: this volume.
Whitfield, M., and Turner, D.R., 1980, Theoretical studies of the chemical speciation of lead in sea water, in: "Lead in the Marine Environment", M. Branica and Z. Konrad, eds., Pergamon Press, Oxford.

DISCUSSION: H.W. NURNBERG

F.H. FRIMMEL   Have you got any information about possible
                      interference from your oceanic humics on the surface
                      of your H.M.D.E., and, if yes, what can you say about
                      the concentration range?

H.W. NURNBERG   Adsorption interferences occur but not for con-
                      centrations of marine humics below 0.8 to 1.2 mg/l.

R.F. VACCARO   Can you tell the biologist anything about the
                      relative toxicities of chemical species such as
                      $CdCl_2$ vs $CdSO_4$, or, $PbCl_2$ vs $PbCO_3$? Also, can you
                      advise the biologist as to how to conduct a meaning-
                      ful experiment in this regard?

H.W. NURNBERG   The voltammetric measurement can not give
                      immediately biological answers. It determines and
                      identifies the metal species in the dissolved phase.
                      From this, one might derive conclusions on interac-
                      tions with interfaces (see Förstner and Salomons,
                      this volume). Such research ultimately will have
                      relevance in an indirect manner for biologists.

NEUTRON ACTIVATION ANALYSIS APPLIED TO SPECIATE TRACE ELEMENTS

IN FRESHWATER

E. Orvini, M. Gallorini, M. Speziali

C.N.R. Centre for Radiochemistry
and Activation Analysis
27100 Pavia, Italy

INTRODUCTION

The evaluation of the eco-toxicological consequences of heavy metals on the biosphere requires an assessment of their biological effects. This involves the determination of the different ionic species and the various chemical forms of each element. For many elements, the usual electrochemical methods for analyzing different chemical species do not meet the required sensitivity; ppm or even ppb concentration ranges represent often the actual content of many matrices. For many elements such as V, Cr, As and Se, one of the most sensitive methods of trace determination (and one which goes down to ppb level) is neutron activation analysis or NAA. Unfortunately, this method gives information only on the total number of nuclei present regardless of their outer structure and chemical forms. In this work, some examples are given of the feasibility of NAA for the determination of trace amounts of different chemical species of some heavy metals, when coupled to specific separation procedures.

VANADIUM

Chemical forms of dissolved vanadium in freshwater

The analytical method developed for the separation and the analysis of different chemical forms of vanadium in freshwater is given in Figure 1, and includes:
    a)   the filtration of the sample through a glass fibre disk;
    b)   the separation of the different vanadium chemical species by ion-exchange chromatography on Chelex 100 and Dowex 1X8

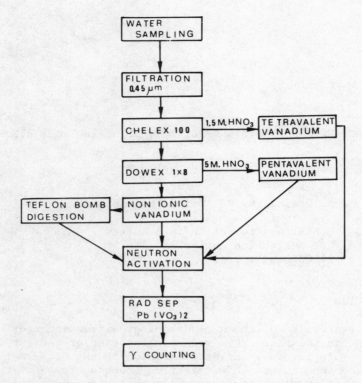

Figure 1. Flow chart of the method used to speciate vanadium.

           columns connected in sequence (ion-exchange system) with selective elutions of the different adsorbed vanadium compounds;
c) neutron activation analysis (NAA) of vanadium in the different chromatographic fractions containing the different vanadium species.

The chemical yield of the complete procedure was checked using a ^{48}V radiotracer under experimental conditions which were as near as possible to those existing in nature (as shown by Orvini et al., 1979). The separation scheme was proven to be effective in the pH range of 4 to 7.

Experiments on freshwater samples

The analytical method has been applied to the analysis of many natural water samples collected at six different sites in northern Italy. Water samples were collected in teflon vessels, stored separately in polyethylene bottles, and, 500 ml aliquots of each were submitted to the procedure illustrated in Figure 1. The induced ^{52}V

($T_{1/2}$ = 3.77 min), from the $^{51}V(n,\gamma)^{52}V$ reaction, was precipitated as $Pb(VO_3)_2$ within 3.5 min after the end of irradiation for the quantitative measurement of the vanadium concentration. The standards for NAA consisted of solutions containing 50 ng of NH_4VO_3. To check the influence of the vanadium content of the reagents (blank) on the determination of the vanadium in the water samples, the separation steps were repeated with deionized water alone. Standards and blanks were irradiated at the same time as the samples, submitted to radiochemical separation and counted in the same manner as for freshwater samples.

Determination of the vanadium chemical species in freshwater samples

Table 1 reports the summary of the concentration data for the neutron activation analysis of the vanadium, in the fractions 1, 2 and 3, corresponding to complexed tetravalent and pentavalent vanadium (Figure 1) obtained from ion-exchange chromatography of water samples collected at six different sites. Each value refers to at least three single determinations. The following conclusions can be drawn.

a) The vanadium from meteoric water collected from an unpolluted area in the inner Alps region was very low in fraction 1.
b) Fraction 1 from all the waters from the city of Pavia, the Ticino River upstream and downstream of Pavia, the Po River, and spring water shows the presence of complexed vanadium in amounts ranging from 14% to 45% of the total vanadium.
c) Vanadium was detected in a similar proportion in fractions 2 and 3 (with a slightly higher value in fraction 2) from the city of Pavia, the Ticino River north and south of Pavia, the Po River and meteoric waters. The proportion between the two fractions is different in the case of the spring water.

CHROMIUM (III) AND CHROMIUM (VI) IN WATER

For this evaluation, neutron activation analysis was coupled to ion-exchange chromatography using AG 50 W-X4 and Ag 1-X8 resin beds connected in sequence, as shown in Figure 2.

Table 1. Determination of vanadium chemical species in freshwater.

Sample	pH	Amounts of Vanadium (μg/l)			Total Vanadium (μg/l)	
		F1 (complexed)	F2 (tetravalent)	F3 (pentavalent)	F1+F2+F3	measured
Ticino river a	6.8	0.210	0.240	0.231	0.681	0.646
Ticino river b	6.7	0.117	0.341	0.270	0.728	0.703
Po river	7.4	0.108	0.070	0.062	0.240	0.225
Tap water	7.4	0.160	0.392	0.231	0.783	0.708
Meteoric water	6.9	0.05×10^{-3}	4.4×10^{-3}	3.1×10^{-3}	7.5×10^{-3}	7.0×10^{-3}
Spring water	6.8	0.054	0.103	0.226	0.383	0.374

a: upstream of Pavia   b: downstream of Pavia

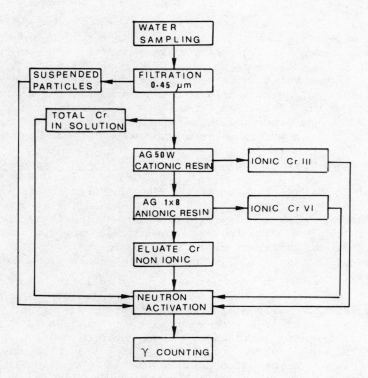

Figure 2. Flow chart of the method used to speciate chromium.

Isolation of single chromium species

Measured aliquots (up to 1000 ml) of water were passed through an ion-chromatographic column system. The water pH was adjusted to 5.5 with a few drops of 2M HNO_3. The chromatographic system consists of two columns, of AG 50 W-X4 cationic and AG 1-X8 anionic resins, connected in sequence. The AG 50 W-X4 resin (H^+ form) was cleaned by repeated washing (approximately 5 bed volumes) with a 1M HNO_3 solution; this procedure was followed by repeated rinsing (10 bed volumes) with distilled deionized water until the pH of the eluate was 5.5. The AG 1-X8 resin (originally in the Cl^- form) was treated with 2N NH_4OH solution and washed with deionized water until pH 5.5. The quartz columns were filled to a height of 2.5 cm with the resins. At the operative pH of 5.5, the AG 50 resin fixes all of the Cr(III) ions in solution, while the Cr(VI) chromate ions pass on to the next AG 1 column, where they are completely retained. The operative flow rate of the separation gave an optimum result at 0.8 ml/min. The eluate was concentrated to nearly 2 ml, by mild evaporation (at 60°C) in a teflon beaker on an electric heater, and then transferred into a quartz vial. Non-ionic chromium salts should be present in this eluate. The total chromium content was determined separately, by analyzing the residue obtained through evaporation at 60°C of 1000 ml of filtered water sample.

Neutron activation analysis

The different chromium fractions were evaluated quantitatively using an instrumental neutron activation analysis. The whole quartz columns and the Cr(III) and Cr(VI) fractions fixed on the resins were heat-sealed at both ends. The suspended matter on the glass fiber filter, the total dissolved chromium fraction and the non-ionic chromium solution were sealed in Specpure quartz vials for irradiation. Blank resins and all chromium fractions, together with primary Cr(VI) standards, were irradiated for 30 hours in the central thimble facility of the TRIGA reactor of the University of Pavia at a flux of approximately 9×10^{12} n cm^{-2} sec^{-1}. After a cooling period of one week, the vials were washed externally with concentrated nitric acid and counted directly on a Ge(Li) detector (10% efficiency; 2.0 KeV resolution on the 1332 KeV ^{60}Co line) connected to a 4096 channel computerized LABEN spectrometer. For the quantitative chromium evaluation, the ^{50}Cr (n,γ)^{51}Cr reaction and the 320 KeV gamma photopeak from ^{51}Cr were utilized.

Chemical yields and method control

The chemical yields of each separation step were thoroughly checked using ^{51}Cr radiotracer solutions, as demonstrated by Orvini et al. (1980). The key points checked were:

a) quantitative retention of ionic Cr(III) on the AG 50 W-X4 resin;
b) complete passage of Cr(VI) ions through the AG 50 W-X4 resin;
c) quantitative retention of Cr(VI) ions on the AG 1-X8;
d) that complexing agents such as Cl$^-$, HPO$_4^{2-}$, EDTA, oxalic acid and citric acid do not influence the behaviour of the chromium ions on the resin system.

The method was then thoroughly checked by analyzing some synthetic water samples spiked with known amounts of Cr(III) and Cr(VI) ionic salts. The operational conditions under which the overall method was controlled were very close to those of most naturally occurring waters. The operative pH of 5.5 does not affect the Cr(III)/Cr(VI) equilibria during the separation time. The method was proven to be effective in the speciation of chromium in actual surface waters also in the presence of complexing agents (such as SO$_4^{2-}$, HPO$_4^{2-}$, Cl$^-$, oxalic acid, citric acid and EDTA). As shown by the Istituto di Ricerca sulle Acque del CNR (1974), only high concentrations of these agents, much over the range usually occurring in nature, do actually affect the results. The chromium contents of blanks (resins, quartz tubes, filters and deionized distilled water) have always proved to be less than 50 nanograms, which is the sensitivity of the method under the experimental conditions used. The interference contribution from the $^{54}Fe(n,\alpha)^{51}Cr$ reaction was checked and it proved to be negligible in the irradiation position used. The method was applied to the Cr(III), Cr(VI) speciation in selected natural waters; the results obtained on two river sites are reported in Table 2. The different ratios, Cr(III)/Cr(VI), may reflect different origins of the chromium salts at the collecting points.

ARSENIC (V) AND ARSENIC (III) IN WATER

A method is reported for the determination of As(III), As(V) and total inorganic arsenic using neutron activation analysis coupled to selective-hydrides generation as shown in Figure 3. The overall analytical procedure for arsenic speciation includes a filtration of the water samples through a glass disk (0.45 μm), the separation of different As(III) and As(V) dissolved species using selective hydride evolution and a collection on $AgNO_3$ filter traps, as demonstrated by Orvini et al. (1981a). The single arsenic species collected on different traps are analyzed by neutron activation. The method has been proven to be reliable in solving the arsenic speciation problem (through analyses of water samples spiked with different arsenic compounds in the range of 20 ppb). Measured aliquots of the filtered water (up to 200 ml) were transferred in a standard 250 ml filter flask for arsine generation. The pH of the solution was buffered to 4.5-5.0 (with sodium acetate - acetic acid buffer) and the temperature was adjusted to 60°C with an electric heater. Five ml of 5% $NaBH_4$ solution were slowly injected into the flask, during continuous stirring. In a reaction time of about 6-10 minutes, all the As(III) present is converted to arsine; it is stripped from the solution along with the evolved hydrogen and recovered on the $AgNO_3$ trap. The pH at the end of this period is about 5.5. After this time, at the open loop of the trap, a water vacuum pump is inserted and the two-way valve is opened to flush the reaction vessel and recover the residual arsine. The flask is removed from the heater and cooled down to room temperature. Then, the $AgNO_3$ trap is disconnected and a new one is applied for the second step of arsine generation from As(V). To this aim, 5 ml of concentrated HCl are injected, through the syringe injector, to adjust the pH to 1. Then, new 5% $NaBH_4$ solution (10 ml)

Table 2. Chromium speciation in natural waters (μg/l).

Sample	pH	Cr(III) A	Cr(VI) B	Non-ionic Chromium C	Total	Total A+B+C	Suspended particles 0.45 μm
Po river	7.4	0.83 ± 0.025	0.71 ± 0.025	< 0.05	1.45 ± 0.045	1.54 ± 0.025	5.33 ± 0.15
Ticino river	6.8	0.20	3.56 ± 0.17	0.34 ± 0.01	3.69 ± 0.015	4.10 ± 0.15	3.56 ± 0.17

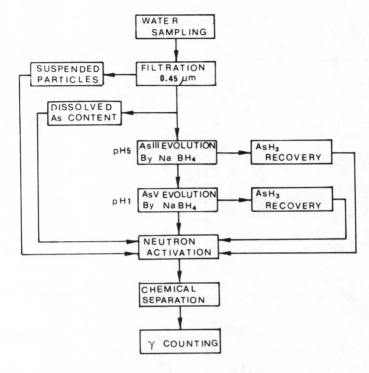

Figure 3. Flow chart of the method used to speciate arsenic.

is injected slowly, and, in 10 minutes all the As(V) compounds are reduced to arsine, stripped from the solution and fixed on the $AgNO_3$ trap.

Neutron activation analysis

Entire $AgNO_3$ traps were heat-sealed at both ends and irradiated for 10 minutes in the central thimble facility of the TRIGA reactor of the University of Pavia at a flux of approximately 9×10^{12} n cm^{-2} sec^{-1}. After irradiation, the trap vials were washed externally with concentrated nitric acid, then opened and transferred into the reservoir of a tin dioxide column (4 cm high, 0.8 cm i.d.). To dissolve the Ag compounds and recover the arsenic, 10 ml of 7N HNO_3 were used. The trap and the column were washed twice with 10 ml portions of 7N HNO_3. All the activity due to ^{110m}Ag passes through, while the ^{76}As remains on the column. The columns were dismantled and counted on a Ge(Li) gamma spectrometer (10% efficiency, 2.0 KeV resolution on the 1332 KeV ^{60}Co line) connected to a 4096 channel computerized LABEN spectrometer. For the quantitative evaluation of arsenic, the ^{75}As $(n,\gamma)^{76}$As reaction and the 560 KeV gamma photopeak from ^{76}As were utilized.

Chemical yields and method control

The chemical yields at each separation step were checked with As-spiked solutions in the range of 20 ppb. The key points checked were:
a) quantitative evolution and recovery of As(III) under the experimental conditions (pH 5.0);
b) negligible evolution of As(V) under the same conditions (pH 5.0);
c) quantitative evolution and recovery of As(V) under the experimental conditions (pH 1).

The overall method was then checked thoroughly by doing analyses of some synthetic water samples spiked with known amounts of fresh As(III) and As(V) ionic solutions.

Results and discussion

The operative conditions at which the overall method was controlled are very close to those of most naturally occurring waters. The operative pH of 5.0 does not affect the As(III)/As(V) equilibria during the short separation time. The arsenic contents of the blanks ($NaBH_4$ filters, distilled deionized water, quartz wool and polyethylene vials) have been proven to be lower than the sensitivity of the method under the experimental conditions used (5 ng). The method was applied to an As(III)/As(V) speciation in selected natural waters; two rivers were investigated and the results are reported in Table 3. The different ratios, As(III)/As(V), may reflect different origins of the arsenic salts at the sampling points. The high arsenic content in the suspended fraction of these waters confirms that a high fraction of arsenic in natural river water may be tied up in suspended materials such as algae, sediments, plankton, bacteria, coagulated particles and etc., as shown by Robertson (1974).

SELENIUM (IV) AND SELENIUM (VI) TRACES IN NATURAL WATERS

A procedure is described for the determination of single species

Table 3. Arsenic speciation in natural waters (µg/l).

Sample	pH	As(III) A	As(V) B	Total A+B	Total	Suspended particles 0.45 µm
Po river	7.4	0.71 ± 0.035	0.43 ± 0.04	1.14 ± 0.04	1.23 ± 0.03	2.89 ± 0.12
Ticino river	6.8	1.04 ± 0.04	1.86 ± 0.14	2.90 ± 0.13	3.04 ± 0.03	4.18 ± 0.14

of selenium compounds in natural waters using neutron activation
analysis, as reported by Orvini et al. (1981b). The method involves
the separation of single selenium species, with a selective reduction
of selenite to elemental selenium and with a specific absorption of
selenate on an ionic retention medium, as shown by Robberecht and Van
Grieken (1980). Also, the selenium contents in suspended matter, in
colloidal metallic selenium and in the non-ionic fraction are evaluated in actual natural water samples, following the scheme of Figure
4 which includes:

a) filtration of water samples through a glass fiber filter
   disk (0.45 μm);
b) separation of different selenium species (colloidal, IV, VI,
   non-ionic);
c) neutron activation analysis of the single fraction.

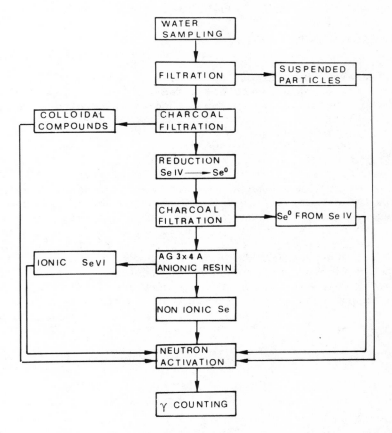

Figure 4. Flow chart of the method used to speciate selenium.

Chemical yields and method control

The chemical yields of each separation step were thoroughly checked using ^{75}Se radiotracer solutions. The key points checked were:

a) the quantitative retention of metallic selenium reduced from ionic Se(IV) on activated charcoal (at a pH lower than 5, the retention was proven to be quantitative);
b) that Se(VI) ions were completely eluted from the activated charcoal (the acid pH does not affect the retention);
c) the quantitative retention of Se(VI) ions on AG 3-X4A resin;
d) that complexing agents such as Cl^- and HPO_4^{2-} do not modify the behaviour of selenium ions on the resin and charcoal systems.

The method was then thoroughly checked by doing analyses of some synthetic water samples spiked with known amounts of Se(IV) and Se(VI) ionic salts. A comparison of the selenium quantities added and found demonstrated the method's reliability.

Results and discussion

The operational conditions under which the overall method was controlled are very close to those of most naturally occurring waters. The operative pH of 5.0 does not affect the Se(IV)/Se(VI) equilibria during the separation time. The method was proven to be effective in the speciation of selenium in actual surface waters also in the presence of complexing agents such as SO_4^{2-}, Cl^- and HPO_4^{2-}. Only high concentrations of these agents, much over the range usually occurring in nature, have an effect on the results. The Se contents of blanks (resins, quartz vials and tubes, filters, and deionized distilled water) have always proven to be less than 10 nanograms, which is the sensitivity of the method under the experimental conditions used. The method was applied to the Se speciation in surface waters; two samples from sewage flowing into the Ticino River and one sample from the Po River were analysed. The results are reported in Table 4 and they confirm the applicability of the method. The results for the two rivers appear to show some peculiar situations at the collecting points; however, the results seem to confirm that the selenate content in polluted waters is lower than is the selenite, as reported by Robberecht and Van Grieken (1980). The Se content in the actual "suspended particles" fractions was always less than the sensitivity limit.

Table 4. Determination of selenium chemical species in freshwater (µg/l).

	pH	Colloidal metallic Se A	Se(IV) B	Se(VI) C	Non-ionic Se D	Total Se A+B+C+D	Total dissolved Se
Sewage 1 to Ticino river	6.8	10.58 ± 0.81	7.5 ± 0.4	3.22 ± 0.40	7.79 ± 0.81	29.1 ± 2.6	25.2 ± 2.1
Sewage 2 to Ticino river	6.8	10.20 ± 0.90	7.5 ± 0.6	5.78 ± 0.6	9.18 ± 1.0	32.6 ± 3.1	26.3 ± 2.6
Po river	7.4	0.1 ± 0.01	0.4 ± 0.04	0.3 ± 0.03	0.05 ± 0.01	0.8 ± 0.08	0.5 ± 0.05

Total values by independent determination on filtered water

REFERENCES

Istituto di Ricerca sulle Acque del CNR, 1974, "Quaderno 24 - Elementi e criteri per la definizione del livello di accettabilità delle acque di scarico", Multigrafica Editrice, Roma.

Orvini, E., Delfanti, R., Gallorini, M., and Speziali, M., 1981a, Determination of trace amounts of arsenic (V) and arsenic (III) in natural waters by neutron activation analysis, in: "Anal. Proc.", (Chemical society. Analytical division), London, 21(6).

Orvini, E., Lodola, L., Gallorini, M., and Zerlia, T., 1981b, Determination of selenium (IV) and selenium (VI) traces in natural waters by neutron activation analysis, in: "Proc. Int. Conf. on Heavy Metals in the Environment", Amsterdam, in press.

Orvini, E., Lodola, L., Sabbioni, E., Pietra, R., and Goetz, L., 1979, Determination of the chemical forms of dissolved vanadium in freshwater as determined by ^{48}V radiotracer experiments and neutron activation analysis, Sci. Total Environm., 13:195.

Orvini, E., Zerlia, T., Gallorini, M., and Speziali, M., 1980, Determination of chromium (VI) and chromium (III) traces in natural waters by neutron activation analysis, Radiochem. Radioanal. Lett., 43:173.

Robberecht, H.J., Van Grieken, R.E., 1980, Sub-part-per-billion determination of total dissolved selenium and selenite in environmental waters by X-ray fluorescence spectrometry, Anal. Chem., 52:449.

Robertson, D.E., 1974, B.N.W.L. - SA Report 5174, Battelle Pacific Northwestern Laboratories, Richland, Washington, 99352.

DISCUSSION: E. ORVINI, M. GALLORINI AND M. SPEZIALI

M. SMIES   In your selenium determinations, the separate analyses of Se(IV) and Se(VI) appear to systematically overestimate Se relative to the determination of total Se. Could you expand on this?

E. ORVINI   The differences range between 15 and 20%. This may be attributed to the blank contribution which, in each single determination, is lower than the sensitivity limit. However, it may be relevant when added to the actual selenium content in each fraction.

E. STEINNES   In two of your separation schemes you feed effluents, from a 450 nm membrane filtration, directly into an ion-exchange column. In such a case, your column may contain not only ionic species but also colloidal particles which have penetrated the filter.

E. ORVINI   Yes, you are right. In fact, soon we intend to introduce a specific filtration step so as to obtain information relating directly to the colloidal fraction.

Y.K. CHAU   I wonder where the organic compounds of trace elements will fit into the separation schemes! For example, there is methylarsonic acid and dimethylarsinic acid in fresh waters; such compounds could be reduced to their corresponding arsines by $NaBH_4$ at the pH for As(III) reduction. Similarly, there are organic compounds of Se and Cr to be considered.

E. STEINNES   As a supplement to your examples on the applications of NAA to speciation studies, I wish to present a table which indicates the feasibility of NAA as an element determination technique in trace element speciation studies on natural surface waters.

A Table on the feasibility of NAA.

- A. Not applicable, or, sensitivity not adequate.

  Be, B, F, Ti, Ni, Sr, Cd, Sn, Pb

- B. Sensitivity adequate but other techniques applicable.

  Mn, Fe, Cu, Zn, As, Se, Hg

- C. Elements for which NAA appears superior to alternative techniques.

1) transition metals
   V, Cr, Co, Sc

2) halogens
   Cl, Br, I

3) alkali metals
   Rb, Cs

4) rare earth elements
   La, Ce, Sm, Eu, Tb, Dy, Ho, Yb, Lu

5) other rare elements
   Sb, Ba, Hf, Ta, W

6) radioactive elements
   Th, U

TRACE ELEMENT SPECIATION IN SURFACE WATERS:

INTERACTIONS WITH PARTICULATE MATTER

U. Förstner

Technical University Hamburg-Harburg
Environmental Engineering Division
Eissendorfer Strasse 38
D-2100 Hamburg 90, West Germany

and

W. Salomons

Delft Hydraulics Laboratory
Haren Branch
c/o Institute for Soil Fertility
P.O. Box 30003
9750 RA Haren (Gr.), The Netherlands

INTRODUCTION

   The uptake of trace metals by organisms occurs chiefly in the dissolved phase. However, it is important to note that, with regard to the concentration and availability of such trace substances, the interactions with solid phases must be considered along with the associated mechanical and chemical processes such as bioturbation, sorption, diffusion and mobilization (Figure 1).

   Suspended matter, originating from the weathering and erosion of soils and rocks as well as from anthropogenic sources, consists of a variety of components including clay minerals, carbonates, quartz, feldspars and organic solids. These components are usually coated with hydrous manganese and iron oxides or by organic substances, which to a large extent affect the interaction processes between solids and dissolved metals.

   Depending on internal conditions (the composition and adsorption characteristics of the suspended matter) and external conditions

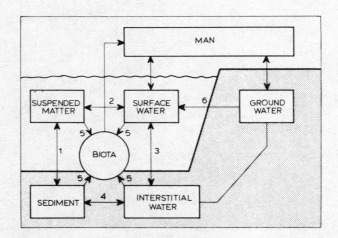

Figure 1. Interactions between abiotic and biotic reservoirs in the aquatic environment (Salomons and van Pagee, 1981) 1 - erosion, dredging, bioturbation, sedimentation; 2 - adsorption, desorption; 3 - diffusion, compaction, bioturbation; 4 - mobilization, adsorption; 5 - accumulation by organisms; 6 - river bank infiltration

(for example, pH, Eh, the presence of natural and man-derived organic and inorganic ligands), a redistribution of trace elements takes place during transport and deposition of suspended matter, particularly during diagenesis of the resulting sediment. The recycling of mineralized organic matter and the pore-fluid transfer processes are essential components in the nutrient and pollutant dynamics of aquatic systems (Lerman, 1977).

PROCESSES AFFECTING THE BINDING OF DISSOLVED TRACE ELEMENTS ON SEDIMENTS

The redistribution of trace elements over the solid and dissolved phases may be the result of direct precipitation and/or adsorption processes. The forms of sorption most commonly encountered in connection with solute transport in soil and sediment are adsorption, chemisorption and ion exchange (Travis and Etnier, 1981). Mechanistically, there is no difference between "sorption" and "coprecipitation" (Jenne, 1976).

The process of "sorption" consists of several single mechanisms (Lieser, 1975; Salomons and Förstner, 1982, in preparation) and these are described below.

Physical adsorption on the external surface of a particulate is based on van der Waals forces of the relatively weak ion-dipole or dipole-dipole interactions (ca. 1 kcal/mole). Additional reactions can occur with physical sorption on the particle's inner surface or in pores (for example, capillary condensation within the pores or inclusions of molecules or ions that snugly fit into the pore system). Representative solid substances include iron oxides, aluminum hydroxides, clay minerals and molecular sieves (zeolites are an example of the latter).

Chemical adsorption is characterized by the formation of chemical associations between ions or molecules from solution and the surface of particles. Such "surface bonds" occur, for example, during hydrolytic adsorption of ions from solution, for the case of ions which could be involved in a condensation reaction with OH groups on the surface of certain (Si, Fe, Al, Mn) compounds, for example,

$$-\overset{|}{\underset{|}{Si}}-OH \ + \ HO-\underset{OH}{Fe}-OH \ \rightarrow \ Si-O-\underset{OH}{Fe}-OH \ + \ H_2O.$$

Sorption based on <u>ion exchange</u> is a chemical process in which negative or positive charges in the mineral lattice (such as in clay minerals) are compensated for by ions possessing opposite charges, which, more or less hydrated in the inner layers, are exchangeable by ions from the solution.

In a general sense, a sorptive-desorptive reaction between solution and solid phases can be a kinetic one (in which the relative amount of solute in the solution and in the solid matrix changes with time), or it can be an equilibrium situation in which the above relationship is attained rapidly and thereafter remains constant. The term "adsorption isotherm" is frequently used in a general sense to include the totality of processes acting to establish equilibrium; however, a good fit of sorption data to a particular "adsorption isotherm" does not in itself constitute proof of any <u>specific</u> sorptive mechanism (Travis and Etnier, 1981).

The surface charge of particulates in natural waters, which may arise from chemical reactions at the surface, by lattice imperfections at the solid surface (or by isomorphous replacements within the lattice), and by adsorption of surfactant ions (Stumm and Morgan, 1981), is strongly pH-dependent (Figure 2a). With regard to the adsorption of metals on oxide surfaces, the onset of the adsorption is characteristic for each metal and the adsorption increases sharply over a small pH-interval (Figure 2b). This means that a small shift in pH in surface waters, as may occur in lakes, causes a sharp increase or decrease in dissolved metal levels.

Various theories have been proposed to describe and interpret the adsorption of metal ions on hydrous oxide surfaces. The summary below has been presented by Stumm et al. in 1976 (see also Stumm et al., 1980):

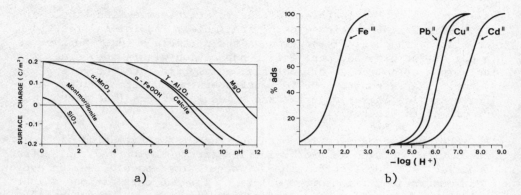

Figure 2. a) Effect of pH on surface charge of inorganic particles (from Stumm and Morgan, 1981, p. 613).
The authors note that these simplified curves are based on results by different investigators whose experimental procedures are not comparable and may depend upon solution variables other than pH.
b) Adsorption of metal ions on amorphous silica as a function of pH (from Stumm and Morgan, 1981; after Schindler et al., 1976).

(1) the Gouy-Chapman-Stern-Graham model which accounts for specific and electrostatic adsorption;
(2) the adsorption-hydrolysis model which postulates that the adsorption of hydrolyzable metal ions is directly related to the presence of hydrolyzed species;
(3) the ion-solvent interaction model which considers coulombic, solvation and specific chemical energy interactions as the ion approaches the interface, and which implies that a lowering of the ionic charge of the metal species (such as by hydrolysis) decreases the ion-solvent interaction (which represents a barrier to close approach of multiply-charged ions to the surface);
(4) the ion exchange model according to which cations, upon adsorption on the hydrous oxide surface groups ($\equiv$Me-OH), replace protons;
(5) the surface complex formation model in which the hydrous oxide surface groups ($\equiv$Me-OH, or $=$Me$\underset{HO}{\overset{OH}{\diagup}}$) are treated similarly to the amphoteric functional groups in polyelectrolytes as complex forming species.

Since <u>surface complex formation constants</u> have been determined, such as for the oxide/water interface (Davis and Leckie, 1979), humic substances (Perdue, 1979) and metal-multiligand systems (MacCarthy and Smith, 1979), models can be devised to evaluate the competition between the sites of soluble ligands and those of surfaces of metal ions (Morel et al., 1973; Lu and Chen, 1977; Jenne, 1979). The surface formation constants show the same trend in stability as the corresponding solute complex formation constants (Stumm et al., 1980; Dugger et al., 1964; Schindler et al., 1976). In practice, however, sorption processes are highly complicated interactions of various mechanisms and substrates. For example, in waters rich in organic matter, minerals may be solubilized by the combined processes of complexation and reduction (Theis and Singer, 1974); reincorporation of metals into the sediment involves the mechanisms of adsorption, flocculation, polymerization and precipitation (Jonasson, 1977). The latter processes are particularly significant in the estuarine mixing zone (Sholkovitz et al., 1978; Sholkovitz and Copland, 1981). Typical processes of binding of dissolved trace elements onto solid phases are described in the following sections.

Direct Precipitation of Metal Compounds

Direct precipitation of metal compounds may take place when external factors change. These precipitation barriers commonly arise because of changes in pH, in oxidation potential, or in concentrations of precipitating substances. Causes for precipitation barriers include interactions of water with solids, mixing with other waters, and loss or addition of gases on emergence of groundwater at the surface (Rose et al., 1979). Major types are (see Perel'man, 1967, as cited by Rose et al., 1979):

(a) oxidation type (iron- and manganese- oxides or native sulfur precipitated by oxidation of reducing solutions and usually caused by water emergence at the surface or a flow of reducing water out of a swamp);
(b) reducing type (U, V, Cu, Se, and Ag precipitated as metals or lower-valency oxides by reduction of oxidizing water and usually caused by an encounter with organic matter or a mixing with reducing waters or gases);
(c) reducing sulfide type (Fe, Cu, Ag, Zn, Pb, Hg, Ni, Co, As, and Mo are precipitated as sulfides by reduction of oxidizing sulfate waters, usually by the action of sulfate-reducing bacteria) (U, V, and Se may also be precipitated and the causes are the same as for the reducing type, but they require the presence of dissolved sulfate);
(d) sulfate carbonate type (Ba, Sr, and Ca precipitated by increased sulfate or carbonate as a result of the mixing of waters, the oxidation of sulfide, or passage into carbonate rock);
(e) alkaline type (Ca, Mg, Sr, Mn, Fe, Cu, Zn, Pb, Cd, and other elements precipitated by increased pH, usually caused by the interaction of acid waters with carbonates or silicate rocks, or its mixing with alkaline waters);
(f) adsorptive type (adsorption or coprecipitation of ions on accumulations of Fe-Mn-oxides, clays, and organic materials)

(the cations of transition metals and those with high valence tend to be more strongly adsorbed than anions and low-valency cations).

If relevant stability constants are known, the equilibrium concentration of a dissolved metal, its inorganic speciation and its stable solid compounds can be calculated. The solubility curves for cadmium and zinc in Figure 3a, taking into account the range of carbonate content normally encountered in inland waters (10^{-3} to 10^{-2} mole dissolved CO_2 per litre; from Hem, 1972), show a minimum at pH values of 9.3. Assuming pH values to be 7 to 8 in normal river waters (under oxidizing conditions) it has been calculated that between 100 and 1000 µg/l zinc (and 5 to 100 µg/l cadmium) may be present in the dissolved state. In the system, $Zn + S + CO_2 + H_2O$ (Figure 3b, from Hem, 1972), where the zinc concentration is 10^{-5} molar (together with 10^{-3} moles of dissolved CO_2 and sulfur), three solid phases are possible: the sulfide, carbonate and hydroxide of zinc. In aerated waters, zinc carbonate is the stable phase if the pH falls below 8.3;

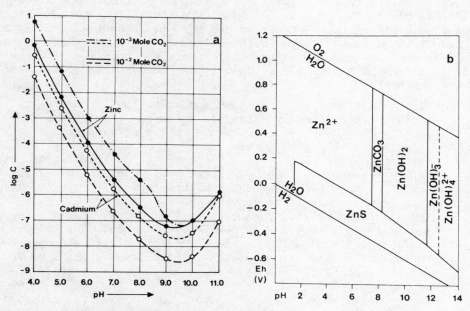

Figure 3. Solubilities of zinc and cadmium as a function of pH (Figure 3a) and fields of stability of solids and dissolved zinc species in the system $Zn + CO_2 + S + H_2O$ at 25° C and 1 atm (Figure 3b) in water (from Hem, 1972).
    (a) Total dissolved carbon dioxide species, 10^{-3} and 10^{-2} mole/l, resp.; ionic strength of 0.0; Log C in moles Zn or Cd/l.
    (b) Dissolved zinc activity, 10^{-5} mole/l; dissolved carbon dioxide and sulfur species, 10^{-3} mole/l.

above a pH of 8.3 a precipitation of zinc hydroxide occurs. The
solution of this type of chemical equilibrium calculation is usually
accomplished by the use of readily-accessible computer programs
(for a review, see Nordstrom et al., 1979).

As there is a lack of data concerning most geochemically-important solid-solution reactions, kinetic factors cannot readily be
assessed (Stumm and Morgan, 1981, p. 230). For irreversible reactions
in a closed system, there is no equilibrium state and such reactions
progress toward complete exhaustion of that reactant which is stoichiometrically limiting (Stumm and Morgan, 1981, p. 113). Among
irreversible reactions which are slow (half-times of several minutes
or longer), in certain aquatic environments, are metal-ion oxidations
(e.g., $Mn^{+2} + O_2$ at pH 8$\sim$10^{10} sec), oxidation of sulfides, various
metal ion polymerizations (e.g., aluminum ion) and aging of hydroxide precipitates. However, some of these reactions can be accelerated greatly by biological catalysis.

Metal Sorption on Fe/Mn Oxyhydrates

For the sorption of trace elements in aquatic systems, the more
important sites generally are comprised of thermodynamically metastable phases (Jenne, 1976) which exhibit extensive isomorphic substitution (Jones and Bowser, 1978). This is mainly true for iron
and manganese oxyhydrates which can be present in X-ray-amorphous
forms, microcrystalline forms and, in more "aged" crystalline particles, as coatings. Formation of hydrous Fe/Mn oxides occurs where
groundwater is discharged to the surface through springs, at the
thermocline regions in eutrophic lakes and fjords (where the anoxic
hypolimnion water is in contact with oxidized sediment), at the point
where neutralization of acidic waters takes place (e.g., at the
junction of rivers exhibiting different pH values and at the mixing
zones of river water with seawater). A further site, particularly
for the deposition of Mn-oxide, is the surface of carbonate minerals
where there is a microzone of higher pH (Lee, 1975). Accumulation
of manganese takes place after the deposition of sediments in the
oxidized surface layer.

The shapes of the concentration-depth profiles of iron and
manganese, in both marine and freshwater environments, are modified
with respect to the position of the redox-cline and its stage of
development in a seasonally anoxic system (Davison, 1981 a, b).
Most of the iron which reaches a sediment is permanently incorporated
whereas manganese is re-released (Wedepohl, 1980). An interesting
phenomenon is the enrichment of manganese (Sundby et al., 1981)
and also, to some extent, of iron and phosphorous (Salomons and
Gerritse, 1980) in surface layers of estuarine sediments. This
enrichment causes a redistribution of the trace metals in the estuarine circulation pattern. The erosion of the estuarine sediment
surface layer and landward transport may enable an increased adsorp-

tion onto freshly precipitated manganese hydroxides (Salomons, 1980). Furthermore, if they are flushed out of the estuarine circulation pattern, their trace metals will contribute to the flux into the oceanic sediments (Sundby et al., 1981). It is these differentiations under changing redox conditions which determine the relative accumulation of trace metals in Fe/Mn concretions in both marine and freshwaters. In deep-sea nodules, Mn, Co, Mo and Tl are concentrated more than one hundred-fold relative to their crustal abundances (Ni, Ag, Ir and Pb are concentrated from fifty- to one hundred-fold while Cu, Zn, Cd, W and Bi are concentrated ten- to fifty-fold) (Cronan, 1976).

Practical evidence of specific sorption effects of trace elements on hydrous oxides has been provided by the study of freshwater Fe/Mn-concretions (Callender and Bowser, 1976). The majority of the nodule material shows concentric bands of alternating Fe-rich (lighter) and Mn-rich (darker) laminations; minor transition metal cations such as copper, cobalt, nickel, and zinc are associated with the manganese phase, whereas elements with aqueous anionic charge such as silica, phosphate and arsenic are related to the iron lamina.

Adsorption of cadmium, zinc, copper, lead, silver, arsenic, selenium, and chromium onto amorphous iron oxyhydroxides and γ-alumina was studied by Leckie and coworkers (1980) experimentally as a function of adsorbent and adsorbate concentrations, solution composition, and pH. Fractional adsorption of dissolved metal increases abruptly in a narrow pH range; the pH of adsorption increases in the order Pb< Cu< Zn< Cd< Ag and in the order SeO_4<CrO_4< SeO_3< AsO_4 under otherwise identical conditions (similar to the adsorption "edges" of metals on amorphous silica as shown in Figure 2b). Competition experiments indicate that, in many cases, different metal ions preferentially adsorb to different groups of sites. Therefore, the adsorption of one metal ion often may have only a small effect on the adsorption of a second ion (Leckie et al., 1980). These experiments also have shown that ligands which form dissolved complexes with metal ions can either increase or decrease metal adsorption. Chloride, sulfate, and ammonia generally decrease fractional adsorption of cadmium at a given pH; on the other hand, cadmium thiosulfate complexes adsorb more strongly on oxides with sufficiently positive surface charge (e.g., FeOOH,γ-Al_2O_3 in Figure 2a) than do free aquo complexes. Trace elements, in both anionic and cationic form, can potentially be removed from very dilute solutions utilizing either a preformed solid, $Fe(OH)_3$, or by precipitating the solid in situ (Leckie et al., 1980).

The adsorption of metal ions on manganese hydroxides has been studied in detail (Murray et al., 1968; Murray, 1975). The affinity of metals for such a surface follows the order Mg<Ca<Sr<Ba<Ni<Zn<Mn< Co. The adsorption isotherm of cobalt was used to explain the enrichment of cobalt in suspended matter from the Black Sea (Murray, 1975). Nickel, copper and especially cobalt exhibit marked specific adsorption as evidenced by the finite adsorption of these ions at the zero point of charge (Murray et al., 1968). The pH of onset of adsorption of metals on iron hydroxides in artificial estuarine solutions follows the order Cu<Zn<Mn and it occurs at pH levels well below the hydrolysis of the cations in the bulk solution. The adsorption isotherms for copper are independent of salinity, whereas those for Zn and Mn show an increase in the pH of the adsorption edge with increase in salinity. It is suggested that the major cations, calcium and magnesium, are probably co-adsorbed and competition from these species for adsorption sites increases with increasing salinity (Millward and Moore, 1981). Recent experiments by Laxen and Sholkovitz (1981) under natural conditions show that significant adsorption of the metals Cd, Cu and Pb (but not Ni and Ca) is possible onto hydrous ferric material formed in a natural water. However, not all of the metal is available for adsorption, possibly as a result of a competitive complexation with humic acid (see section on competition for sorption sites and ligands).

Metal Sorption on Organic Substances

Organic surfaces for metal sorption could arise in three possible ways (see review by Hart, 1982):

(1) from organisms such as bacteria and algae;
(2) by the breakdown of plant and animal material and by the condensation of lower-molecular-weight organics (this material would characteristically have a high molecular weight and would possess many of the properties of polyelectrolytes and colloidal materials);
(3) by lower-molecular-weight organic matter being sorbed onto clay or metal oxide substrates (Davis, 1980; Davis and Gloor, 1981; Tipping, 1981).

Although the difference between these three surface types is not well understood, with respect to trace metal uptake, there is general agreement that at least one major binding mechanism involves salicylic entities; other stronger binding entities (such as peptides) may also be present in some systems (Hart, 1981). Data given by Hunter (1980) and by Tipping (1981) suggest that at least part of the organics adsorbed onto the particulate matter in natural waters has carboxylic acid and phenolic functional groups available for binding with traces.

The hydrogen ion concentration, or pH, is probably the single most important factor influencing metal adsorption onto both inorganic and organic surfaces. Experiments performed by Nelson et al. (1981) indicate that the effect of pH on the adsorption of copper and cadmium by _bacteria_ contrasts with the pH adsorption "edges" for these metals on inorganic surfaces. Whereas, on oxide surfaces heavy metal adsorption increases from near zero to near 100% over a narrow pH range of about two units (see Figure 2b), the highly attenuated pH adsorption "edge" for bacteria in Figure 4 may represent the greater

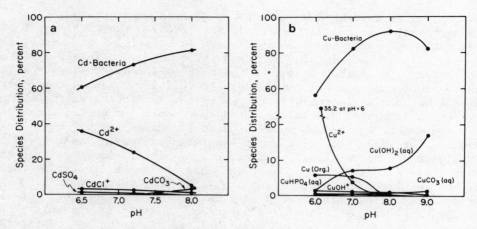

Figure 4. Metal adsorption on bacteria (Nelson et al., 1981). Speciation-distribution diagram for cadmium (4a) and copper (4b) as a function of pH.

range over which protonation of surface (and, to a much lesser extent, probably intercellular) functional groups occurs (Nelson et al., 1981). The adsorption maximum of copper in the vicinity of pH 8 (Figure 4b) appears to be unique to the biological system; copper speciation above pH 7 is strongly influenced by formation of the soluble $Cu(OH)_2$ (aq.) complex which becomes the dominant soluble species above neutral pH values. The data given by Nelson et al. (1981) suggest that the equilibrium distribution of metals between bacterial solids and solutes is mediated primarily by physical-chemical factors and is not influenced greatly by active biological transport processes.

Since the bulk of soluble organic matter has characteristics similar to those of fulvic acid, it is not surprising that a number of studies have reported conditional formation constants (^+K) for trace metal complexes (in natural waters) closely parallelling those for trace metal fulvates, both in terms of absolute values and pH dependency (Hart, 1981, 1982). The trace metal adsorption capacity of organic matter is generally between that for metal oxides and clays (Guy and Chakrabarti, 1976).

The sorption of a mixture of eleven metals on humic acid was studied by Kerndorf and Schnitzer (1980); it was not possible to find correlations between the affinities of these metals to sorb on humic acid and their atomic weights, atomic numbers, valencies and crystal and hydrated ionic radii. However, it was possible to describe the adsorption with an empirical formula.

Competition for Sorption Sites and Ligands

The sorption of metals on sediments is complicated because the various organic and inorganic ligands in solution and the various components in the sediments all compete for complexation of the trace metals. With regard to competing ligands in solution, two situations can be envisaged (Davis and Leckie, 1978) as follows.

(1) The complexing ligand (for instance, chloride) is not adsorbed onto the sedimentary components, and thus it competes with surfaces for coordination of the metal ion. Increased concentration of the ligand causes a decrease in adsorption as exemplified by the chlorinity increase in estuaries (Figure 5). It should be noted, however, that the concentration of the suspended matter may also increase (turbidity maximum, erosion of sediments) causing an increase in adsorption despite a rise in chlorinity.

(2) The complexing ligand is adsorbed onto surfaces. If this is the case, the adsorption of metal ions can be significantly enhanced. This appears to be the case for EDTA (Davis and Leckie, 1978) and NTA (Elliott and Huang, 1979). In this way, the speciation of trace metals in solution determines the particulate metal speciation.

Apart from competition between surfaces and ligands in solution, the various phases in the sediment (internal factors) also compete for complexation of the trace metals. Oakley et al. (1981) studied, in model systems, the competition between bentonite clay, iron and manganese hydroxides and humic acids for complexation of metals. Their model predicted that the clay fraction is a major sink for copper and cadmium in the artificial system used.

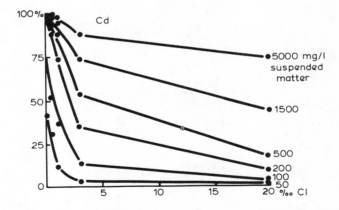

Figure 5. Adsorption of cadmium (5 µg/l) on sediment from the Rhine River: influence of chlorinity and suspended matter concentration (from Salomons, 1980).

Characteristic differences in solid/solution interactions of trace elements were determined in experiments on sand filter columns which were fed with river water containing added doses of humic acids (6 mg/l DOC) and dissolved metals (1 mg Cu/l + 10 mg Zn/l, respectively; Förstner et al., 1979). After 10 hours, the copper concentrations in the filter outlet had increased significantly, whereas no increase in the zinc concentrations was observed over the whole 24-hour measuring period (Figure 6). Anodic stripping voltammetry determinations showed that zinc in the inflowing solution is basically labile-bound. A nearly total elimination of zinc during passage through the column (Table 1) can be explained as a result of an active interaction with the Fe/Mn hydroxides of the filter particles (which were analyzed by sequential chemical leaching). The copper component, on the other hand, is in a more stable association with the complexing (organic) ligands, and remains partially unaffected by the physicochemical interactions with the filter material.

In the natural environment, the individual sedimentary components may act as nucleation centers for the deposition of iron and manganese hydroxides (Aston and Chester, 1973), and humic materials may

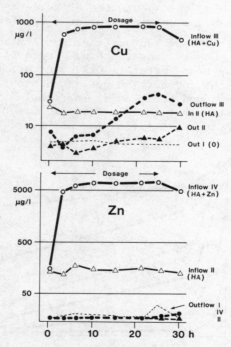

Figure 6. Sand filter column experiments with Ruhr River Water (Column I) and humic acid (II), copper (III) and Zinc (IV) additives; concentrations of copper and zinc in the inflowing and outflowing waters (Forstner et al., 1979).

Table 1. Filter column experiments - anodic stripping voltammetry data (conditions after 30 hours; see Figure 5). Förstner et al. (1979).

	Humic acid/Cu-dosage			Humic acid/Zn-dosage		
	total µg/l	% labile[a]	% bound[b]	total µg/l	% labile	% bound
Inflowing water	400	2%	98%	3500	98%	2%
Outflowing water	29	20%	80%	13	60%	40%

[a] labile metal measured at pH 4.7, acetate buffer
[b] bound = total - labile

adsorb on clay minerals or iron hydroxides (Tipping, 1981). There is considerable evidence that the surface characteristics of particles, immersed in natural seawater and estuarine waters, are lost and that new surface properties are acquired which are consistent with the formation of a macromolecular organic film (Neihof and Loeb, 1974; Loeb and Neihof, 1977; Hunter and Liss, 1979). In this way, the sorption properties of sedimentary components are changed, making it difficult to apply laboratory results to actual conditions in nature unless the type and properties of the surface film are known (see reviews by Hart, 1981, 1982).

Recently, Balistrieri et al. (1981) combined field observations on the trace metal scavenging activity of sinking particulate matter in the marine environment with theoretical surface chemistry. The equilibrium constants obtained were compared with those that define metal interactions in laboratory systems on typical metal oxides and organic compounds (Figure 7a, b). The comparison indicated that the metal/particulate interactions closely resemble the interactions between organic compounds and metals. The authors propose, therefore, that scavenging is the result of metal interactions with organic coatings present on the oceanic particles. The data thus far obtained seem to suggest that organic-metal interactions play an important role in the initial distribution of a trace metal between the solid and the soluble phase, either through competition or enhanced adsorption of organic-metal species or through films of organic matter on sedimentary particles. Subsequent diagenetic effects (aging of metastable phases, decay of organic matter, dissolution of minerals), however, may cause a redistribution of trace metals.

Diagenetic Reactions Involving Metal-Rich Particulates

Decomposition of organic matter, which is mediated by microorganisms ("catabolic" processes), generally follows a definite succession in sediments depending upon the nature of the oxidizing

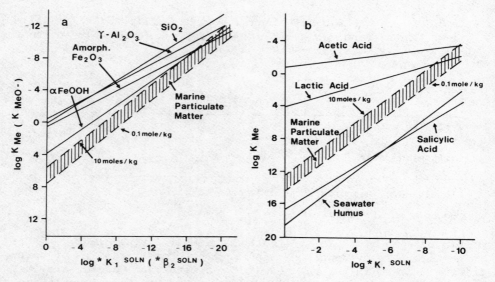

Figure 7. A comparison of the equilibrium constants defining metal-marine particle interactions with some reference metal oxide compounds (7a) and some reference organic compounds (7b). From Balistrieri et al. (1981).

agent: the successive events are oxygen consumption (respiration), nitrate reduction, sulfate reduction, and methane formation. The environments corresponding to these processes are, respectively: "oxic", "post-oxic", "sulfidic", and "methanic" (Berner, 1980). For the recognition of early diagenetic environments (Table 2; Berner, (1981), MnO_2-type minerals constitute a good indicator of truly oxic conditions. Post-oxic (weakly reducing) conditions, involving successively the reduction of nitrate, manganese and iron, are characterized by the presence of Fe^{2+}-Fe^{3+}-silicates such as glauconite, the absence of sulfide minerals and only minor contents of organic matter (authigenic Fe^{2+}-Fe^{3+}-silicates, siderite and vivianite are stable neither in the oxic nor in the sulfidic environment). Sulfidic and methanic (both strongly reducing) environments typically contain significant amounts of organic matter, in addition to either newly formed or pre-existing (from former diagenetic processes) sulfidic iron minerals such as pyrite, mackinawite and greigite. Methane, vivianite and siderite form more readily in non-marine sediments than in marine deposits (Emerson and Widmer, 1978), due to an initially lower sulfate concentration in the water.

The effects of these transformations on the distribution of trace metals can be evaluated both from pore water analyses and from mineral concentrates of typical diagenetic phases. Froelich et al. (1979) have shown that, as dissolved O_2 reaches zero with depth,

Table 2. Geochemical classifications of sedimentary (diagenetic) environments (Berner, 1981). "C" refers to concentrations in moles per liter. Total sulfide is represented by H_2S.

Environment	Characteristic phases
I. Oxic ($C_{O_2} = 10^{-6}$)	hematite (Fe_2O_3); goethite (FeOOH); MnO_2-type minerals, no organic matter.
II. Anoxic ($C_{O_2} \leq 10^{-6}$)	
A. Sulfidic ($C_{H_2S} \geq 10^{-6}$)	pyrite, marcasite (FeS_2); rhodochrosite ($MnCO_3$); alabandite (MnS); organic matter.
B. Nonsulfidic ($C_{H_2S} \leq 10^{-6}$)	
1. Post-oxic	glauconite and other Fe^{+2}-Fe^{+3}-silicates (also siderite, vivianite, rhodochrosite); no sulfide minerals; minor organic matter.
2. Methanic	siderite ($FeCO_3$); vivianite ($Fe_3(PO_4)_2 \cdot 8H_2O$); rhodochrosite; earlier formed sulfide minerals; organic matter.

dissolved Mn^{2+} from MnO_2 reduction simultaneously begins to build up in concentration. Rhodochrosite solubility controls the concentration of the Mn^{2+} of pore fluids in both lacustrine (Callender et al., 1974) and marine (Holdren et al., 1975) sediments. Later, at greater sediment depth, the process of iron reduction to dissolved Fe^{2+} begins, thus indicating the persistence of iron oxides, but not manganese oxides, under anoxic conditions (Berner, 1981). Under some lower redox conditions, $Fe(OH)_3$ and siderite might be expected to coexist (Jones and Bowser, 1978). Due to the scarcity of organic matter in these oxic environments, the phase equilibria of trace metals are mainly affected by pH-conditions and inorganic water constituents. The build-up of ammonia and DOC (Figure 8) in pore waters may lead to significant complexing of Cu^{2+} and consequently to increased copper

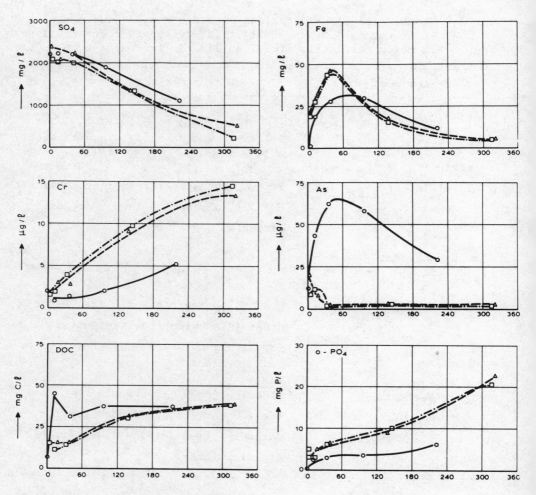

Figure 8. Changes in the composition of pore waters in original freshwater (□), brackish water (Δ) and marine (o) sediments dumped into a marine environment over a 360 day period (Kerdijk and Salomons, 1982).

levels (Leckie and Davis, 1979). In stronger reducing environments, the trace metal distribution should be controlled primarily by the formation of (highly insoluble) sulfides; at lower H_2S concentrations, it should be controlled by the stability of carbonate and phosphate phases. However, metal complexation by organic ligands in pore fluids effectively competes with inorganic water constituents. Significant enrichment of trace metals in pore waters of sediment samples from southern California basins, Saanich Inlet (British Columbia) and Loch Fyne (Scotland) has been explained by complexation effects with organic substances (Brooks et al., 1968; Presley et al., 1972; Duchart et al., 1973). Recent water analyses on the Woronora River of New South Wales (Australia) by Batley and Giles (1980), including speciation studies, indicate an increased solubility of "bound" species of lead and copper in the anoxic interstitial water (mangrove sediment) compared with oxic surface and interstitial waters; this should be the result of organic complexation, whereas the increase in "labile" metal species is most probably associated with sulfide species (Table 3).

The extent of the influence of organic complexes on metal solubilization depends on a competition with hydrogen sulfide, which is simultaneously produced in the reduced part of the sediment (compare graphs in Figure 8 - sulfate decrease and DOC increase). It can be expected that most of the H_2S is precipitated as sulfides (compare the simultaneous sulfate decrease and iron decrease in Figure 8). If the metal-chelating compounds migrate upward, they will mobilize the metals in the oxidizing uppermost layer of the sediment. In such a dynamic system, some of the metals are chelated and complexed before they meet the sulfidic environment (Hallberg, 1978).

The overall effects of early diagenetic mineral formation on the relative enrichment of heavy metals in recent lake deposits can be seen from examples in Table 4 (adapted from Förstner, 1981); these data, however, should not be generalized to cover all cases, since

Table 3. Trace metal species in interstitial and surface waters from the Woronora River, New South Wales, Australia (Batley and Giles, 1980).

	Cadmium		Lead		Copper	
	Labile[a] µg/l	Bound[b] µg/l	Labile µg/l	Bound µg/l	Labile µg/l	Bound µg/l
Surface water	0.11	0.09	0.28	0.32	0.74	1.15
Interstitial water I 0-8 cm (sand sediment)	0.15	0.10	0.28	0.30	0.68	1.05
Interstitial water II (mangrove sediment)	0.29	0.11	0.50	0.60	1.35	1.80

[a] labile metal measured at pH 4.7, acetate buffer

[b] bound = total minus labile

Table 4. Effects of early diagenetic mineral phases on the relative enrichment (compared to bulk sediment) of heavy metals in recent lake deposits (after Förstner, 1981).

Mineral phase	Lake example(s)	Effect (examples)
Carbonates		
Rhodochrosite	Lake Michigan[1]	+ to ++ heavy metals
Siderite	Black Sea[2], Birket Ram (Israel)[3]	+ Mn, $\pm$ heavy metals
Sulfides		
Hydrotroilite	Lake Constance[4]	+ Ni, Cu, Cr; $\pm$ Pb, Zn
Pyrite	Black Sea[5]	++ Cu, Ni, Co; $\pm$ Pb, Zn, Mn
	Lake Kivu[6]	++ Ni
Phosphates	Lake Geneva[7]	+ Fe, Co, Cu, Zn, Pb (?)
Silicates		
Nontronite	Lake Malawi[4]	+ Co, Mn; $\pm$ Zn, Pb, Cu

[1]Callender et al. (1974); [2]Stoffers and Müller (1978); [3]Singer and Navrot (1978); [4]Förstner (1981); [5]Volkov and Fomina (1974); [6]Degens et al. (1972); [7]Jaquet (in press)

the separation of phases is often not satisfactory. Incorporation into pyrite minerals seems to provide a distinct mechanism of enrichment of transition metals (note examples from Black Sea sediments). Studies of coarse-grained aggregates of unstable iron sulfide of the hydrotroilite type (FeS·H_2O), separated by sieving from the fine-grained matrix taken from Lake Constance surface sediments (Förstner, 1982), indicate an approximate six-fold enrichment of iron in the concentrate compared with the composition of the surrounding sediment. These studies also showed that, when compared to the sediment matrix, Cu, Co, and Cr were two times greater and Ni was 1.5 times greater; on the other hand, Pb, Zn and Mn did not undergo a significant increase. Generally, the accumulation of metals in the early diagenetic mineral phase listed here seems to be distinctly smaller than in Fe/Mn oxyhydrates.

The early diagenetic reactions have been studied experimentally in the field (Kerdijk and Salomons, 1982). Three large pits (80 x 30 x 6m) were dug in the Rhine estuary below the water table and were filled with freshwater (Waalhaven), brackish water (Botlek) and marine (Maasmond) sediment. The consolidation rate and the changes in pore water composition were studied over a one-year period (Figure 8). The changes in the concentrations with time can be divided into three different categories:

(1) the concentrations show a decrease (this is the case for sulfate in all cases, nickel in the sediments from the Maasmond, and arsenic in the Botlek and Waalhaven sediments);
(2) the concentrations show a continuous increase (this is the case for O-PO$_4$, NH$_4$, silicon, Cr, BOD$_5$ and DOC);
(3) the concentrations show a maximum and then afterwards decrease (this is the case for iron and manganese in all cases, copper in sediments from the Botlek and Waalhaven sediments, and cadmium in the Maasmond sediment).

A rapid increase in iron and manganese concentrations is observed in the first few weeks (Figure 8e); during this period no sulfate reduction takes place as is shown by the constant levels of sulfate. The electron acceptors for the degradation of the organic matter are apparently oxygen and nitrate. The low oxygen levels cause a reduction of iron and manganese and a subsequent appearence of Fe^{2+} and Mn^{2+} in the pore waters. Sulfate concentrations start to decrease after about 2 months. The decrease is most rapid in the sediments with the highest organic matter contents (Botlek and Waalhaven) but is slower in the Maasmond sediment. With the decrease in sulfate concentrations, and the production of sulfide, the iron and manganese levels start to decrease, possibly as the result of the formation of sulfidic minerals. Silicon and ortho-phosphorous concentrations show a continuous increase in their concentrations. The phosphorous release probably takes place as a consequence of the breakdown of the organic matter; about 10-20% of the total phosphorous in the sediments is organically bound (Salomons and Gerritse, 1981). The behaviour of arsenic differs strongly between the three sediment types. In the Botlek and Waalhaven sediment the concentrations decrease continuously, whereas in Maasmond sediments, the same type of behaviour as for iron is found. Chromium shows a continuous increase in its concentrations; this is probably due to its occurrence in the organic fraction of sediments (Salomons and Förstner, 1980). Furthermore, no sulfides of chromium appear to be stable under aquatic conditions.

SUMMARY AND SUGGESTIONS FOR FURTHER RESEARCH

Particulates play a most important role in the distribution of trace metals between the dissolved and solid phases because they offer sites for adsorption. The interactions are complicated because the various sedimentary surfaces and the inorganic and organic ligands in solution all compete for complexation of the metals (Figure 9). Some complex interactions are as follows.

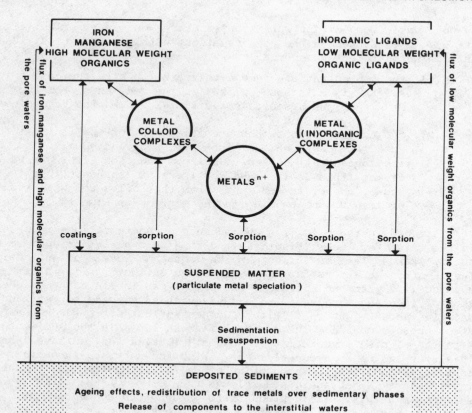

Figure 9. Summary of major processes and mechanisms in the interactions between dissolved and solid metal species in surface waters.

(1) The speciation of the metals in solution may enhance adsorption if the ligands themselves are adsorbed onto mineral surfaces, or, the speciation may prevent adsorption if the complexing ligands form strong complexes (with metals) which are not adsorbed onto mineral surfaces.

(2) The speciation of the solid phases, particularly with respect to the nature of their surfaces, determines the extent of the sorption, although mineral particles themselves are able to adsorb trace metals. The particles are often covered with hydrous iron and manganese oxides and/or organic films which have a great ability to remove trace metals from solutions. The nature of these coatings is still subject to investigations; for a complete under-

standing of the metal behaviour in aquatic systems more studies are needed.
(3) Aging effects and other diagenetic processes taking place after deposition cause a redistribution of trace metals over the various components of the sediments; the available data suggest that these processes cause a stronger binding to the sedimentary particles. However, for a better understanding more research is needed on these kinetic changes and redistributions.
(4) Sediments indirectly influence the speciation of dissolved metals because they are a substrate for biogeochemical transformations which result in the formation of methylated species of some trace metals. Additionally, diagenetic changes cause an enrichment of dissolved organic matter (DOM) in the interstitial waters. The DOM undoubtedly influences the speciation of the interstitial trace metals and, after a subsequent release of the DOM to overlying waters (by bioturbation, compaction and erosion), it also influences the speciation of metals there.

In particular, an understanding of the rapid fluctuation conditions in estuaries with regard to effects on trace metal behaviour and the role of the deposited sediments requires an integrated approach using both laboratory studies and biogeochemical field studies. A more detailed understanding of these processes is necessary to deal adequately with the effects of disposal of both non-radioactive and radioactive wastes in the aquatic environment.

ACKNOWLEDGEMENTS

We would like to thank D. Godfrey who helped prepare the English version of the manuscript. The support of NATO for travel and accomodation during the workshop, "Trace Element Speciation in Surface Waters and its Ecological Implications", is highly appreciated.

REFERENCES

Aston, S.R., and Chester, R., 1973, The influence of suspended particles on the precipitation of iron in natural water, Estuarine Coastal Mar. Sci., 1:225.
Balistrieri, L., Brewer, P.G., and Murray, J.W., 1981, Scavenging residence times of trace metals and surface chemistry of sinking particles in the deep ocean, Deep-Sea Res., 28A:101.
Batley, G.E., and Giles, M.S., 1980, A solvent displacement technique for the separation of sediment interstitial waters, in: "Contaminants and Sediments, Vol. 2", R.A. Baker, ed., Ann Arbor Science Publ., Ann Arbor, Michigan.
Berner, R.A., 1980, "Early Diagenesis: A Theoretical Approach", Princeton University Press, Princeton, N.J.

Berner, R.A., 1981, A new geochemical classification of sedimentary environments, J. Sedim. Petrol., 51:359.

Brooks, R.R., Presley, B.J., and Kaplan, I.R., 1968, Trace elements in the interstitial waters of marine sediments, Geochim. Cosmochim. Acta, 32:397.

Callender, E., and Bowser, C.J., 1976, Freshwater ferromanganese deposits, in: "Handbook of Strata-Bound and Stratiform Ore Deposits, Vol. 7", K.H. Wolf, ed., Elsevier, New York.

Callender, E., Bowser, C.J., and Rossmann, R., 1974, Geochemistry of ferromanganese and manganese carbonate crusts from Green Bay, Lake Michigan, Trans. Am. Geophys. Un., 54:340.

Cronan, D.S., 1976, Manganese nodules and other ferromanganese oxide deposits, in: "Chemical Oceanography, Vol. 5", J.P. Riley and R. Chester, eds., Academic Press, London.

Davis, J.A., 1980, Adsorption of natural organic matter from fresh water environments by aluminum oxide, in: "Contaminants and Sediments, Vol. 2", R.A. Baker, ed., Ann Arbor Science Publ. Ann Arbor, Mich.

Davis, J.A., and Gloor, R., 1981, Adsorption of dissolved organics in lake water by aluminum oxide: Effect of molecular weight, Environ. Sci. Technol., 15:1223.

Davis, J.A., and Leckie, J.O., 1978, Effect of adsorbed complexing ligands on trace metal uptake by hydrous oxides, Environ. Sci. Technol., 12:1309.

Davis, J.A., and Leckie, J.O., 1979, Speciation of adsorbed ions at the oxide/water interface, in: "Chemical Modeling in Aqueous Systems", E.A. Jenne, ed., Am. Chem. Soc. Symp. Ser. 93, Washington, D.C.

Davison, W., 1981a, Supply of iron and manganese to an anoxic lake basin, Nature, 290:241.

Davison, W., 1981b, Transport of iron and manganese in relation to the shapes of their concentration-depth profiles, in: "Proc. 2nd Int. Symp. Interactions between Sediments and Freshwater", P. Sly, ed., Junk Publ., The Hague, in press.

Degens, E.T., Okada, H., Honjo, S., and Hathaway, J.C., 1972, Microcrystalline sphalerite in resin globules suspended in Lake Kivu, East Africa, Miner. Deposita. 7:1.

Duchart, P., Calvert, S.E., and Price, N.B., 1973, Distribution of trace metals in the pore waters of shallow water marine sediments, Limnol. Oceanogr., 18:605.

Dugger, D.L., Stanton, J.H., Irby, B.N., McConnel, B.L., Cunnings, W.W., and Maatman, R.W., 1964, The exchange of twenty metal ions with the weakly acidic silanol group of silica gel, J. Phys. Chem., 68:757.

Elliott, H.A., and Huang, C.P., 1979, The adsorption characteristics of Cu(II) in the presence of chelating agents, J. Colloid & Interface Sci., 70:29.

Emerson, S., and Widmer, G., 1978, Early diagenesis in anaerobic lake sediments. II. Thermodynamic and kinetic factors controlling the formation of iron phosphate, Geochim. Cosmochim. Acta, 42:1307.

Förstner, U., 1981, Recent heavy metal accumulations in limnic sediments, in: "Handbook of Strata-Bound and Stratiform Ore Deposits, Vol. 9", K.H. Wolf, ed., Elsevier, Amsterdam.

Förstner, U., 1982, Accumulation phases for heavy metals in limnic sediments, in: "Proc. 2nd Int. Symp. Interactions between Sediments and Freshwater", P. Sly, ed., Junk Publ., The Hague, in press.

Förstner, U., Nähle, C., and Schöttler, U., 1979, Sorption of heavy metals in sand filters in the presence of humic acids. Summaries of papers of the International Symposium on Artificial Groundwater Recharge, Dortmund (F.R.G.), May 14-18, 1979, Institute of Water Research, Dortmund, 61/1-3.

Froelich, P.N., Klinkhammer, G.P., Bender, M.L., Luedtke, N.A., Heath, G.R., Cullen, D., Dauphin, P., Hammond, D., Hartmann, B., and Maynard, V., 1979, Early oxidation of organic matter in pelagic sediments of the eastern equatorial Atlantic: suboxic diagenesis, Geochim. Cosmochim. Acta, 43:1075.

Guy, R.D., Chakrabarti, C.L., 1976, Studies of metal-organic interactions in model systems pertaining to natural waters, Can. J. Chem., 54:1600.

Hallberg, R., 1978, Metal-organic interaction at the redoxcline, in: "Environmental Biogeochemistry and Geomicrobiology, Vol. 3", W.E. Krumbein, ed., Ann Arbor Science Publ., Ann Arbor, Michigan.

Hart, B.T., 1981, Trace metal complexing capacity of natural waters: A review, Environ. Technol. Lett., 2:95.

Hart, B.T., 1982, Uptake of trace metals by sediments and suspended particulates: A review, in: "Proc. 2nd Int. Symp. Interactions between Sediments and Freshwater", P. Sly, ed., Junk Publ., The Hague, in press.

Hem, J.D., 1972, Chemistry and occurrence of cadmium and zinc in surface water and groundwater, Water Resour. Res., 8:661.

Holdren, G.R., Bricker, O.P., and Matisoff, G., 1975, A model for the control of dissolved manganese in the interstitial waters of the Chesapeake Bay, in: "Marine Chemistry in the Coastal Environment", T.M. Church, ed., Am. Chem. Soc. Symp. Ser. 18, p. 364.

Hunter, K.A., 1980, Microelectrophoretic properties of natural surface-active organic matter in coastal seawater, Limnol. Oceanogr., 25:807.

Hunter, K.A., and Liss, P.S., 1979, The surface charge of suspended particles in estuarine and coastal waters, Nature, 282:823.

Jenne, E.A., 1976, Trace element sorption by sediments and soils - sites and processes, in: "Symposium on Molybdenum, Vol. 2", W. Chappell and K. Petersen, eds., Marcel Dekker, New York.

Jenne, E.A., ed., 1979, "Chemical Modeling in Aqueous Systems - Speciation, Sorption, Solubility and Kinetics", ACS Symposium Series 93, American Chemical Society, Washington, D.C.

Jonasson, I.R., 1977, Geochemistry of sediment/water interactions of metals, including observations on availability, in: "The Fluvial Transport of Sediment-Associated Nutrients and Contaminants", H. Shear and A.E.P. Watson, eds., IJC/PLUARG. Publ., Windsor, Ontario.

Jones, B.F., and Bowser, C.J., 1978, The mineralogy and related chemistry of lake sediments, in: "Lakes - Chemistry, Geology, Physics", A. Lerman, ed., Springer-Verlag, New York.

Kerdijk, H.N., and Salomons, W., 1982, Importance of early diagenetic reactions for the dumping of dredged material, in preparation.

Kerndorf, H., and Schnitzer, M., 1980, Sorption of metals on humic acid, Geochim. Cosmochim. Acta, 44:1701.

Laxen, D.P.H., and Sholkovitz, E.R., 1981, Adsorption (co-precipitation) of trace metals at natural concentrations on hydrous ferric oxide in lake water samples, Environ. Technol.Lett.,2:561.

Leckie, J.O., and Davis, J.A., 1979, Aqueous environmental chemistry of copper, in: "Copper in the Environment. Part I: Ecological Cycling", J.O. Nriagu, ed., Wiley, New York.

Leckie, J.O., Benjamin, M.M., Hayes, K., Kaufman, G., and Altmann, S., 1980, Adsorption/coprecipitation of trace elements from water with iron oxyhydroxides, Final Report. Research Project 910-1/CS-1513. Prepared for Electric Power Research Institute, Palo Alto, California.

Lee, G.F., 1975, Role of hydrous metal oxides in the transport of heavy metals in the environment, in: "Heavy Metals in the Aquatic Environment", P.A. Krenkel, ed., Pergamon Press, Oxford.

Lerman, A., 1977, Migrational processes and chemical reactions in interstitial waters, in: "The Sea, Vol. 6", E.D. Goldberg, I.N. McCave, J.I. O'Brien and J.H. Steele, eds., Wiley, New York.

Lieser, K.H., 1975, Sorption mechanisms, in: "Sorption and Filtration Methods for Gas and Water Purification", M. Bonnevie-Svendsen, ed., NATO Advanced Study Institute, Series E, Vol. 13.

Loeb, G.I., and Neihof, R.A., 1977, Adsorption of an organic film at the platinum-seawater interface, J. Mar. Res., 35:283.

Lu, C.S.J., and Chen, K.Y., 1977, Migration of trace metals in interfaces of seawater and polluted surficial sediments, Environ. Sci. Technol., 11:174.

MacCarthy, P., and Smith, G.C., 1979, Stability surface concept. A quantitative model for complexation in multiligand mixtures, in: "Chemical Modeling in Aqueous Systems", E.A. Jenne, ed., Am. Chem. Soc. Symp. Ser. 93, Washington, D.C.

Millward, G.E., and Moore, R.M., 1982, The adsorption of Cu, Mn and Zn by iron oxyhydroxide in model estuarine solution, Water Res., 16:901.

Morel, F., McDuff, R.E., and Morgan, J.J., 1973, Interactions and chemostasis in aquatic chemical systems: role of pH, pE, solubility and complexation, in: "Trace Metals and Metal-Organic Interactions in Natural Waters", P.C. Singer, ed., Ann Arbor Science Publ., Ann Arbor, Michigan.

Murray, D.J., Healy, T.W., and Fuerstenau, D.W., 1968, The adsorption of aqueous metal on colloidal hydrous manganese oxide, Am. Chem. Soc. Adv. Chem. Ser. 79, p. 74.

Murray, J.W., 1975, The interaction of metal ions at the manganese dioxide-solution interface, Geochim. Cosmochim. Acta, 39:505.

Neihoff, R.A., and Loeb, G.I., 1974, Dissolved organic matter in sea water and the electric charge of immersed surfaces, J. Mar. Res., 32:5.

Nelson, P.O., Chung, A.K., and Hudson, M.C., 1981, Factors affecting the fate of heavy metals in the activated sludge process, J. Wat. Pollut. Control Fed., 53:1323.

Nordstrom, D.K., and eighteen co-authors, 1979, A comparison of computerised chemical models for equilibrium calculations in aqueous systems, in: "Chemical Modeling in Aqueous Systems", E.A. Jenne, ed., Am. Chem. Soc. Symp. Ser. 93, Washington, D.C.

Oakley, S.M., Nelson, P.O., and Williamson, K.J., 1981, Model of trace-metal partitioning on marine sediment, Environ. Sci. Technol., 15:474.

Perdue, E.M., 1979, Solution thermochemistry of humic substances, in: "Chemical Modeling in Aqueous Systems", E.A. Jenne, ed., Am. Chem. Soc. Symp. Ser. 93, Washington, D.C.

Presley, B.J., Kolodny, Y., Nissenbaum, A., and Kaplan, I.R., 1972, Early diagenesis in a reducing fjord, Saanich Inlet, British Columbia. II. Trace element distribution in interstitial water and sediment, Geochim. Cosmochim. Acta, 36:1073.

Rose, A.W., Hawkes, H.E., and Webb, J.S., 1979, "Geochemistry in Mineral Exploration", Academic Press, London.

Salomons, W., 1980, Adsorption processes and hydrodynamic conditions in estuaries, Environ. Technol. Lett., 1:356.

Salomons, W., and Förstner, U., 1980, Trace metal analysis on polluted sediments. Part II: Evaluation of environmental impact, Environ. Technol. Lett., 1:506.

Salomons, W., and Gerritse, R.G., 1981, Some observations on the occurrence of phosphorous in recent sediments from western Europe, Sci. Total Environm., 17:37.

Salomons, W., and Van Pagee, J.A., 1981, Prediction of NTA levels in river systems and their effect on metal concentrations, in: "Proc. Int. Conf. Heavy Metals in the Environment", Amsterdam, in press.

Schindler, P.W., Fürst, B., Dick, R., and Wolf, P.U., 1976, Ligand properties of surface silanol groups. Surface complex formation with Fe^{3+}, Cu^{2+}, Cd^{2+}, Pb^{2+}, J. Colloid & Interface Sci., 55:469.

Sholkovitz, E.R., and Copland, D., 1981, The coagulation, solubility and adsorption properties of Fe, Mn, Cu, Ni, Cd, Co and humic acids in a river water, Geochim. Cosmochim. Acta, 45:181.

Sholkovitz, E.R., Boyle, E.A., and Price, N.B., 1978, The removal of dissolved humic acids and iron during estuarine mixing, Earth Planet. Sci. Lett., 40:130.

Singer, A., and Navrot, J., 1978, Siderite in Birket Ram Lake sediments. Abstr. 10th Int. Congr. Sedimentology, July 9-14, 1978, Jerusalem, p. 616.

Stoffers, P., and Müller, G., 1978, Mineralogy and lithofacies of Black Sea sediments - Leg 42B Deep Sea Drilling Project, in: Initial Reports of the Deep-Sea Drilling Project, 42 (2), by D.A. Ross, Y.P. Neprochnov et al., U.S. Govt. Printing Office, Washington, D.C., p. 373.

Stumm, W., and Morgan, J.J., 1981, "Aquatic Chemistry, 2nd Edition", Wiley, New York.

Stumm, W., Hohl, H., and Dalang, F., 1976, Interaction of metal ions with hydrous oxide surfaces, Croat. Chem. Acta, 48:491.

Stumm, W., Kummert, R., and Sigg, L., 1980, A ligand exchange model for the adsorption of inorganic and organic ligands at hydrous oxide interfaces, Croat. Chem. Acta, 53:291.

Sundby, B.M., Silverberg, N., and Chesselet, R., 1981, Pathways of manganese in an open estuarine system, Geochim. Cosmochim. Acta, 45:293.

Theis, T.L., and Singer, P.C., 1974, Complexation of iron(II) by organic matter and its effect on iron(II) oxygenation, Environ. Sci. Technol., 8:569.

Tipping, E., 1981, The adsorption of aquatic humic substances by iron hydroxides, Geochim. Cosmochim. Acta, 45:191.

Travis, C.C., and Etnier, E.L., 1981, A survey of sorption relationships for reactive solutes in soil, J. Environ. Qual., 10:8.

Volkov, I.I., and Fomina, L.S., 1974, Influence of organic material and processes of sulfide formation on distribution of some trace elements in deep-water sediments of the Black Sea, Am. Ass. Petrol. Geol. Mem., 20:456.

Wedepohl, K.H., 1980, Geochemical behaviour of manganese, in: "Geology and Geochemistry of Manganese, Vol. I", I.M. Varentsov and G. Grasselly, eds., Akademiai Kiado, Budapest.

DISCUSSION: U. FORSTNER AND W. SALOMONS

G.G. LEPPARD   Have organic films on particles been well-characterized and, if so, is there any kind of film which is clearly of great importance?

U. FORSTNER   Although numerous experimental studies have investigated the adsorption properties of trace metals with organic ligands, an extrapolation to natural waters is still difficult because many natural waters contain organic substances and colloids of unknown composition to complicate these systems (see Sholkovitz and Copland, 1981). This problem, in particular, pertains to those surface films whose characterization is even more difficult than that of dissolved organic species. Hart (1982, in press) has given a review of the methodological aspects of surface organic films (electron microscopy, Raman microprobe, photoelectron spectroscopy and, most important, electrophoresis measurements). Using the latter method, Hunter (1980) showed that well-characterized solid surfaces, when freshly exposed to seawater, became covered with a tenacious film of organic matter and, with respect to your question, he showed that carboxylic acid and phenolic functional groups were the major ionizable functional groups in this organic film.

R. BONIFORTI   What do you think about the hypothesis that the most important properties of marine suspended matter are established by organic coatings? Also, do you think that this process may be different in importance in deep sea water with respect to coastal water?

U. FORSTNER   In extension of Dr. Leppard's question, one has to stress the importance of the surface organic films. Tipping (1981) has shown that even concentrations as low as 0.6 mg/L may be sufficient to cover the surface of any particulate matter present. Data from Balistrieri and co-workers (1981) suggest that the scavenging component of deep-ocean particles has predominantly an organic nature; several questions, however, are raised regarding the stability of organic material throughout the water column. Although there should be differences between coastal marine and deep sea particles with respect to the ratio of organic and inorganic constituents, as well as regarding the composition of the organic films, no data are presently available to answer the question of the possible differences as regards metal adsorbancy. In the estuarine environment, there is the characteristic mechanism of the coagulation of river-borne, Fe-humic colloids (Boyle, Edmond and Sholkovitz, Geochim. Cosmochim. Acta, 41:1313, 1977). In the open ocean, the distribution of most metals is controlled by biological uptake at the surface and regeneration from sinking biogenic debris deeper in the water column as shown, for example, from the covariance of cadmium with phosphate (Boyle, Sclater and Edmond, Nature, 263:42, 1976). A large fraction of the Mn, Co and Ni currently being deposited on the ocean floor may be transported from the continental margins to the deep ocean; once elements are transported down to the deep ocean, a certain fraction of each element is remineralized and dissolved back into the water column, while another fraction is incorporated into pelagic clays and Mn-nodules (Li, Geochim. Cosmochim. Acta, 45:1659, 1981).

H.W. NURNBERG — The demonstrated tendency towards a release of adsorbed heavy metals from suspended particulates with increasing salinity will be a most important event determining the fate of such metals in the estuarine depths towards the open sea. The redissolved metals can be taken up to a significant extent by the phytoplankton. In this context, the speciation by organic ligands will be favorable in principle for accumulation. Whether significant uptake occurs, however, will depend on kinetic barriers.

U. FORSTNER — I generally agree with this interpretation of the data. One should, however, admit that no information is presently available on the fate of cadmium once it is released from particulate matter; perhaps there is a mechanism of fixation on colloidal iron and manganese compounds, or perhaps there is a re-adsorption on organic matter.

M. BRANICA — How does the concentration of heavy metals in an estuary depend on the very pronounced precipitation of DOC in estuarine conditions?

U. FORSTNER — The precipitation of DOM, as well as the simultaneous precipitation of iron (and manganese), provide new sites for adsorption of dissolved metals. However, a competition with chloride ions will take place for the complexation of dissolved trace metals. If the precipitation takes place at high chlorinities, then less adsorption will take place compared with precipitation at low chlorinities, (Salomons, 1980). At present it is difficult to distinguish, using field data, between removal of dissolved trace metals though (1) adsorption on organic matter or through (2) adsorption on iron/manganese oxyhydrates.

E.K. DUURSMA — For about 30 (oceanic) marine sediments, the metal "speciation" in these sediments showed phenomena such as strontium being completely exchangeable. Also, for cadmium this was shown for a few sediments with the exchange going from solid to liquid. Other metals were less to much less exchangeable over a long period of time. The techniques used were: (1) leaching with NH_4 acetate/acetic acid; (2) exchange of radionuclides from water to sediment. My question is "How is cadmium bound to fresh water sediments"?

U. FORSTNER   Since the topic of chemical leachability is not included in our contribution, we should add a few remarks with regard to methods (for a compilation, see Salomons and Förstner, 1980). For the estimation of the environmental impact of metal-contaminated solid materials (such as dredged substances, sewage sludges and polluted soils), it is necessary to determine the metal fractions which are either released directly at the solid/water interface or which can be active in middle-term chemical reactions. Based on our experience and an evaluation of the literature, we have suggested a standard extraction procedure consisting of four steps: (1) exchangeable cations were determined after treatment with 1 M ammonium acetate; (2) readily-reducible and carbonate-associated metals were extracted with acidified hydroxylamine hydrochloride at pH 2; (3) organically-bound metals were partly released after extraction with acidified hydrogen peroxide, followed by an extraction with ammonium acetate to remove any readsorbed metal ions; (4) total analysis ($HF/HClO_4$) is performed on the material remaining from the first three extractions. From the data we obtained on both weakly and strongly polluted sediments, two points seem to be particularly important. First, for all metal-containing examples, a clear decrease of the residual component is shown with increasing overall metal concentrations. Such data suggest that the surplus of metal contaminants introduced into the aquatic system from man's activities usually exists in relatively unstable chemical forms and should, therefore, be predominantly accessible for short term processes, including biological uptake. Secondly, there is a marked difference between the low extractability in the ammonium acetate and the acidified hydroxylamine steps of the metal pair, copper/lead, and the good extractability of the pair, zinc/cadmium, indicating a much better availability of the latter elements. Data from a Rhine River sediment sample, containing 9 mg Cd/kg, show approximately 30% in exchangeable form and about 40% in easily reducible/carbonate associations. The high percentage of exchangeable cadmium is unique among the metals studied by us.

IS ANALYTICALLY-DEFINED CHEMICAL SPECIATION THE ANSWER WE NEED TO UNDERSTAND TRACE ELEMENT TRANSFER ALONG A TROPHIC CHAIN?

Renato Baudo

CNR - Istituto Italiano di Idrobiologia
Largo V. Tonolli, 50/52 - 28048 Verbania, Italy

In the last few years, a large amount of research on trace element emission and dispersion through aquatic ecosystems has underlined the need for more detailed studies on the function and role of these elements in the various ecological compartments. Following the well-known cases of injuries to man from Hg and Cd, all "heavy metals" and related trace elements have been regarded as powerful toxicants, potentially dangerous even for human health. Still more recently, with the advent of an ever increasing mass of data, it has been recognized that the impact of these elements on aquatic environments, with regard to undesirable effects on aquatic life, is mostly controlled by their physico-chemical status rather than by their "total" concentration in water.

The most challenging problem is, at present, the definition of analytical tools enabling us to recognize and to quantify the ecotoxicological risks of the different chemical species of several trace elements. New chemical methods, and improvements in the analytical instrumentation used for the study of chemical speciation in aquatic systems, are continually being presented; however, there is a risk that this effort might be a mere scientific exercise, leading only to contradictory results, if the target is not more precisely defined.

In this paper, in reviewing the available literature, I shall try to reveal some obscure areas needing future attention in the much-debated question of trace element transfer along an aquatic food chain. Knowing that many essential elements are homeostatically regulated, we can start from the assumption that <u>the mechanisms determining the toxicity and the uptake of these elements are not necessarily the same</u>; this means that the two problems may have different solutions and thus have to be treated separately.

Since the early works on trace element transfer along the food chain (prompted mainly for the health protection of fish food consumers), terms such as "bioaccumulation", "bioconcentration" and "biomagnification" have become part of scientific terminology. Unfortunately, the widespread use of these terms has so broadened their meaning that they are now often improperly and misleadingly used. If this paper is to be correctly understood, I should define what these words mean to me:

- <u>Bioaccumulation</u> refers to the ability of an organism to accumulate a chemical from its environment by every means (Isensee et al., 1973), and it can be subdivided into "direct bioaccumulation" (from air, water and sediments) and "indirect bioaccumulation". Bioaccumulation is therefore a synonym of absorption-uptake, as defined by Laties (1959) (a general term describing ion and salt penetration of whatever kind into cells, tissues or organs), and does not imply that active uptake mechanisms have to be involved.

- <u>Bioconcentration</u> is the accumulation of a chemical residue in organisms by transportation of the chemical through the gills or other membranes; the term excludes accumulation by the consumption and absorption of contaminated food (Veith et al., 1979).

- <u>Bioconcentration factor</u> (or concentration factor or transfer factor) is a constant of proportionality between the concentration of the chemical in the organism and in the water. This definition requires that the concentration in the organism be linearly related to concentration in water (Veith et al., 1979), and should be based on the wet weight of the whole organism referring to the same weight unit of water.

- <u>Biomagnification</u> is the building up of residues via food chains, and applies to chemicals that have the ability to move through the food chain resulting in a higher concentration of the chemical at each trophic level (Ellgehausen et al., 1980).

According to the definitions above, we find in the literature many improper applications of these terms. For example, the comparison of concentration factors of various ecological compartments, regardless of the existence of a true proportionality with concentration in water, has been used to "prove" the biomagnification of some trace elements. On the contrary, in the light of the most recent research, it is now accepted that at least Ag, Al, As, B, Ba, Be, Cd, Cr, Co, Cu, Fe, Li, Mn, Mo, Ni, Pb, Pu, Ru, Sb, Se, Sr, Tl, V and Zn do not correspond with the definition of "biomagnified elements" (compare, for examples, Gommes and Muntau, 1975; Foley et al., 1978; Papadopoulou et al., 1978; Phillips and Russo, 1978; Williams and Giesy, 1978; Amiard-Triquet and Saas, 1979; Heyraud and Cherry, 1979; Stoeppler and Brandt, 1979; Stoeppler and Nürnberg, 1979; Spear and Pierce,

1979; Li, 1980). In some recent experimental works (Gächter and Geiger, 1979), no appreciable increase along the trophic chain was demonstrated even for mercury: Jernelöv and Lann (1971) suggested that the high levels of mercury often found in top-predator fish are partly due to a direct uptake of the element from water, noting that, through their food, the predators reach a concentration of the same magnitude as that of their prey.

Nevertheless, apart from deciding on the correct expression to be used and admitting the implicit interest for human health, the transfer of all trace elements along a food chain *a priori* cannot be neglected. In fact, according to Brooks and Rumsby (1965), the pathways for the uptake of these elements in aquatic biota can be depicted as follows:

- ingestion of particulate suspended material;
- ingestion of elements via their preconcentration in food material;
- uptake from water (complexing of metals by coordinate linkage with appropriate organic molecules, incorporation of metals into physiologically important systems, and uptake by adsorption/exchange onto external surfaces).

For autotrophic organisms, disregarding the much-debated question of relative uptake from water and sediments by rooted aquatic macrophytes, only the third pathway works.

Laties (1959) recognized four kinds of chemical uptake for plants:

(1) - entry into the apparent free space;
(2) - adsorption-exchange;
(3) - adsorption;
(4) - carrier transport.

Besides these, two additional mechanisms, leakage and diffusion-exchange, are in fact controlling the concentrations inside the plant mass.

The usual pre-treatment steps of plant samples do not make it possible to eliminate the effect of particle trapping and bacterial epiphytic growth that, according to Patrick and Loutit (1977), may account for a significant fraction of the trace element amounts improperly referred to as "plant" content. This means that these additional two ways of false absorption have also to be considered in comparing the concentrations of producers and consumers.

For heterotrophic organisms, the trace element uptake generally follows a curvilinear relationship, with an initial linear uptake that levels off, becoming asymptotic with the duration of exposure; often the accumulation is reported to be inversely related to the concentration of the trace element at the start of the exposure. This explains why most of the experimental work done in the laboratory, using concentrations of a chemical which are many times higher than what is usually found in natural environments, means little or nothing for predictive interpretations.

The relative importance of absorption from water and food has long been investigated (mostly using a radiotracer in controlled experiments) without reaching a conclusive statement, neither for a single element nor for a selected animal species. Nevertheless, many authors have tried to depict the pattern of trace element absorption (both in plants and animals) by elaborating mathematical models (Table 1).

Some of these take into consideration the exposure time, the efficiency of the chemical residue transfer through the trophic chain, the relationships with metabolic mechanisms such as oxygen passage across the gills and the food assimilation rate; however, all have in common the fundamental consideration that the direct transfer from water to the organism is related to the concentration of the element in water.

Unfortunately, even if the importance of the chemical speciation of the trace elements for determining their "bioavailability" could be set aside in the past, none of such models have tried to relate the internal concentration to some specific form of the chemical in the external medium.

Recently, at least for some elements, the need for quantification of the chemical species involved in understanding absorption has already been proven: George et al. (1976) found that in the mussel, transferrin, ferritin and $Fe(OH)_3$ were taken up directly in gills and gut by pinocytosis, with different efficiencies, kinetics and clearance; Kimura et al. (1979) measured the rates of uptake, the turnover and the concentration factors of organic-complexed and ionic forms of Co by mussels; Von Heidemarie (1979) studied the influence of Co chemical forms on accumulation in Nereis diversicolor; Boudou et al.(1979) investigated the transfer of various mercury compounds along a freshwater food chain; Buchanan et al. (1980) speculated that Fe accumulation in spatangoid echinoderms derived from oxidative deposition of a soluble ferrous salt ingested in reducing conditions; van der Putte et al. (1981) found that pH-dependent variations of the forms of Cr(VI) determine different patterns of accumulation and elimination in rainbow trout; and so on.

So, the question is not "if" but "how" chemical studies on speciation could help in the interpretation of the various pathways of the overall absorption process (Figure 1). Let us examine some of these pathways according to the nature of the uptake.

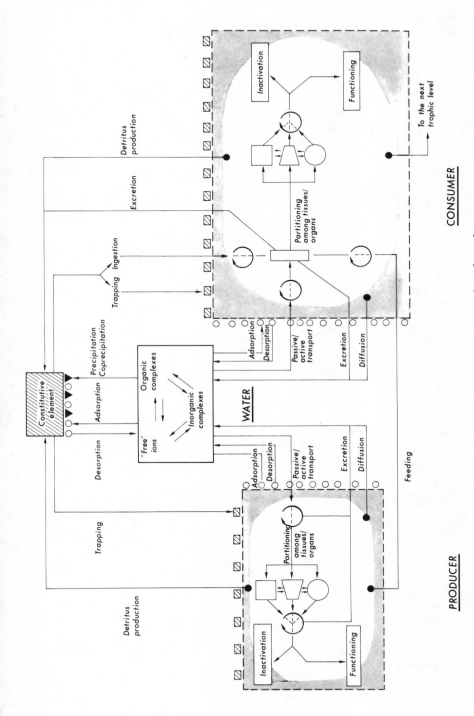

Figure 1. Pathways of absorption and related processes.

Table 1. Models of chemical uptake in aquatic biota.

Hg in *Esox lucius* (Fagerstrom et al. 1975)

$$\frac{\delta X(t,u)}{\delta u} = p_1^{(t,u)} \cdot W(t,u)^b + p_2^{(t,u)} \cdot \frac{\delta W(t,u)}{\delta u} - p_3^{(t,u)} \cdot X(t,u) \cdot W(t,u)^{b-1}$$

Hg in *Perca flavescens* (Norstrom et al. 1976)

$$\frac{dP}{dt} = \frac{e_{pw} C_{pw}}{e_{ox} C_{ox} q_{ox}} \left(a_{lr} W^g + b \frac{dW}{dt} \right) + \frac{e_{pf} C_{pf}}{e_f} \cdot \left[a_{lr} W^g + (b+1) \frac{dW}{dt} \right] - k_{cl} p W^n$$

Hg in *Lepomis macrochirus* (Curtis et al. 1977)

$$Q_t = \frac{UC}{b+g} \left[1 + \frac{b}{m} + \frac{g}{k} + \left(\frac{b}{b+g-m} + \frac{g}{b+g-k} - 1 \right) e^{-(b+g)t} - \left(\frac{b}{m} + \frac{b}{b+g-m} \right) e^{-mt} - \left(\frac{g}{k} + \frac{g}{b+g-k} \right) e^{-kt} \right]$$

Cs in a food chain (Aoyama et al. 1978)

$$\frac{dX_2}{dt} = k_{02} X_0 + \frac{k_{01}}{k_{10}} X_0 \cdot k_{12} \cdot (1 - e^{-k_{10}t}) - \left[k_t \left(\frac{C}{b + e^{-at}} \right)^{-B} + \frac{a \cdot e^{-at}}{b + e^{-at}} \right] X_2(t)$$

B in *Lemna minor* (Glandon and Mc Nabb 1978)

$$U = \left\{ \frac{[B_{t_1}] M_{t_1} - [B_{t_0}] M_{t_0}}{(M_{t_1} - M_{t_0})/2} \right\} / \Delta t$$

Co in *Mytilisepta virgatus* (Kimura et al. 1979)

$$\frac{dQ_t}{dt} = u \cdot S_t - b \cdot Q_t$$

Hg in *Serranus cabrilla* (Radoux and Bouquegneau 1979)

$$C_t = C_{ss} (1 - e^{-kt})$$

Toxicant in a trophic level (Jørgensen 1979)

$$\frac{dC_n}{dt} = MY_n (C_{n-1} \cdot YT_n - C_n \cdot YF_n) + C_n (RESP_n - EXC_n) + UT_n \cdot TOX_n$$

Zn in *Ascophyllum nodosum* (Seip 1979)

$$z_i = \left[u_i C_w^q - s_i \left(\frac{Z_i}{N_i + C_i} - C_w \right) \cdot N_i \right] \cdot \Delta t$$

Zn in *Procambarus acutus acutus* (Giesy et al. 1980)

$$\frac{dQ_a}{dt} = Q_f \cdot k_{fa} + Q_w \cdot k_{wa} - Q_a \cdot k_{aw}$$

Cu in *Salmo gairdneri* (Dixon and Sprague 1981)

$$C = antilog (1.27 + 0.70 E + 1.15 T - 0.09 E^2 - 3.13 T^2 + 0.55 TE + 0.18 E^3 + 8.85 T^3 - 5.29 T^2 E - 2.37 E^2 T$$

Table 1. (Legend)

X body burden of Hg, W body weight, b exponent relating metabolic rate to body weight, P_1 P_2 uptake parameters, P_3 excretion parameters, t time, u age

dP/dt rate of uptake, e_{pw} efficiency of uptake from water, C_{pw} concentration in water, e_{ox} efficiency of oxygen transfer across the gills, C_{ox} concentration of oxygen in water, q_{ox} caloric equivalent of inspired oxygen, a_{lr} metabolic rate coefficient, W fish weight, g body weight exponent, b proportionality constant, e_{pf} efficiency of uptake from food, C_{pf} concentration in food, e_f efficiency of caloric uptake from food, k_{cl} clearance coefficient, p body burden of Hg, n exponent ranging from -0.2 to -0.8

Q_t whole body concentration at time t, U fraction per hour absorbed by a fish from water that passes over its gills, C concentration in water, b g k m transfer and/or elimination rates from three fish compartments (three parts of the fish body)

X_2 concentration in a predator fish, k_{02} transfer coefficient from water to the predator fish, X_0 concentration in water, k_{01} transfer coefficient from water to the prey, k_{10} transfer coefficient from the prey to water, k_{12} transfer coefficient from prey to predator, $C/(b + e^{-at})$ change in weight of the predator fish, a b C B k constants

$[B_t]$ concentration at time t, M_t plant mass at time t, Δt time interval, $(M_{t_1} - M_{t_0})/2$ average standing crop biomass over the time interval

Q_t concentration in the mussel, t time, u rate of uptake per day, S_t concentration in water, b turnover rate = 0.693 / biological half-time

C_t Hg concentration at time t, C_{ss} asymptotic or steady state concentration, k coefficient corresponding to 0.693 / theoretical biological half-time

C_n concentration of the toxicant, nth trophic level, MY_n growth rate of nth trophic level, YT_n yield factor of toxicant, YF_n yield factor of feed, $RESP_n$ respiration rate at nth trophic level, EXC_n rate of excretion, UT_n uptake rate of toxicant, TOX_n concentration of toxicant in water

z_i amount of Zn in age class i taken up during the time interval Δt, u_i uptake rate in age class i, s_i secretion rate from age class i, Z_i accumulated amount of Zn over time, N_i biomass of age class i, C_i initial concentration in algae, C_w concentration in water, q constant

Q_f mass of metal in food, Q_a mass of metal in animal, k_{fa} coefficient of uptake from food, k_{wa} coefficient of uptake from water, k_{aw} coefficient of elimination from animal

C whole body Cu concentration, E = (log level of Cu exposure) - 1.924, T = log (duration of exposure + 10) - 1.276

UPTAKE FROM WATER

Much research on direct element uptake from water has often suggested that adsorption is an important route of absorption for many metals and at various trophic levels: Fe on Mytilus edulis (Hobden, 1969); Cu, Zn, Cd, Pb, Cr, Ni, Fe and Mn in shrimp and fish (Horowitz and Presley, 1977); Co, Cd, Zn, Ni, Cu and Mn on bacteria and yeasts (Norris et al., 1979); Zn in Lemanea fluviatilis (Harding and Whitton, 1981); V in Mytilus galloprovincialis (Miramand et al., 1980); V in Lysmata seticaudata and Carcinus maenas (Miramand et al., 1981). Moreover, adsorption plays an important role in various hypotheses: the inverse relationship between the body level of a chemical and the body size of an organism may be due to the different surface/volume ratios (Ünlü and Fowler, 1979); the inverse relationship between metal accumulation rates and salinity, or hardness, may involve competition for binding sites on cell surfaces (Zitko and Carson, 1976; Phillips, 1977; McFarlane and Franzin, 1980; Miramand et al., 1980, 1981); the differences in accumulation trends between metal-tolerant and non-tolerant populations may be explained by a difference in the number of binding sites on the external surface of an organism (Brown, 1976, 1977; Hall, 1980, 1981).

Finally, Steinberg and Herrmann (1980) have suggested that adsorption works also for organically-bound metal, and Leppard et al. (1977) have speculated on the role in contact cation exchange of the electron-opaque microscopic fibrils found on the surface of some lake algae and bacteria, and also found free in the water column. All these experiences support the feeling of Bourg (1979) that the importance of the solid-solution interface (including those of living organisms) has been too often undervalued. Moreover, these sets of facts represent the first confirmation that the relationships between trace element absorption and chemical speciation could be noticeably different from the relationships governing the toxic effects of trace elements.

As regards the other possible ways of direct uptake of trace metals by biota from water, the relationships with chemical speciation have been suggested often but seldom tested experimentally: Evans (1980) tried to relate Cu accumulation in crayfish to ionic Cu concentration in water, and concluded that "the form(s) of copper which are biologically accumulated by the crayfish remain unknown" because cupric ion concentration alone does not account for the total amount of metal absorbed; Luoma and Jenne (1976) and Neff et al. (1978), studying the heavy metal availability to benthos, came to the conclusion that absorption in benthic organisms is not related to the amount of metal present in the soluble form. Moreover, the attempts to correlate particular forms of metals, as determined by various chemical extractants, to bioavailability were not successful; this indicates the need for new, more sensitive, analytical methods that would make it possible to study the uptake-elimination kinetics of the different chemical species.

At present (Baudo, 1980), the studies dealing with chemical speciation in aquatic environments are based mainly on various combinations of the following techniques:

- electrochemical methods (including element-specific electrodes and polarographic-voltammetric methods);
- ion-exchange and chelation methods;
- gel-filtration and ultrafiltration.

Apart from the limitations intrinsic to each technique, which have been discussed fully in other chapters of this volume, all the techniques suffer from two major criticisms:

- they define the different chemical species of an element according to some chemical parameter (such as mobility or strength of binding) which has a positive chemical meaning, but which has not proven to be the parameter on which aquatic organisms base their own discriminatory capacity;
- they are applied to the chemical equilibria of aqueous solutions which had been derived by subtracting the biotic components from an initial medium of interest (by filtration, centrifugation, dialysis, ...) before starting the analysis (due to the mutual interactions between organisms and their surroundings, the equilibria in the microzone around them may be more or less shifted, by a physical separation process, from those observed in the whole solution).

Given that trace element uptake necessarily has to overcome the problem of a passage through a physical barrier (that is, passage through an internal or external biological surface), I believe that more research with specially-adapted techniques, based on dialysis, could be very useful (Laube et al., 1979). In addition, future studies may be helped greatly by new methods of surface analysis. In fact, from the early work of Hercules et al. (1973), where electron spectroscopy (ESCA) was claimed to be not suitable in biomedical or environmental applications as a result of its poor sensitivity, this technique has been greatly improved so that, recently, it has been used successfully to trace the uptake and bonding of some elements within aquatic organisms (George et al., 1976, 1978, 1980; Buchanan et al., 1980; Martoja et al., 1980). This powerful technique (or possibly other related methods of surface analysis) may be applied to the study of the mechanisms involved in trace element absorption and, in combination with the definition of their chemical status in the external medium, it may constitute a means for individuation of the true bioavailable fraction of these elements. Another possible line of research which is based on the use of radiotracers could be utilized appropriately for the understanding of some basic mechanisms (for example, those governing the turnover of the element in the organisms and within the external medium), but, it suffers from the severe limitation of being suitable only for use in laboratory experiments; it is very difficult to extrapolate such results to natural conditions.

UPTAKE THROUGH FOOD

Even in studies devoted to elucidating the trace element transfer along a food chain, surprisingly few researches have considered both the total body concentration and the localization of the elements in the various parts of an organism. In my opinion, a knowledge of the relative distribution of a chemical among the various tissues and organs is of fundamental importance in understanding its fate within an organism and in elucidating the action of eventual internal detoxifying mechanisms. It is, however, quite useless to trace the transfer along a trophic chain if such a tracing effort is not connected to information dealing with the nutritional value of the various parts of the prey for its predator. For example, what can we learn of the fate of siliceous, calcareous, chitinous, cartilaginous and bony structures during digestion? And what is their importance relative to the whole body in storing trace elements?

Let us consider that the chemical form in which an element is stored (in food) is supposed to influence the accumulation efficiency of the consumer (see, for example, Phillips and Russo, 1978, for Hg; Ünlü, 1979, and Ünlü and Fowler, 1979, for As; Chau et al., 1980, for Pb; Chou et al., 1978, for Cd) and the partitioning among the various internal tissues and organs. This leaves the field open for the study of trace element speciation within an organism; certainly, such a study will present a very hard task to be faced in the near future, but, at the present moment, it seems that the instruments to be used are already available (Van Loon, 1979). As an example, Chau et al. (1979) have experimented satisfactorily with the interfacing of gas chromatographic instruments with an atomic absorption spectrophotometer to separate and detect various organic forms of trace elements. When one realizes that other separation techniques, such as high pressure liquid chromatography (HPLC), are now being extensively investigated, I believe that the solution of the analytical problems in such studies is only a matter of time provided that they really attract our interest.

CONCLUSION

In conclusion, as must be evident from this presentation, I have made up my mind that, due to the complexity of the problem, only very sophisticated techniques can enable us to reach some conclusive statements on the mechanism of trace element absorption in aquatic biota. Nevertheless, this is not what the ecologist really needs; he needs a powerful but simple analytical tool which could make it possible to check quickly and precisely the potential hazard of a chemical in an ecosystem. I hope, after enough information about basic mechanisms has been gathered, that someone will discover just such a simple analytical tool (for example, something based on kinetic analysis - see Brezonik, 1974). Assuming that an enzyme-like, element-specific compound could be identified for every trace element of ecotoxicological concern, its functioning in natural water samples would be related to the actual bioavailability of the element.

REFERENCES

Amiard-Triquet, C., and Saas, A., 1979, Modalités de la contamination de deux chaînes trophiques dulçaquicoles par le cobalt 60. II. Contamination simultanée des organismes par l'eau et la nourriture, Water Air Soil Pollut., 12:141.

Aoyama, I., Inoue, Y., and Inoue, Y., 1978, Experimental study on the concentration process of trace element through a food chain from the viewpoint of nutrition ecology, Water Res., 12:831.

Baudo, R., 1980, Chemical speciation of trace elements in the aquatic environments: a literature review, Mem. Ist. Ital. Idrobiol., 39: in press.

Boudou, A., Delarche, A., Ribeyre, F., and Marty, R., 1979, Bioaccumulation and bioamplification of mercury compounds in a second level consumer, Gambusia affinis. Temperature effects, Bull. Environ. Contam. Toxicol., 22:813.

Bourg, A., 1979, Spéciation chimique des métaux traces dans les systèmes aquatiques. Importance de l'interface solide-solution, J. Français Hydrol., 10:159.

Brezonik, P.L., 1974, Analysis and speciation of trace metals in water supplies, in: "Aqueous-Environmental Chemistry of Metals", A.J. Rubin, ed., Ann Arbor Science Publ., Ann Arbor, Michigan.

Brooks, R.R., and Rumsby, H.G., 1965, The biogeochemistry of trace element uptake by some New Zealand bivalves, Limnol. Oceanogr., 10:521.

Brown, B.E., 1976, Observations on the tolerance of the isopod Asellus meridianus Rac. to copper and lead, Water Res., 10:555.

Brown, B.E., 1977, Uptake of copper and lead by a metal-tolerant isopod Asellus meridianus Rac., Freshw. Biol., 7:235.

Buchanan, J.B., Brown, B.E., Coombs, T.L., Pirie, B.J.S., and Allen, J.A., 1980, The accumulation of ferric iron in the guts of some spatangoid echinoderms, J. Mar. Biol. Ass. U.K., 60:631.

Chau, Y.K., Wong, P.T.S., Bengert, G.A., and Kramar, O., 1979, Determination of tetraalkyllead compounds in water, sediments, and fish samples, Anal. Chem., 51:186.

Chau, Y.K., Wong, P.T.S., Kramar, O., Bengert, G.A., Cruz, R.B., Kinrade, J.O., Lye, J., and van Loon, J.C., 1980, Occurrence of tetraalkyllead compounds in the aquatic environment, Bull. Environ. Contam. Toxicol., 24:265.

Chou, C.L., Uthe, J.F., and Zook, E.G., 1978, Polarographic studies on the nature of cadmium in scallop, oyster, and lobster, J. Fish. Res. Board Can., 35:409.

Curtis, E.H., Beauchamp, J.J., and Blaylock, B.G., 1977, Application of various mathematical models to data from the uptake of methyl mercury in bluegill sunfish (Lepomis macrochirus), Ecol. Model. 3:273.

Dixon, D.G., and Sprague, J.B., 1981, Copper bioaccumulation and hepatoprotein synthesis during acclimation to copper by juvenile rainbow trout, Aquat. Toxicol., 1:69.

Ellgehausen, H., Guth, J.A., and Esser, H.O., 1980, Factors determining the bioaccumulation potential of pesticides in the individual compartments of aquatic food chains, Ecotoxicol. Environ. Safety, 4:134.

Evans, M.L., 1980, Copper accumulation in the crayfish (Orconectes rusticus), Bull. Environ. Contam. Toxicol., 24:916.

Fagerström, T., Kurtén, R., and Åsell, B., 1975, Statistical parameters as criteria in model evaluation: kinetics of mercury accumulation in pike Esox lucius, Oikos, 26:109.

Foley, R.E., Spotila, J.R., Giesy, J.P., and Wall, C.H., 1978, Arsenic concentrations in water and fish from Chautauqua Lake, New York, Env. Biol. Fish., 3:361.

Gächter, R., and Geiger, W., 1979, MELIMEX, an experimental heavy metal pollution study: Behaviour of heavy metals in an aquatic food chain, Schweiz. Z. Hydrol., 41:277.

George, S.G., Pirie, B.J.S., and Coombs, T.L., 1976, The kinetics of accumulation and excretion of ferric hydroxide in Mytilus edulis (L.) and its distribution in the tissues, J. Exp. Mar. Biol. Ecol., 23:71.

George, S.G., Pirie, B.J.S., Cheyne, A.R., Coombs, T.L., and Grant, P.T., 1978, Detoxication of metals by marine bivalves: an ultrastructural study of the compartmentation of copper and zinc in the oyster, Mar. Biol., 45:147.

George, S.G., and Pirie, B.J.S., 1980, Metabolism of zinc in the mussel, Mytilus edulis (L.): A combined ultrastructural and biochemical study, J. Mar. Biol. Ass. U.K., 60:575.

Giesy, J.P., Bowling, J.W., and Kania, H.J., 1980, Cadmium and zinc accumulation and elimination by freshwater crayfish, Arch. Environ. Contam. Toxicol., 9:683.

Glandon, R.P., and McNabb, C.D., 1978, The uptake of boron by Lemna minor, Aquat. Bot., 4:53.

Gommes, R., and Muntau, H., 1975, La distribution de quelques métaux lourds (Zn, Cu, Cr, Ni, Mn, Co) dans la zone littorale des bassins sud et de Pallanza du lac Majeur, Mem. Ist. Ital. Idrobiol., 32:245.

Hall, A., 1980, Heavy metal co-tolerance in a copper-tolerant population of the marine fouling alga, Ectocarpus siliculosus (Dillw.) Lyngbye, New Phytol., 85:73.

Hall, A., 1981, Copper accumulation in copper-tolerant and non-tolerant populations of the marine fouling alga, Ectocarpus siliculosus (Dillw.) Lyngbye, Botanica Mar., 24:223

Harding, J.P.C., and Whitton, B.A., 1981, Accumulation of zinc, cadmium and lead by field populations of Lemanea, Water Res., 15:301.

Hercules, D.M., Cox, L.E., Onisick, S., Nichols, G.D., and Carver, J.C., 1973, Electron spectroscopy (ESCA): use for trace analysis, Anal. Chem., 45:1973.

Heyraud, M., and Cherry, R.D., 1979, Polonium-210 and lead-210 in marine food chains, Mar. Biol., 52:227.

Hobden, D.J., 1969, Iron metabolism in Mytilus edulis. II. Uptake and distribution of radioactive iron, J. Mar. Biol. Ass. U.K., 49:661.

Horowitz, A., and Presley, B.J., 1977, Trace metal concentrations and partitioning in zooplankton, neuston, and benthos from the South Texas Outer Continental Shelf, Arch. Environ. Contam. Toxicol., 5:241.

Isensee, A.R., Kearney, P.C., Woolson, P.C., Jones, G.E., and Williams, V.P., 1973, Distribution of alkyl arsenicals in model ecosystem, Environ. Sci. Technol., 7:841.

Jernelöv, A., and Lann., H., 1971, Mercury accumulation in food chains, Oikos, 22:403.

Jørgensen, S.E., 1979, Modelling the distribution and effect of heavy metals in an aquatic ecosystem, Ecol. Model., 6:199.

Kimura, Y., Honda, Y., and Katsurayama, K., 1979, Comparative uptake and elimination of radiocobalt in organic complexed and ionic forms by mussel, Mytilisepta virgatus, J. Radiat. Res., 20:291.

Laties, G.G., 1959, Active transport of salt into plant tissue, Annu. Rev. Plant Physiol., 10:87.

Laube, V., Ramamoorthy, S., and Kushner, D.J., 1979, Mobilization and accumulation of sediment-bound heavy metals by algae, Bull. Environ. Contam. Toxicol., 21:763.

Leppard, G.G., Massalski, A., and Lean, D.R.S., 1977, Electron-opaque microscopic fibrils in lakes: their demonstration, their biological derivation and their potential significance in the redistribution of cations, Protoplasma, 92:289.

Li, W.K.W., 1980, Cellular accumulation and distribution of cadmium in Isochrysis galbana during growth inhibition and recovery, J. Plankton Res., 2:283.

Luoma, S.N., and Jenne, E.A., 1976, Estimating bioavailability of sediment-bound trace metals with chemical extractants, in: "Trace Substances in Environmental Health - $\underline{X}$", D.D. Hemphill, ed., Univ. of Missouri Press, Columbia, Missouri.

Martoja, M., Tue, V.T., and Elkaïm, B., 1980, Bioaccumulation du cuivre chez Littorina littorea (L.) (Gastéropode Prosobranche): signification physiologique et écologique, J. Exp. Mar. Biol. Ecol., 43:251.

McFarlane, G.A., and Franzin, W.G., 1980, An estimation of Cd, Cu and Hg concentrations in livers of northern pike, Esox lucius, and white sucker, Catostomus commersoni, from five lakes near a base metal smelter at Flin Flon, Manitoba, Can. J. Fish. Aquat. Sci., 37:1573.

Miramand, P., Guary, J.C., and Fowler, S.W., 1980, Vanadium transfer in the mussel Mytilus galloprovincialis, Mar. Biol., 56:281.

Miramand, P., Guary, J.C., and Fowler, S.W., 1981, Uptake, assimilation, and excretion of vanadium in the shrimp, Lysmata seticaudata (Risso), and the crab, Carcinus maenas (L.), J. Exp. Mar. Biol. Ecol., 49:267.

Neff, J.W., Foster, R.S., and Slowey, J.F., 1978, "Availability of Sediment-adsorbed Heavy Metals to Benthos with Particular Emphasis on Deposit-feeding in Fauna", U.S. Dept. Comm. Natl. Tech. Inf. Service AD/A-061, 152:286 pp.

Norris, P.R., and Kelly, D.P., 1979, Accumulation of metals by bacteria and yeasts, Dev. Ind. Microbiol., 20:299.

Norstrom, R.J., McKinnon, A.E., deFreitas, A.S.W., 1976, A bioenergetics-based model for pollutant accumulation by fish. Simulation of PCB and methylmercury residue levels in Ottawa River yellow perch (Perca flavescens), J. Fish. Res. Board Can., 33: 248.

Papadopoulou, C., Zafiropoulos, D., Hadjistelios, I., Vassilaki-Grimani, M., and Yannopoulos, C., 1978, Trace elements in pelagic organisms and a pelagic foodchain of the Aegean Sea, IVes Journées Etud. Pollutions, Antalya, C.I.E.S.M., 231-232.

Patrick, F.M., and Loutit, M.W., 1977, The uptake of heavy metals by epiphytic bacteria on Alisma plantago-aquatica, Water Res., 11:699.

Phillips, D.J.H., 1977, The common mussel Mytilus edulis as an indicator of trace metals in Scandinavian waters. I. Zinc and cadmium, Mar. Biol., 43:283.

Phillips, G.R., and Russo, R.C., 1978, "Metal Bioaccumulation in Fishes and Aquatic Invertebrates: A Literature Review", (U.S.) EPA-600/3-78-103, 116 pp.

Radoux, D., and Bouquegneau, J.M., 1979, Uptake of mercuric chloride from sea water by Serranus cabrilla, Bull. Environ. Contam. Toxicol., 22:771.

Seip, K.L., 1979, A mathematical model for the uptake of heavy metals in benthic algae, Ecol. Model., 6:183.

Spear, P.A., and Pierce, R.C., 1979, "Copper in the Aquatic Environment: Chemistry, Distribution, and Toxicology", Publ. Environ. Sec., Ottawa, Canada, No. NRCC 16454, 227 pp.

Steinberg, C., and Herrmann, A., 1980, Utilization of Dissolved Metal Organic Compounds by Freshwater Microorganisms, Paper presented at 21st. SIL Congress, Kyoto, Japan, August, 1980.

Stoeppler, M., and Brandt, K., 1979, Comparative studies on trace metal levels in marine biota. II. Trace metals in krill, krill products, and fish from the Antarctic Scotia Sea, Z. Lebensm. Unters. Forsch., 169:95.

Stoeppler, M., and Nürnberg, H.W., 1979, Comparative studies on trace metal levels in marine biota. III. Typical levels and accumulation of toxic trace metals in muscle tissue and organs of marine organisms from different European Seas, Ecotoxicol. Environ. Safety, 3:335.

Ünlü, M.Y., 1979, Chemical transformation and flux of different forms of arsenic in the crab Carcinus maenas, Chemosphere, 5:269.

Ünlü, M.Y., and Fowler, S.W., 1979, Factors affecting the flux of arsenic through the mussel Mytilus galloprovincialis, Mar. Biol., 51:209.

van der Putte, I., Lubbers, J., and Kolar, Z., 1981, Effect of pH on uptake, tissue distribution and retention of hexavalent chromium in rainbow trout (Salmo gairdneri), Aquat. Toxicol., 1:3.

Van Loon, J.C., 1979, Metal speciation by Chromatography/Atomic spectrometry, Anal. Chem., 51:1139A.

Veith, G.D., DeFoe, D.L., and Bergstedt, B.V., 1979, Measuring and estimating the bioconcentration factor of chemicals in fish, J. Fish. Res. Board Can., 36:1040.

Von Heidemarie, K., 1979, Die Belastung von Wattenorganismen im Elbe-Aestuar durch Radionuklide: Versuche zur Kontamination von Nereis diversicolor O.F. Müller mit Co-57, Arch. Hydrobiol., Suppl., 43:265.

Williams, D.R., and Giesy, J.P., 1978, Relative importance of food and water sources to cadmium uptake by Gambusia affinis (Poeciliidae), Environ. Res., 16:326.

Zitko, V., and Carson, W.G., 1976, A mechanism of the effects of water hardness on the lethality of heavy metals to fish, Chemosphere, 5:299.

DISCUSSION: R. BAUDO

M. BRANICA — The answer to your "title question" is yes and such an answer receives very strong backing.

R. BAUDO — Yes! We need to have analytically-defined chemical speciation to understand trace element transfer along a trophic chain.

Y.K. CHAU — I wish to comment that organolead compounds (such as lead alkyls) are accumulated mainly in a specific fatty tissue. Also, I believe that a mission-oriented speciation method can be developed to suit the needs of the biologist.

G.E. BATLEY — If, by kinetic analysis, you are referring to the catalytic methods of the Yatsimirskii group, I can make some comments based on my experiences with these techniques. While reaction rates will be dependent on chemical form, there are many interferences from other elements. Thus, any speciation scheme based on these methods will require some preliminary separation stages.

R. BAUDO — I see you are not against the basic idea.

H.W. NURNBERG — Asking for a simple, reliable, rapid, low-cost, specific, universal, analytical technique to measure the bioavailability of a trace metal in a natural water is asking for a miracle. However, the problem

H.W. NURNBERG (continued)   can be tackled using existing analytical methods if the natural water sample is first separated into its major measureable components and if the resultant data are considered as information bits for the whole system. This is a feasible and sound approach. Regarding your request for the new technique, a kinetic analysis certainly will <u>not</u> be an optimal solution.

R. BAUDO   The sense of my proposal was that we need a simple method just for testing.

P. VALENTA   The use of kinetic methods in natural water systems is hopeless. Chemical kinetics depend much more on the parameters of the medium than on the chemical equilibrium and, also, the reaction path itself can be changed inadvertently. These methods can be used for one substance only in very precisely controlled conditions.

R. BONIFORTI   You have presented a distribution pattern of lead in fish. In your opinion, what is the effect of biological variability on that distribution pattern?

R. BAUDO   Data from the analysis of biological materials is generally not too precise; such data has a high relative standard deviation. For the research that I presented, this may be due to the inaccuracy of the methods used or to a subdivision of the organisms analyzed without a proper knowledge of the metabolic mechanisms determining the distribution within a given organism of the trace element under consideration.

LABORATORY AND FIELD APPROACHES TO ENVIRONMENTAL EFFECTS
MONITORING WITH EMPHASIS ON SOME MICROBIAL-HEAVY METAL RELATIONS

Ralph F. Vaccaro

Woods Hole Oceanographic Institution
Woods Hole, Massachusetts 02543

INTRODUCTION

   Over the past 30 years an increasingly rigid rationale has
evolved concerning what constitutes acceptable environmental
impacts from the release of society's wastes. This trend, particularly pronounced with regard to the oceans, has in turn generated an unfulfilled demand for more discerning effects monitoring
to disclose prevailing levels of environmental stress. The per-
plexing shortfall regarding environmental effects monitoring is
somewhat analogous to that of an employee paid to rid a certain
section of beach of cans, bottles and other debris only to be
abruptly confronted with the additional task of counting all brown
sand particles.

   Current environmental concern places significant emphasis on
community structure and population interaction, a viewpoint
eminently consistent with the classical definition of a biological
ecosystem which stresses both population and substrate interactions. Proponents of the ecosystem approach to environmental
quality standards hold that man's critical interests are best
served by a conscientious regard for earth's renewable resources
and that any impact on community function is inadmissible. The
consequential need to examine entire ecosystems, as well as
individual organisms, is a prime reason for today's unprecedented
need for more responsive environmental monitoring.

   Additional motivation for establishing more effective environmental monitoring stems from rather recently recognized epidemio-
logical concerns relating to preventive medicine. Formerly,
environmental protection and surveillance centered on public

health criteria adopted to rid human populations of infectious diseases. Modern preventive medicine must also contend with a massive variety of insidious chemicals and discrete viruses capable of imparting deleterious genetic, birth and cellular defects. A lengthy list of 129 environmental pollutants of concern, recently distributed by the U.S. Environmental Protection Agency, clearly reflects this complication. Presumably, each chemical species listed is of sufficient environmental or public health concern to warrant independent in situ evaluation.

Futher impetus for improved marine monitoring originates from the polarization of opinions concerning the validity of assimilative capacity analyses as applied to the oceans. Proponents of this concept hold that the oceans offer a finite capacity for waste assimilation in a manner which offers scientific, social and economic advantages over land or atmospheric disposal (Goldberg, 1981). An opposing and more conservative opinion stresses the inadequacy of the informational base available to construct reliable assimilative capacity models, hence the strong likelihood of unreliable conclusions (Kamlet, 1981). Particular concern is expressed for chronic, long-lived toxins, known to accumulate in the oceans, to which no adverse effects have, as yet, been assigned. Whether the impact of chronic pollutants in the oceans is indeed negligible or whether there has been a failure to make critically relevant measurements remains unanswered (Lewis, 1980).

The challenge for improved monitoring has already resulted in a number of highly intensive efforts by multidisciplined teams of scientists studying community response to chemical stress. Several in depth, highly complex, experimental studies have generated much useful information regarding biomass, species diversity, productivity, tolerance acquisition and trophic interactions. However, as yet, no novel and economically attractive monitoring procedures, commensurate with today's need, have become established.

A comprehensive effort in aquatic environmental monitoring entails hydrographic, chemical and biological information which uniquely describes a specific location. Only analytical chemical information can provide definitive evidence regarding the presence and time-space distribution of a specific chemical combination. In practice, however, the chemical effort often becomes unwieldy because of the overwhelming variety of chemical species having environmental significance. Also, the excessive costs of modern analytical instrumentation and its demands for operational expertise can be an added deterrent. Another inevitable analytical limitation concerns the inability of chemical techniques to anticipate obscure interactions between a particular contaminant and other natural water constituents which in combination may exert a more (or less) than additive effect. Finally, routine chemical analyses rarely provide relevant information on chemical speciation which is often of critical importance with regard to a particular biological impact.

Whereas an exclusively analytical chemical approach cannot predict the ecological consequences of a pollutant event, biological monitoring is notably deficient in terms of its ability to identify the specific chemical agent responsible for a particular biological disruption. The major, and unique contribution of biological environmental monitoring is that it records the _net_ effect of all existing variables, including salient contributions from unsuspected sources.

BIOLOGICAL ENVIRONMENTAL MONITORING

The most demanding charge of modern ecological monitoring is to evaluate levels of potency and permanency regarding anthropogenic perturbations which affect natural communities. The underlying rationale recognizes that increasingly higher levels of perturbants progressively affect biological tolerance, function and survival. In a utilitarian sense, the above consequences of stress provide biological signals which can be exploited experimentally for assay purposes. Here a representative test organism acts as the indicator of an end-point for titrating the potency of a particular pollutant.

The standard bioassay technique endorsed by the regulatory agencies typically relies on the mortality of test organisms, as observed in the laboratory, to assess a potentially toxic situation. Such observations are notoriously unreliable for discerning impacts on surviving resident populations. Survival studies, however, are much more appropriate when used in the laboratory to screen for potential toxins or to monitor effluent quality at known point sources of pollution. A further inevitable criticism of the standard bioassay is the uncertainty which arises from the necessary extrapolation of laboratory results to infinitely more complicated natural situations.

Bioassay techniques which address sublethal as opposed to lethal effects are clearly more responsive to the modern demand for effects monitoring. Indeed, such observations, even at the laboratory level, are more appropriate, in that they often encourage confirmatory efforts in the field. Too often, however, a pronounced inability to establish linkage between laboratory findings and specific _in situ_ abnormalities is encountered. Yet, assessments of sublethal effects do appear to offer the best hope for a systematic approach to more stringent effects monitoring, providing representative multi-levels of biological organization are addressed. Such observations should extend to entire groups of organisms and should include their status in terms of standing crops, production, species diversity and general trophic organization. Corresponding consideration should be given to the degree of biological variation which prevails within unstressed, but otherwise comparable, control communities.

EFFECTS OF POLLUTANTS ON ORGANISMS

As an organism's surroundings become increasingly more hostile, its ability to carry on maintenance metabolism becomes progressively disorganized, and ultimately survival depends on its ability to substitute alternative pathways to sustain vital functions. A report from the National Academy of Sciences (NAS, 1971) identifies four levels of biological organization wherein such effects have been recorded. These are shown in Table 1 which also compares the time required to complete representative observations (after Capuzzo, 1981).

BIOCHEMICAL AND CELLULAR RESPONSES

Observations on biochemical effects are ideally suited for a rapid appraisal of environmental contamination, especially when cellular irregularities can be coupled to altered physiological performance. An enzyme complex that appears to offer considerable promise for this purpose is the mixed function oxygenase (MFO) of mammals, fish and invertebrates whose activity is enhanced by many foreign organic compounds, including the polycyclic aromatic hydrocarbons. For example, fish from petroleum-polluted areas are known to accelerate hydroxylase activity in liver and gill tissue as compared with the same species in nonpolluted areas (Payne,

Table 1. Response levels to pollutants in marine ecosystems (adapted from NAS, 1971; Capuzzo, 1981).

Level	Biological organization	Time required for study
I	Biochemical-Cellular	minutes - hours
II	Whole Organism	hours - months
III	Population	days - years
IV	Community	weeks - decades

1976). Thus, MFO activity has been specified as a highly useful source of information on both short and long term petroleum effects.

A second intracellular mechanism recommended for measuring chemical stress is the adenylate energy charge (AEC) which addresses the fractionation of metabolic energy from an organism's adenine nucleotide pool. Cellular energy is stored as adenosine triphosphate (ATP) whose precursors are adenosine monophosphate (AMP) and adenosine diphosphate (ADP). The AEC is calculated from

$$AEC = \frac{ATP + 1/2\ ADP}{ATP + ADP + AMP}.$$

In multicellular organisms or microbial cultures, AEC values can be indicative of physiological condition and growth state. Healthy growing cells display an index range of 0.5 - 0.75 in environments unimpacted as compared with indices of less than 0.5 for organisms irreversibly stressed. Provided that acceptable and reproducible baselines are established, results appear to be consistent for a broad variety of species, the response times are not unreasonably long, and the presence of abnormal indices appears to correlate with subsequent physiological and behavioral changes (Giesy et al., 1978; Ivanovici, 1980).

The index approach to interpretations of environmental status has become extremely popular within recent years in the United States and Europe. At present, more than 80 candidate techniques, in various stages of development, are under official review as possible vehicles for addressing quality of life and environmental damage (Ott, 1978).

WHOLE ORGANISM RESPONSES

Whole organisms which selectively concentrate and react to foreign entities are frequently used to assess biological effects at a particular trophic level. An impressive number of reports and contributions on this subject have appeared within the past decade (National Academy of Sciences, 1971, 1980; Marine Technology Society, 1974; Ontario Ministry of the Environment, 1975; American Soc. for Testing and Materials, 1977; McIntyre and Pierce, 1980).

A classic example of the whole organism approach to environmental monitoring is the use of bivalve molluscs to integrate and record pollution levels in coastal marine waters (Holden, 1973; Butler, 1973). An ongoing effort in the U.S. which utilizes the whole organism approach is the Mussel Watch program, the subject of a recent report by the U.S. National Academy of Sciences (NAS, 1980).

Bivalves such as Mytilus edulis and Mya arenaria typify marine organisms known to accumulate a variety of pollutants at body concentrations which far exceed ambient seawater levels. Significant concentration increases have been recorded for heavy metals, transuranic elements, petroleum and halogenated hydrocarbons. Frequently the bioaccumulation process permits successful analyses of a pollutant whose concentration in seawater is too low for routine analytical detection. A recent Mussel Watch publication emphasizes the extent of variability revealed by bivalve surveillance in U.S. coastal waters for the four classes of perturbants mentioned above (Goldberg et al., 1978).

Prolonged periods of stress affect the general physiology of an organism and lead to important departures in the rates of respiration, excretion, and feeding efficiency. Since each of the above is intimately concerned with an organism's growth potential, growth retardation can be a convenient indication of stress. A useful physiological index which addresses growth potential is known as "the scope for growth".

Scope for growth has been defined as the caloric difference between the energy value of the food consumed by an animal and the value of all energy appropriations from its food intake for purposes other than growth (Winberg, 1960; Warren and Davis, 1967). In practice, caloric equivalents of food intake (C), excretion (U), respiration (M) and gamete production (F) are measured, and scope for growth is calculated as:

$$\text{Scope} = C - U - M - F \tag{1}$$

which can be converted to the following index descriptive of growth efficiency.

$$\text{Efficiency} = \frac{C - (U+M+F)}{C}. \tag{2}$$

Scope-for-growth measurements are often appropriate when direct growth measurements are unduly complicated or lack sufficient sensitivity to reveal stress-induced variation. The scope for growth approach has been successfully applied to both fish (Warren and Davis, 1967) and molluscs (Bayne et al., 1975, 1978; Gilfillan, 1975; Gilfillan et al. 1976; Phelps and Galloway, 1980) to quantify whole organisms-stress relations.

No single physiological measurement can be expected to demonstrate environmental stress under all circumstances. In practice, a suite of indices to enhance the chances of mutually supportive information from two or more measurements is preferable. A case in point is the variety of stress indices originally considered for Mussel Watch, shown in Table 2. Together, they demonstrate the need for a significant amount of prior basic knowledge concerning bivalve metabolism and stress (Bayne et al., 1980; NAS, 1980).

The use of whole organisms as environmental sentinels is subject to many of the same deficiencies which plague other forms of environmental monitoring. Field observations of stress, unduly influenced by the presence of unrecognized pollutants, will inevitably generate unwarranted conclusions. Low tolerance of an organism to sublethal effects may be advantageous in that it not only promises increased monitoring sensitivity but also enhances the possibility of assigning general inhibitory responses to a particular source (Bayne, 1980). Finally, the ability of an organism subjected to chronic sublethal stress to display increasingly greater tolerance levels, while not incompatible with the sentinel organism approach to environmental monitoring, must be taken into account.

POPULATION RESPONSES INCLUDING AQUATIC BACTERIA

To illustrate the impact of a pollutant at the population level is highly demanding and often impossible. Population responses are notoriously non-specific and tend to be obscured by the

Table 2. Condition indices for evaluating stress in bivalve molluscs (after Bayne, 1980; NAS, 1980).

Type Index	Parameter
Euphoric	- Flesh weight vs shell length or volume. - Growth, direct or scope measurements. - Carbohydrate, lipid, protein fractionation. - Gametogenesis
Biochemical	- Adenylate energy ratio. - Lysosomal enzyme integrity. - Amino acid ratios.
Pathological	- Parasite burden. - Mucous excretion. - Granulocytomas. - Neoplasms - Hemocytic infiltrations.

natural patchiness which characterizes all aquatic environments. Uncertainty is heightened when the search is for long-term population effects associated with the chronic presence of a pollutant.

The following types of information on population structure have been recommended for monitoring purposes (Capuzzo, 1981).

1. Species abundance and distribution
2. Age representation
3. Growth of prominant age groups
4. Reproductive and recruitment potentials
5. Pathological syndromes.

The most complete information currently available on the variability of natural marine populations relates to commercially harvested species whose distribution and persistence is the prominent concern of fisheries management. A comparable depth of information is rarely available for economically unattractive species which constitute the bulk of the remaining populations. Typically, it becomes necessary for field studies to rely on more uncertain sampling strategies, such as a comparison of perturbation levels along a pollution gradient.

For laboratory studies at the population level, mixed species of naturally occurring bacteria can be a very useful source of test cells for assay purposes. These microscopic, unicellular organisms possess an extremely favorable surface-to-volume ratio along with highly efficient and versatile enzyme systems. Such advantages are notably prevalent in even the most common heterotrophic bacteria which are ideally suited for microbial assays.

Rates of heterotrophic bacterial activity, which are readily measurable via observations of radio-labeled ^{14}C glucose uptake, under standard conditions, can be advantageously exploited for bioassay information. Test cells are easily obtainable from the mixed population which spontaneously develops in raw water during 18-20 hour incubations at room temperature. Bioassay results are recorded by plotting, as a percentage, the relation between the average ^{14}C uptake for a series of different waste concentrations vs that of a waste-free control. The resulting ^{14}C uptake patterns can be used to quantify sublethal microbial effects as the effective concentration of a pollutant which accounts for a 50 percent reduction in uptake behavior (EC-50). Our Woods Hole laboratory has used the heterotrophic microbial assay to assess the effects of both heavy metal and dissolved organic pollutants on marine bacterial populations.

As is generally accepted, the potential toxicity of a metal is dominated more by the nature and abundance of coexisting complexing ligands than by the total metal concentration (Barber,

1973). Natural ligands lessen the impact of metals on aquatic systems when they act as nature's analogues of synthetic chelators, such as ethylenediaminetetraacetic acid (EDTA) or nitrilotriacetic acid (NTA) (Erickson, 1972; Chau, 1973; Lewis et al., 1973; Sunda and Guillard, 1976).

Relevant Woods Hole experiments have addressed the effects of EDTA on copper-induced inhibition of the ^{14}C glucose uptake response by a copper-sensitive, marine bacterial population (Gillespie and Vaccaro, 1978). Here, membrane-filtered (0.22 µm), Nantucket Sound surface water (ambient copper concentration undetermined) which had been irradiated by ultra-violet light (450 watts, 3 hrs) was the common seawater employed. Results are shown in Figure 1, wherein the midpoint of each of the segments showing maximum change within each sigmoid response curve was designated the saturation point for experimentally-added EDTA (Davey et al., 1973). Thus the saturation value of 3 µg Cu· liter^{-1} corresponded to 0.0 µM EDTA and was taken as the blank measurement after chelate degradation by the ultra-violet radiation, hence the inherent copper tolerance of the test bacterial population.

When applied to a natural seawater sample of known origin, the saturation point can provide an estimate of its potential to chelate and detoxify a subsequent copper incursion. Similarly, the total amount of copper present (i.e., ambient plus added copper) at the saturation point provides an estimate of the total chelation capacity provided in a given water sample (Gillespie and Vaccaro, 1978).

The above analysis has also been used by us to intercompare the apparent copper complexation capacities at various oceanic locations. Pacific Ocean water taken from Saanich Inlet, British Columbia, consistently demonstrated complexation capacities ranging from 10-15 µgCu· liter^{-1}. As shown in Figure 2a, comparable analyses, conducted on Atlantic samples selected to provide a range of variation, have demonstrated copper capacities of 3 and 7 µgCu· liter^{-1} respectively for the waters of the remote Sargasso Sea and nearby Vineyard Sound, off Woods Hole. In contrast, a much greater complexation capacity of 32 µgCu· liter^{-1} characterized a nearby low-salinity, highly eutrophic water of a marine marsh contiguous with Nantucket Sound (Figure 2b).

Another useful application of the ^{14}C bacterial assay technique concerns changes in heavy metal tolerance by microbial populations. In this case, ^{14}C uptake comparisons are made on bacterial populations which have been preconditioned by prior exposure to different regimes of heavy metal concentration (Gillespie and Vaccaro, 1981). Evidence described below strongly suggests that the previous heavy metal history of a particular

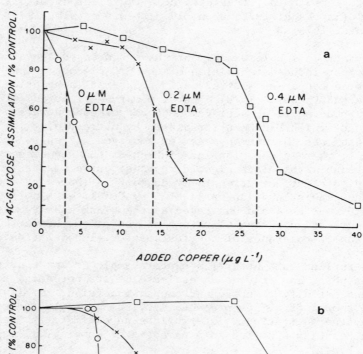

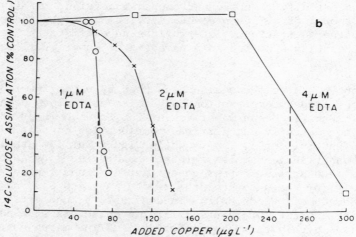

Figure 1. Effects of EDTA on assimilation of ^{14}C glucose by isolate no. 5 in UV-treated seawater containing increasing concentrations of added copper (after Gillespie and Vaccaro, 1981).

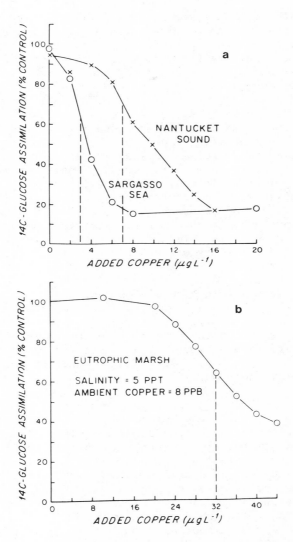

Figure 2. Bioassay measurements of copper chelation capacities of natural waters from the Sargasso Sea, Nantucket Sound and a eutrophic salt marsh (dashed lines represent saturation point of natural chelators - after Gillespie and Vaccaro, 1981).

water mass can also have an important influence on the stress levels which prevail at a particular time.

To evaluate tolerance acquisition, primary cultures comprising six replicate suspensions of mixed natural marine bacteria in seawater were prepared; two for copper additions and two for mercury additions. The remaining two primary cultures which received no added copper or mercury were the designated controls. Copper concentrations added were 0, 25 and 50 µg liter^{-1} while mercury enrichments were 0, 0.25 and 2.5 µg liter^{-1}. All six primary cultures were stored in darkness at 5°C for a period of six months before being used as an inoculum for secondary cultures representative of different degrees of previous heavy metal exposure. Ultimately, on a predetermined time scale, the metal concentrations of the secondary cultures were restored to levels originally provided for the primary cultures. While still under the influence of the six renewed metal concentrations, changes in ^{14}C uptake patterns, obtained under reproducible conditions and based on about 15 above-ambient concentrations of the appropriate heavy metal, were observed. The final experimental range provided for both copper and mercury was 0 - 250 µg·liter^{-1} of added metal.

Results relating to the development of copper tolerance are shown in Figure 3(a,b,c) and comparable information for mercury appears in Figure 4 (a,b,c). Interpretation of these results is facilitated by comparing the number of instances wherein copper or mercury additions had little or no effect on ^{14}C uptake (i.e., when the uptake response approached 100 percent of a control excluded from metal addition).

Regarding the copper series, the greatest metal inhibition corresponded to the culture developed from cells stored for a six month period in the absence of added copper (Figure 3a). Subsequent exposure of this culture to as little as 15 µg·liter^{-1} copper caused a 50 percent decrease in ^{14}C uptake. Later, as the ambient copper concentration was increased to 5 and still later to 25 µg·liter^{-1} copper, the concentration of copper required for a 50 percent reduction in uptake showed a corresponding increase to 22 and 200 µg·liter^{-1}. Finally, a two day exposure at 50 µg·liter^{-1} produced a population which was essentially unaffected by a copper concentration as high as 250 µg·liter^{-1}.

In marked contrast to the above behaviour was that noted for the culture which had been stored for six months in the presence of 50 µg·liter^{-1} copper. Descendents of these cells required well over 50 µg·liter^{-1} copper to decrease ^{14}C uptake by 50 percent (Figure 3c). Later, this culture developed even greater resistance to copper when its ambient copper concentration was increased. Ultimately a copper concentration as high as 100 µg·

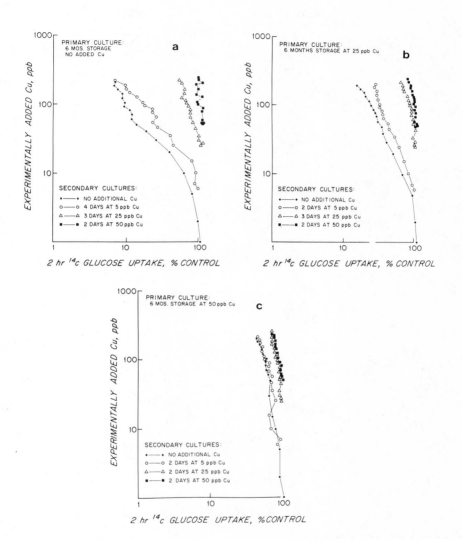

Figure 3. The effect of previous copper exposure on copper tolerance of a marine heterotrophic bacterial community after storage in excess of six months with: (a) no added Cu; (b) 25 µgl^{-1}Cu; and (c) 50 µgl^{-1}Cu (after Gillespie and Vaccaro, 1981).

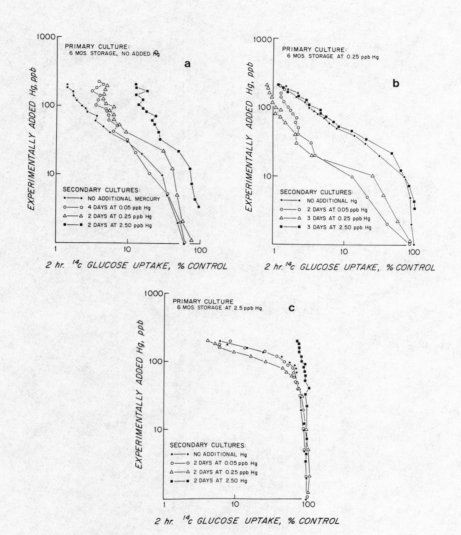

Figure 4. The effect of previous mercury exposure on mercury tolerance of a marine heterotrophic bacterial community after storage in excess of six months with: (a) no added Hg; (b) 0.25 $\mu g l^{-1}$ Hg; and (c) 2.5 $\mu g l^{-1}$ Hg (after Gillespie and Vaccaro, 1981).

liter^{-1} caused only a 10 percent reduction in ^{14}C assimilation. Similar analyses of ^{14}C utilization for the intermediate copper concentration of 25 μg· liter^{-1} resulted in uptake patterns intermediate between the above extremes (Figure 3b).

A similar sequence of changes in ^{14}C uptake patterns has been observed for mercury (Figure 4 - a, b, c) except that an intermediate effect from storage at 0.25 μg· liter^{-1} mercury was not so clear-cut. Nevertheless, in the extreme case, the bacterial survivors of a mercury exposure of 2.5 μg· liter^{-1} generated a population whose mercury tolerance increased to a point where no appreciable inhibition was observed from a concentration as high as 100 μg· liter^{-1} mercury (Figure 4c). A logical expansion of tolerance studies to include simultaneous observations on additional trophic levels would appear to be an appropriate research target for the near future.

COMMUNITY RESPONSES

Natural communities exhibit highly complicated interactions on the part of member populations which produce a pyramidal flow of chemical energy from primary producers to top predator. Unperturbed, as opposed to stressed, marine communities generally display greater stability with regard to species representation and population densities. Adverse population effects at one or more trophic levels often lead to an unstable community structure via the invasion and success of more tolerant adventitious species. Unambiguous pollution studies addressing entire marine communities inevitably must contend with the vastness of the oceans and a common inability to define precise experimental boundaries (Menzel and Case, 1977). An added and familiar complication is the need to assess the large variety of chemical species of concern in modern oceanic surveillance.

Within the past 10 years, many investigators in North America and in Europe have adopted large scale enclosures of natural waters as a more practical expedient for open ocean pollution research. A 1980 Symposium on Enclosed Experimental Ecosystems, held in Sidney, British Columbia, Canada, was, in effect, an international forum on enclosure research. Aspects of enclosure research discussed included structural and experimental design, duplication of natural communities and trophic effects from specified pollutants (Grice and Reeve, in press). Table 3 identifies some of the more prominent, long-term efforts in environmental research which have relied upon the enclosure format.

Early encouragement for marine pollution studies in North America at the community level was provided by the International Decade of Oceanography (IDOE) Program which led to activation of

Table 3. Prominent examples of enclosure research oriented toward the assessment of environmental effects.

Project Title and Location	Enclosure Type	Volume	Principal Interaction Studied
Plankton Tower Kiel, West Germany	Plastic	30 m^3	Plankton - sediment
*POSER Rosfjorden, Norway	Plastic	30 m^3	Plankton - nutrient
Den Helder Program Den Helder, Netherlands	Plastic	Up to 16 m^3	Plankton - toxicity
Loch Ewe Enclosures Loch Ewe, Scotland	Plastic	100 m^3, 300 m^3	Plankton - toxicity
**MERL Program Narragansett Bay Rhode Island, USA	Fixed Tanks in series	13 m^3	Plankton - toxicity
***CEPEX Program Sidney, B.C. Canada	Plastic	68 m^3, 1300 m^3	Plankton - toxicity

* Plankton Observations with Simultaneous Enclosures Rosfjorden
** Marine Ecosystems Research Laboratory
*** Controlled Ecosystem Pollution Experiment

---

the Controlled Ecosystem Pollution Experiment (CEPEX) in 1973. This multidisciplinary effort was designed to ascertain the response of plankton communities to low levels of pollutants. A review of the CEPEX experimental procedure would seem appropriate here to illustrate how demanding aquatic pollution studies at the community level can be.

The CEPEX approach to a simulated ocean involved sequestering, as test organisms, the entire marine populations of large experimental water columns. The transparent plastic enclosures used were influenced by the earlier designs of many investigators (Goldman, 1962; Antia et al., 1963; Klussman and Inglish, 1968;

McAllister et al., 1971; Strickland and Terhune, 1971). Experimental units termed Controlled Experimental Ecosystems (CEEs) were designed to hold small (68 m^3) and large (ca 1300 m^3) volumes of seawater. The diameter of the smaller enclosures was 2.4 m and their length about 16 m while the larger CEEs were 9.5 m in diameter and about 23 m in length (Figure 5). The CEEs were deployed in Saanich Inlet, British Columbia, Canada.

The enclosure approach to pollution research requires extensive, multidisciplined support to record time-related population events exhibited by a broad variety of biological species. In CEPEX the dominant perturbants used were heavy metals and hydrocarbons. By mid 1981 CEPEX reports had appeared in about 100 publications while an additional 14 manuscripts were in press. Some early findings, only briefly summarized below, appeared in issues of the Bulletin of Marine Sciences 27(1), 1977, and Marine Science Communications 3(4), 1977, devoted to pollution research.

1. Organisms having short generation times, such as single-celled bacteria and phytoplankton, showed the earliest signs of stress but also demonstrated the fastest recoveries.
2. Zooplankton were generally more susceptible to stress in CEEs than in the laboratory.
3. Size plays an important role in an invertebrate's response to stress, both adults and large organisms being respectively less sensitive than juveniles and small organisms.
4. A repetitive sequence of community events was often observed as a common response to a variety of pollutants.
5. Differences in heavy metal concentrations that result in catastrophic, rather than sublethal, consequences cannot always be analytically discriminated.

The amount and intensity of effort required for community-oriented enclosure experiments clearly exceeds, by far, the resources of the typical water quality laboratory. On balance, enclosure research has generated an impressive amount of useful information regarding some prominent pollutant effects on multi-trophic aquatic assemblages. Significantly, however, none of these large multidisciplinary efforts appears to have satisfactorily solved today's environmental monitoring shortfall.

There is general agreement that the existing discrepancy between deliverable and desirable monitoring capabilities undermines the firm foundation needed to establish and sustain important water quality decisions. However, optimistic observers are of the opinion that enough information is already available to provide a reasonably sound basis for predicting environmental consequences in the oceans. The opposing pessimistic view holds that the oceans should be exempted from a more active role in waste

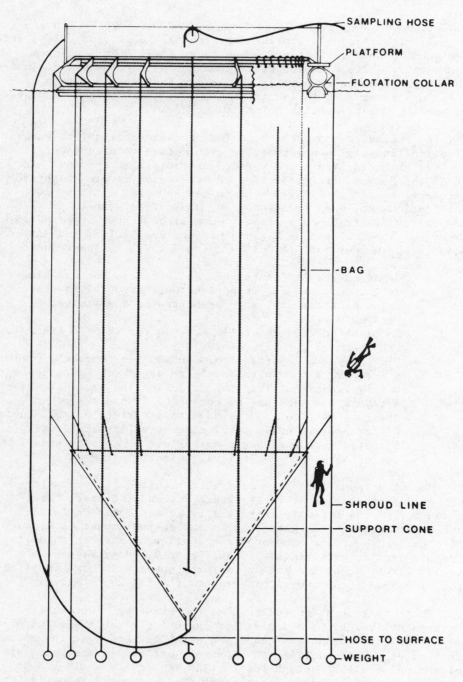

Figure 5. Two-dimensional view of full-size "controlled experimental ecosystem - (CEE)". The total length is 23m, inside diameter is 9.5m, volume capacity is 1300 m^3 (after Gillespie and Vaccaro, 1981).

disposal, pending a much better understanding of all of the attendant environmental consequences. The following contrasting statements summarize the degree of incompatibility between these two extreme points of view.

1. "The determination of an assimilative capacity is a scientific judgement based on the available, but sometimes incomplete, wisdom of the day." .... "Ocean disposal becomes acceptable when land or atmospheric dissemination of wastes becomes scientifically, economically, or socially unjustifiable." (Goldberg, 1981).

2. ".... we cannot prudently rely on a permissive approach based on crude assimilative capacity models in the hope that after-the-fact monitoring and a decades-long response time will ensure that health and the environment are protected." (Kamlet, 1981).

REFERENCES

American Society for Testing and Materials - Committee on Pesticides, 1977, Aquatic Toxicology and Hazard Evaluation, in: "Proc. First Annual Symp. on Aquatic Toxicology", F.L. Mayer and J.L. Hamelink, eds., American Society for Testing and Materials, Philadelphia.

Antia, N.J., McAllister, C.D., Parsons, T.R., Stevens, K., and Strickland, J.D.H., 1963, Further measurements of primary production using a large-volume plastic sphere, Limnol. Oceanogr., 8:166.

Barber, R.T., 1973, Organic ligands and phytoplankton growth in nutrient-rich seawater, in: "Trace Metals and Metal-Organic Interactions in Natural Waters", P.C. Singer, ed., Ann Arbor Science Publ., Ann Arbor, Michigan.

Bayne, B.L., Gabbott, P.A., and Widdows, J., 1975, Some effects of stress in the adult on eggs and larvae of Mytilus edulis L., J. Mar. Biol. Ass. U.K., 55:675.

Bayne, B.L., Holland, D.L., Moore, M.N., Lowe, D.M., and Widdows, J., 1978, Further studies on the effects of stress in the adult on eggs and larvae of Mytilus edulis L., J. Mar. Biol. Ass. U.K., 58:825.

Bayne, B.L., Anderson, J., Engel, D., Gilfillan, E., Hoss, D., Lloyd, R., and Thurberg, F.P., 1980, Physiological techniques for measuring the biological effects of pollution in the sea, Rapp. P.-v. Réun. Cons. Perm. Int. Explor. Mer., 179:88.

Butler, P.A., 1973, Residues in fish, wildlife and estuaries, Pestic. Monit. J., 6:238.

Capuzzo, J.M., 1981, Predicting pollution effects in the marine environment, Oceanus, 24(1):25.

Chau, Y.K., 1973, Complexing capacity of natural water - its significance and measurement, J. Chromatogr. Sci., 11:579.

Davey, E.W., Morgan, M.J., and Erickson, S.J., 1973, A biological measurement of copper complexation capacity of seawater, Limnol. Oceanogr., 18:993.

Davies, J.M., Gamble, J.C., and Steele, J.H., 1975, Preliminary studies with a large plastic enclosure, in: "Estuarine Research, Vol. I", L.E. Cronin, ed., Academic Press, New York.

Erickson, S.J., 1972, Toxicity of copper to Thalassiosira pseudonana in unenriched inshore seawater, J. Phycol., 8:318.

Gilfillan, E.S., 1975, Decrease of net carbon flux in two species of mussels caused by extracts of crude oil, Mar. Biol., 29:53.

Gilfillan, E.S., Mayo, D., Hanson, S., Donovan, D., and Jiang, L.C., 1976, Reduction in carbon flux in Mya arenaria caused by a spill of No. 6 fuel oil, Mar. Biol., 37:115.

Gillespie, P.A., and Vaccaro, R.F., 1978, A bacterial bioassay for measuring the copper-chelation capacity of seawater, Limnol. Oceanogr., 23:543.

Gillespie, P.A., and Vaccaro, R.F., 1981, Heterotrophic microbial activity in experimentally perturbed marine ecosystems, in: "Marine Environmental Pollution, Vol. II, Dumping and Mining", Elsevier Publishing Co., Amsterdam.

Giesy, J.P., Duke, R., Bingham, R., and Denzer, S., 1978, Energy charges in several molluscs and crustaceans: natural values and responses to cadmium stress, Bull. Ecol. Soc. Am., 59:66.

Goldberg, E.D., 1981, The oceans as waste space: the argument, Oceanus, 24(1):2.

Goldberg, E.D., Bowen, V.T., Farrington, J.W., Harvey, G., Martin, J.H., Parker, P.L., Risebrough, R.W., Robertson, W., Schneider, E., and Gamble, E., 1978, The Mussel Watch, Environ. Conserv., 5:101.

Goldman, C.R., 1962, A method of studying nutrient limiting factors in situ in water columns isolated by polyethylene film, Limnol. Oceanogr., 7:99.

Grice, G.D., Harris, R.P., Reeve, M.R., Heinbokel, J.F., and Davis, C.O., 1980, Large-scale enclosed water-column ecosystems: an overview of Foodweb I, the final CEPEX experiment, J. Mar. Biol. Ass. U.K., 60:401.

Grice, G.D., and Reeve, M.R., editors, "Marine Mesocosms: Biological and Chemical Research in Experimental Ecosystems", Springer-Verlag, in press.

Holden, A.V., 1973, International cooperative study of organochlorine and mercury residues in wildlife, 1969-71, Pestic. Monit. J., 7:37.

Ivanovici, A.M., 1980, Adenylate energy charge: An evaluation of applicability to assessment of pollution effects and directions for future research, Rapp. P.-v. Réun. Cons. Perm. Int. Explor. Mer, 179:23.

Kamlet, K.S., 1981, The oceans as waste space: the rebuttal, Oceanus, 24(1):10.

Klussman, W.G., and Inglish, J.M., 1968, Polyethylene tubes for studies of fertilization and productivity, Proc. SEast. Ass. Game Fish Commnrs., 22:415.

Lewis, J.R., 1980, Options and problems in environmental management and evaluation, Helgoländer Wiss. Meeresunters., 33:452.

Lewis, A.G., Whitfield, P., and Ramnarine, A., 1973, The reduction of copper toxicity in a marine copepod by sediment extract, Limnol. Oceanogr., 18:324.

Lund, J.W.G., 1972, Preliminary observations in the use of large experimental tubes in lakes, Verh. Internat. Verein. Limnol., 18:71.

Marine Technology Society - American Petroleum Institute, 1974, "Marine Bioassays", (Proc. Workshop on Marine Bioassays), Sponsored by the U.S. Environmental Protection Agency, the American Petroleum Institute, and the Marine Technology Society, Washington, D.C.

McAllister, C.D., Parsons, T.R., Stephens, K., and Strickland, J.D.H., 1961, Measurements of primary production in coastal sea water using a large-volume plastic sphere, Limnol. Oceanogr., 6:237.

McIntyre, A.D., and Pierce, J.B., editors, 1980, "Biological Effects of Marine Pollution and the Problems of Monitoring", (Proc. ICES Workshop, Beaufort, N.C., 1979), Rapp. P.-v. Réun. Cons. Perm. Int. Explor. Mer, Volume 179: 346 pp.

McLaren, I.A., 1969, Population and production ecology of zooplankton in Ogac Lake, a landlocked fiord on Baffin Island, J. Fish. Res. Board Can., 26:1485.

Menzel, D.W., and Case, J., 1977, Concept and design: Controlled Ecosystem Pollution Experiment, Bull. Mar. Sci., 27:1.

National Academy of Sciences - National Research Council, 1971, "Marine Environmental Quality", A report to the Ocean Science Committee, National Academy of Sciences, Washington, D.C., 107 pp.

National Academy of Sciences - National Research Council, 1980, "The International Mussel Watch", A report to the Environmental Studies Board, Commission on Natural Resources, National Academy of Sciences, Washington, D.C., 248 pp.

Ontario Ministry of the Environment, 1975, Report, Water Resources Branch, Toronto, "Proc. Aquatic Toxicity Workshop", and, from the Limnology and Toxicity Section, "Proc. Second Annual Toxicity Workshop" (mimeo.), G.R. Craig, ed., Ontario Ministry of the Environment, Toronto, Ontario, Canada, 339 pp.

Ott, W.R., 1978, "Environmental Indices - Theory and Practice", Ann Arbor Science Publ., Ann Arbor, Michigan.

Payne, J.F., 1976, Field evaluation of benzo[a]pyrene hydroxylase induction as a monitor for marine pollution, Science, 191:945.

Phelps, D.K., and Galloway, W.B., 1980, A report on the coastal environmental assessment stations (CEAS) program, Rapp. P.-v. Réun. Cons. Perm. Int. Explor. Mer , 179:76.

Schelske, C.L., and Stoermer, E.F., 1971, Eutrophication, silica depletion and predicted changes in algal quantity in Lake Michigan, Science, 173:423.

Strickland, J.D.H., and Terhune, L.D.B., 1961, The study of in situ marine photosynthesis using a large plastic bag, Limnol. Oceanogr., 6:93.

Sunda, W.G., and Guillard, R.R.L., 1976, The relationship between cupric ion activity and the toxicity of copper to phytoplankton, J. Mar. Res., 34:511.

Warren, C.E., and Davis, E.C., 1967, Laboratory studies in the feeding bioenergetics and growth of fish, in: "The Biological Basis of Freshwater Fish Production", S.D. Gerking, ed., Blackwell Publications, Oxford.

Winberg, G.G., 1960, Rate of metabolism and food requirements of fishes, Fisheries Research Board of Canada, Translation Series No. 194, Ottawa, Canada.

DISCUSSION: R.F. VACCARO

R. DE BERNARDI   The use of enclosures to study toxic effects on communities does not, in my opinion, allow us to achieve definitive general results; this is so because the results are greatly dependent on the starting community structure and on the trophic role in such communities played by the less tolerant species. As a consequence of these problems, what we observe is not a real toxic effect per se, but rather it is an end result of the interactions between toxicity effects and trophic relations; disturbed trophic relations, in many instances, may exert the greatest effect. In your opinion, is there a possibility to discriminate between these two components in enclosures? Also, can this research be done using an alternative to enclosures?

R.F. VACCARO   Inherent experimental inequalities in enclosure research must be accepted. I would agree that it is often difficult to distinguish between direct pollution effects and subsequent inter-trophic abnormalities. I think that studies on trophic interactions per se within enclosures represent a useful area for investigation; such studies should be expanded before one attempts a more complete understanding of pollution effects.

M. BRANICA   Your results on ^{14}C glucose uptake in the presence of Cu and added EDTA, as well as the "time adaptation" of your populations to high concentrations of Cu, can be explained by the cellular excretion of complexing and binding materials. A change in the distribution of Cu species can be elucidated by a determination of Cu complexing capacity in your water samples.

R.F. VACCARO   I would agree that a more complete information base on the nature of Cu-adaptation in our experiments might have been obtained, were alternative (chemical) measurement techniques available to us. However, based on our overall background in this experimental area (and the brevity of the contact periods between cells and Cu), we believe that the mechanism being inhibited by Cu is the cell membrane transport system for glucose uptake (in this case, for ^{14}C glucose).

H.W. NURNBERG   At the present state of knowledge, any dumping of metals into the sea should be strongly discouraged. Shallow coastal waters may be severely and quasi-irreversibly altered by such a practice, thus despoiling the coastal zone for future aquaculture uses necessary to the future food supply. Even relatively small enhancements to trace metal concentrations in the oceans should be regarded as potentially great risks; there are no technical means for man to make proper adjustments.

R.F. VACCARO   Currently in the United States there remains much uncertainty regarding the advantages of near-shore, as opposed to off-shore, waste disposal. Both favorable and unfavorable consequences can be cited regarding either method of disposal. One important consideration is that the near-shore areas along continental margins are the most productive areas of the ocean while being also the very areas usually impacted by waste disposal.

M. SMIES   In relation to Dr. Vaccaro's theme, I would like Dr. Ravera to elaborate on a topic which he brought to our attention earlier. Dr. Ravera, can you explain in what way a test concerning reproduction would be different from bioassays measuring the effect of toxicants on other biological parameters?

O. RAVERA   Bioassays present two major problems to the
scientist: (1) while being useful for screening
pollutants, they must be standardized; and (2) they
cannot be used directly to plan the protection of
natural populations. In addition to regular bioassays, we really do need more information regarding the influence of pollutants on the reproduction
of aquatic species, and, we must be careful in our
extrapolation of results obtained from microorganism
bioassays to assessing the multicellular organisms.
One mustn't forget that, in microorganism experiments with pollutants, a resistant strain often
may be produced in a very short time, a phenomenon
which would tend to be restricted to microbes.

R.F. VACCARO   I am in agreement with Dr. Ravera. It is
wise to place more emphasis on (1) the need for
bioassay standardization, (2) the appropriateness of
in situ growth rates, and (3) a consideration of
the tendency for populations subjected to long-
term stress to develop a specific resistance.

INDEX

Acid effects on speciation, see Speciation of trace elements
Alkylation, see Organometals
Bioaccumulation, 36, 130, 148, 152, 156, 160, 164-166, 168-171, 175, 180-181, 184, 197, 211, 246, 276-278, 281-282, 295-296
Bioassay, 19-20, 36, 47, 76, 130, 132, 147, 157, 184, 192, 197-201, 203, 293-294, 298-305, 313-314
Bioavailability, 1-2, 4, 6-7, 17, 21, 28, 30, 36, 105-110, 117, 127, 130, 158, 167, 169, 178, 180-181, 183-186, 192, 196-197, 199, 227, 245, 273, 278, 282-283, 285, 289
Biological effects (on speciation of trace elements and biological control over respeciation) (see also Metallothioneins) (see also Organometals) (see also Speciation of trace elements), 6, 26, 106, 108-111, 124, 127-128, 130, 132, 134, 171, 178-184, 195-202, 205, 227, 251, 254, 271-272, 279
Biological response
  at community and ecosystem levels, 8, 123-124, 129-131, 137, 141, 143-144, 146, 148, 150-152, 156-158, 163, 181, 200, 275-276, 280, 284, 289, 291-294, 302-307, 312
  mortality, see Toxicity
  photosynthesis changes, 128, 130, 140, 144, 146, 197
  productivity changes, 128, 141-142, 150, 152, 196-197, 294, 297
  reproduction changes, 130-131, 151-152, 161, 192, 296-298, 313-314
  secretion and excretion, 19, 26, 106, 108, 114-115, 134, 164, 166-171, 175, 178-179, 181, 195-196, 198, 200, 202, 279, 281, 296-297, 313
  toxic effects of heavy metals, see Toxicity
Cell surface (cell-water interface)
  components, 7, 23, 30, 106, 110, 124, 129, 162-163, 167, 175, 199, 213, 218, 227, 254, 276, 282-283
  functions, 7-8, 18, 30, 106, 110, 160, 162, 167, 227, 278-279, 282, 313
Colloids (see also Fibrils), 7-8, 18-19, 21-28, 31, 38-41, 43-44, 68, 75, 105, 107, 109-113, 115-116, 121, 158, 181-182, 184, 198 239, 241-242, 253, 264, 270-272
Complexation capacity, 4, 26, 61-64, 68, 75-77, 80, 132, 195, 201-202, 219, 225-226, 299

Experimental biological systems (see also Bioassay)
  aquaria, 123-124
  cultures, 2, 19, 113-115, 123-129, 135, 138, 152, 159, 177, 184, 197, 201-202, 295, 298-299, 302-305
  enclosed water columns, general references, 124-125, 138-150, 155-158, 178, 181, 184-185, 305-308, 312
  major ecosystem studies with enclosures
    CEPEX, 138, 140, 148, 151, 181, 306-307
    MELIMEX, 108, 138, 140-141, 147-148, 151, 181
Fibrils, 6, 8, 26, 68, 105, 110-117, 121-122, 167, 282
Heavy metals, see Metallothioneins, see Trace elements of interest in water
Metallothioneins, 8, 31, 163-164, 166, 168, 170-171, 174-175
Methodology and techniques
  for fractionation
    artifacts, 8, 21-22, 37, 39-43, 46-47, 78-79, 111, 122, 124, 235, 240, 283
    centrifugation, 23, 39, 43-44, 77, 115, 283
    chromatography in general, 20, 23, 65, 88-98, 100, 196, 212, 284
    dialysis and electrodialysis, 22, 29, 41-44, 46-47, 76, 78, 198, 201, 283
    electrophoresis, 43-44, 223, 271
    filtration and ultrafiltration, 7-8, 18, 21, 29-30, 39-42, 44, 46, 75-78, 111, 115, 122, 147, 196-198, 202, 204, 226, 231-232, 234-237, 239, 241-242, 283, 299
    gel filtration/chromatography, 22, 44, 47, 76, 283
    ion exchange and resin separations, 23-24, 28-31, 42, 44, 76-78, 98-99, 127, 201, 204, 231-235, 239-240, 242, 283
    multistage separations, 21, 23, 29-30, 39, 44, 115, 231-235, 239, 242, 283
    solvent extraction, 23, 31, 77-79, 91-92, 115, 196, 198-199, 272-273, 282
  for measurement
    bioassays, see Bioassay
    complexation capacity, see Complexation capacity
    ion selective (specific) electrodes, 21, 30, 76, 78, 85, 283
    isotope dilution (see also Bioassay), 36, 129, 131, 135, 232, 235, 240, 272, 278, 283, 298-299
    neutron activation analysis, 6, 36, 44, 77-78, 212, 231-237, 239, 242
    spectroscopy, 23, 29, 44, 76-78, 80, 88-100, 127, 135, 199, 212, 235, 237, 270, 283-284
    titration methods, 26, 57-63, 68, 75-76, 201
    voltammetry, 6, 21, 24-25, 28-30, 44, 49, 54-58, 60-62, 64-65, 68, 75-78, 85, 100, 127, 135, 209, 211-215, 218-220, 223, 225-227, 230, 256-257, 283
Microscopical assessment (see also Fibrils) (see also Toxicity), 112-114, 130, 270, 283
Models, 20-21, 30-31, 54, 57-59, 61, 64-65, 73-74, 76-79, 81, 85-86, 134, 177, 203-205, 210, 217, 220, 223-224, 248, 255, 278,

INDEX

  280-281, 292, 309
Monitoring, 6, 9, 18, 127, 291-293, 295-298, 305, 307, 309
Nutrients, see Trace elements of interest in water
Organometals (carbon to metal bonding)
  alkylation and alkylated metals, 7, 17, 20, 88, 90-94, 96, 98-100,
      103, 109, 161, 163, 165, 174-175, 178-179, 181-186, 192-193,
      242, 265, 289
general references, 20, 87-92, 94-95, 103, 109, 179, 183-184, 192,
      289
Particles
  coatings on, 25, 49, 107, 111, 113-115, 211, 218, 245-247, 249,
      251-253, 256-257, 264, 270-271
  general references, 3-4, 6-9, 18-22, 25, 37-39, 41, 43-44, 49, 71-
      72, 74, 85, 105-107, 109-114, 122, 124, 127, 140, 142, 152,
      155-156, 158, 175, 179, 181-182, 184-185, 205, 214, 217, 226-
      227, 234, 236-240, 245-248, 251, 255-258, 262-265, 271-272,
      277
  scavenging by, see Sedimentation
  transport by (see also Sedimentation), 107, 111, 184, 246, 257,
      264, 271-272
Physical factors affecting speciation, see Speciation of trace
elements
Redistribution of trace elements, see Colloids, see Particles, see
Sedimentation
Respeciation of trace elements in nature, see Speciation of trace
elements
Sedimentation
  general references, 107, 112, 114, 117, 147, 156, 185, 246, 249,
      264, 271
  in relation to scavenging and associated particle surface activity,
      28, 107, 111, 184, 227, 246, 249, 257, 264, 271
Speciation of trace elements
  alkylation, see Organometals
  biological effects on speciation of trace elements and biological
  control over respeciation, see Biological effects, see Metallothioneins, see Organometals
  by complexing substances
    extracellular products, 19, 26, 127, 134, 150, 164, 167, 175,
        195-203, 313
    fulvic acids, 22, 25-26, 38, 76-77, 178, 180, 254
    humics, 22, 25-26, 28, 38, 42, 44, 47, 61, 73, 76-77, 80, 86, 127,
        178, 181, 201, 203-204, 209, 219, 221-223, 230, 248, 253-258,
        271
    hydroxamates, 86, 199-202
    inorganics, 22, 24-25, 27, 38, 50, 52, 54-59, 71-72, 76, 88,
        109, 178, 180-184, 186, 214-217, 219-222, 230, 235, 240, 250,
        252, 254-255, 259-264, 272, 278-279, 284
    organic acids (small molecules such as amino acids), 22, 38, 55-
        56, 59, 76, 79, 95, 98, 127, 166, 195-199, 203, 217-219, 235,
        258

polysaccharides, 114, 121, 175, 197, 199, 203
proteins (see also Metallothioneins), 31, 38, 44, 162, 166, 180, 183-186, 195, 197-198, 219, 221, 297
synthetic chelators (excluding laboratory resins), 27, 58-59, 61, 64-65, 73, 77, 79, 126-127, 196-199, 201, 220-224, 235, 246, 255, 299-300, 313
labile species, 7, 24, 28-29, 49-50, 54-55, 57, 60, 75-76, 78, 214, 217-218, 220-221, 225, 256-257, 261
metallothioneins, see Metallothioneins
mineral particle surface effects, 25, 107, 211, 245, 247-249, 251-253, 256-257, 262, 264, 272
physical factors affecting, 26, 106, 109, 123, 127, 134, 178, 180, 200, 203, 219, 226, 246-257, 259, 272, 278
respeciation in nature (see also Biological effects) (see also Metallothioneins) (see also Toxicity), 7-9, 49, 105-106, 109-110, 128, 164, 170-171, 180, 197-203, 227, 246-249, 252-264, 271-272, 279
speciation schemes, 4, 7, 18, 22, 24, 28-32, 107, 232, 234, 237, 239, 289

Toxicity
assessment, 7, 18-20, 28, 31-32, 36, 71, 110, 123, 129, 131-132, 152, 159-160, 178, 192, 230, 275, 284, 292-293, 295-297, 305-307
bioassay, see Bioassay
detoxification, 117, 160, 162-164, 168-169, 174-175, 180, 183-184, 186, 196-199, 299
measurement of, 19, 110, 124, 128-131, 152, 157-158, 161, 276, 294-305, 313
mechanisms of toxicity and detoxification, 8, 106, 131, 157, 160-166, 169, 175, 179-180, 282, 284, 313
metallothioneins, see Metallothioneins
reduction through organic complexation (see also Metallothioneins) 19, 106, 109, 181, 184, 197-199, 299
susceptibility as a function of prior stress, 129, 137, 140, 146-148, 297, 299, 302-305, 314
tolerance, 110, 129, 137, 152, 160-165, 282, 292-293, 297, 299, 302-305, 312-314
ultrastructural correlates of, 130, 162-164, 166-168, 170, 175, 294

Trace elements of interest in water
aluminum, 5, 15, 25, 40, 76, 108, 162, 202, 247-248, 251-252, 258, 276
antimony, 24, 27, 93, 98, 100, 108, 243, 276
arsenic, 5, 15, 17, 20, 27, 91-94, 96, 98-99, 108, 159, 162-163, 178-179, 184, 231, 236-238, 242, 249, 252, 260, 263, 276, 284
cadmium, 5, 15, 17, 19, 24, 28-29, 31, 49, 55-56, 61-62, 64-65, 72, 76-77, 100, 108, 111, 127, 141, 144, 148-150, 156, 159, 161-171, 175, 180, 183, 197, 204, 209, 215-224, 226, 242, 248-250, 252-255, 261, 263. 271-273, 275-276, 282. 284
chromium, 5, 15, 17, 23-24, 27, 77, 108, 159, 180, 231, 233-236, 242-243, 252, 260, 262-263, 276, 278, 282

INDEX

cobalt, 17, 23-24, 27, 49, 76-77, 108, 126, 159, 168, 202, 243, 249, 252, 262, 271, 276, 278, 280, 282
copper, 5, 15, 17, 19, 23-24, 26, 28-29, 31, 49, 52, 54, 61-62, 68, 72, 76-77, 79-80, 95, 100, 108, 127, 140-144, 146-150, 152, 156, 158-159, 161, 163-171, 180-181, 195-204, 209, 219, 223, 226, 242, 248-249, 252-257, 259, 261-263, 273, 276, 280-282, 299-303, 305, 313
iron, 5, 15, 17, 22, 25-28, 40, 72, 76-77, 80, 86, 106, 108, 115, 117, 127, 159, 165-169, 181, 196-197, 199-202, 242, 245, 247-249, 251-253, 255-256, 258-264, 271-272, 276, 278, 282
lead, 5, 15, 17, 19-20, 24, 28-29, 40, 49, 54, 56, 58-59, 61-62, 64, 72-73, 76-77, 88, 90, 92-93, 96-98, 100, 108, 114, 127, 141, 144, 148-150, 156, 159, 162-163, 165-166, 168, 171, 182, 193, 202, 204, 215-217, 219-224, 226, 242, 248-249, 252-253, 261-262, 273, 276, 282, 284, 289-290
manganese, 5, 15, 17, 22, 24, 27, 41, 76-77, 108, 115, 117, 159, 165-166, 168, 182, 198-199, 242, 245, 247-249, 251-253, 255-256, 258-259, 262-264, 271-272, 276, 282
mercury (including use in measurements), 5, 15, 17, 20, 24-25, 30, 49, 57, 62, 65, 68-69, 72-73, 85-86, 88, 100, 108, 124, 127-128, 135, 141, 144, 148-150, 156, 159, 161-163, 165-166, 168, 174, 178, 181-183, 193, 219, 223, 242, 249, 275, 277-278, 280-281, 284, 302, 304-305
molybdenum, 17, 108, 171, 249, 252, 276
nickel, 17, 24, 49, 76-77, 108, 127, 159, 169, 202, 242, 249, 252-253, 262-263, 271, 276, 282
nitrogen, 5, 15, 17, 157, 197-198, 202, 259, 263
phosphorous, 2, 5, 15, 17, 72, 77, 79, 108-111, 122, 130, 139, 162, 166, 179, 197, 220, 235, 240, 251-252, 254, 259-260, 262-263, 271, 295
selenium, 17, 27, 91, 108, 159, 162-163, 178, 183-184, 231, 238-242, 249, 252, 276
silicon, 5, 15, 17, 25, 38, 40, 159, 183, 247-249, 252, 258-259, 262-263, 284
thallium, 17, 24, 27, 49, 108, 252, 276
tin, 17, 20, 24, 27, 90-94, 97-98, 100, 183, 242
uranium, 108, 114, 243, 249, 296
vanadium, 17, 159, 165, 199, 231-233, 243, 249, 276, 282
zinc, 5, 15, 17, 24, 26, 31, 41, 49, 52, 58-59, 65, 72, 76-77, 108, 126-127, 141, 144, 146, 148-152, 155-156, 159, 162-164, 166-171, 175, 183, 197-199, 218-224, 226, 242, 249-253, 256-257, 262, 273, 276, 280-282
Transport, see Particles, see Sedimentation
Water types
  effluent, 54, 73, 182, 240-241, 293
  estuarine, 1, 28, 111, 181-182, 186, 201, 214, 217, 219-220, 249, 251-253, 255, 257, 262, 265, 271-272
  lacustrine, 1, 8-9, 26, 42, 62, 91, 96-97, 103, 105, 107, 111-115, 117, 121, 138-139, 144, 147, 151, 155-156, 158, 178, 185, 202, 211, 213, 220-221, 223-224, 226, 247, 251, 259-262, 282, 306

marine, 1, 9, 22-24, 26-28, 41-42, 52, 54-56, 58-61, 64-65, 68, 85, 96-97, 103, 122, 135, 140, 142, 152, 167, 178-183, 195-199, 201, 203, 209, 211-226, 230, 251- 252, 257-260, 262, 271-272, 291-292, 294-296, 298-307, 309, 313
pore waters, 3, 18, 217, 219, 246, 258-261, 263-265
riverine, 1, 22, 28, 40-42, 71-73, 96-97, 111, 220, 233, 236, 238, 240-241, 250-251, 255-256, 261, 271, 273
tap water, 71-72, 97, 233